Commutative Algebra

Commutative Algebra

Second Edition

N S Gopala Krishnan

Universities Press

COMMUTATIVE ALGEBRA

UNIVERSITIES PRESS (INDIA) PRIVATE LIMITED

Registered Office
3-6-747/1/A & 3-6-754/1, Himayatnagar, Hyderabad 500 029, Telangana, India
info@universitiespress.com, www.universitiespress.com

Distributed by
Orient Blackswan Private Limited

Registered Office
3-6-752 Himayatnagar, Hyderabad 500 029, Telangana, India

Other Offices
Bengaluru / Chennai / Guwahati / Hyderabad / Kolkata
Mumbai / New Delhi / Noida / Patna

First edition published by Oxford & IBH 1984
Second Edition published by Universities Press 2016
Reprinted 2017, 2023, 2024, 2025

Cover and book design:

ISBN 978 81 7371 978 3

502247

Typeset in Times New Roman at 11pt *by*
Orient BlackSwan

Printed in India by
Replika Press Pvt. Ltd., Kundli Sonepat 131 028, Haryana.

Published by
Universities Press (India) Private Limited
3-6-747/1/A & 3-6-754/1, Himayatnagar, Hyderabad 500 029, Telangana, India

To the memory of my wife, Meena

Contents

Preface

This book is intended as a textbook on Commutative Algebra. In a vast field like Commutative Algebra, there is a variety of choice of topics for a textbook. However, the choice in the present book has been made with the following two aims. The first aim is to present basic results on Commutative Algebra necessary for elementary Algebraic Geometry. The second aim is to introduce homological methods in Commutative Algebra.

Some efforts have been made to make the book self-contained. Chapter 1 introduces projective modules and their elementary properties. It also deals with flat and faithfully flat modules. The second chapter introduces localisation and gives some applications of localisation to projective modules. The third chapter contains basic properties of Noetherian and Artinian rings. The fourth chapter introduces integral extensions and include results such as the going up and going down theorems, normalisation theorem and the finiteness of integral closure. The fifth chapter deals with valuation rings and Dedekind domains and ends with a proof of ramification formula in the finite separable case. The sixth chapter contains basic results on completion such as Artin Rees Lemma, Krull's intersection theorem and Hensel's Lemma. The seventh chapter deals with *Ext* and *Tor* functors and the projective dimension. The eight and ninth chapters deal with dimension theory at some length. Homological properties of Cohen–Macaulay rings and homological characterisation of regular local rings are given. We also give a proof of Cohen's structure theorem on complete local rings. Chapter 10 gives some applications of homology in the solutions of some outstanding conjectures of 1970's. Further developments on these conjectures are not dealt with in this book.

The style of the book is maintained in the form of lecture notes. This will facilitate readers with some mathematical maturity to study the book on their own. Motivations are given at a number of places, examples follow every definition and exercises are given at the end of each section.

I am thankful to Professors S. Kumaresan, R. C. Cowsik and S. A. Katre for taking excellent initiatives in bringing out this book. I am very much thankful to the late Professor Amit Roy who was the main inspiration for this book and specially for his valuable suggestions to include applications of homology in the solutions of some outstanding conjectures. I am also thankful to Professors R. Sridharan, Balwant Singh and Avinash Sathaye for many helpful suggestions.

My thanks are due to Professor Amartya Kumar Dutta and Dr Neena Gupta for their nice suggestions for enhancing the quality of the book. I am also thankful to Professor Parvati Shastri for her suggestions in improving the initial part of the book. I am extremely thankful to Dr Rabeya Basu for the major task of preparing the typescript, incorporating all suggestions and bringing the book to an excellent form. Sheer love for the subject by all the above individuals is the main reason for this book coming out in the present form. I am also thankful to Mr S. R. Gosavi and Mr Juzer Siamwala for typing some of the initial chapters and to Dr Vivek Mohan Mallick for all the technical help.

Writing this book has been the most satisfying experience of my career and nothing will give me more joy than if the book is used by students for learning the subject and also in their future work.

N. S. GOPALA KRISHNAN
B-1001, Kalpak Gulisthan
Perry Cross Road
Bandra (W), Mumbai 400050

CHAPTER

1

MODULES

This chapter is concerned with a preliminary study of modules. The notion of an ideal which arose from number theory is also important in Algebraic Geometry. It is useful to study ideals from a module-theoretic setup, as operations of linear algebra, such as formation of quotients, products and tensor products, are closed for modules but not for ideals.

We then study special types of modules such as projective modules, flat modules and faithfully flat modules. Projective modules play the role of a vector space while studying linear algebra over a general commutative ring. Flat modules are more general than projective modules.

Sometimes, geometric objects defined over a field behave differently over a bigger field. Such questions are best studied by scalar extension of rings using tensor products. Some properties are preserved under flat extension, but faithfully flat extensions are more useful as they have nice descent properties.

1.1 Free modules

A ring R will always mean a commutative ring with unit element 1.

Definition 1.1.1 An *R-module* M is an abelian group M together with a map $R \times M \to M$, given by $(a, x) \mapsto a.x$, satisfying the following conditions:

$$a \,.\, (x + y) = a \,.\, x + a \,.\, y, \quad a \in R, \quad x, y \in M,$$

$$\begin{aligned} (a+b)\,.\,x &= a\,.\,x + b\,.\,x, & a,b \in R, \quad x \in M, \\ a\,.\,(b\,.\,x) &= (ab)\,.\,x, & a,b \in R, \quad x \in M, \\ 1\,.\,x &= x, & x \in M. \end{aligned}$$

We denote $a\,.\,x$ by ax.

Example 1.1.2 (i) Any vector space V over a field K is a K-module.

(ii) Any abelian group G is a $\mathbb{Z}$-module.

Definition 1.1.3 A subset $N \subset M$ is called a *submodule* if N is a subgroup of the abelian group M and $ax \in N$ for all $a \in R$ and $x \in N$.

Example 1.1.4 (i) Any subspace W of a vector space V.

(ii) All polynomials of degree at most n is a submodule of the R-module $R[X]$.

(iii) The modules $\{0\}$ and M are submodules of M, called *improper submodules.*

Definition 1.1.5 Let M and N be R-modules. A map $f : M \to N$ is called a *homomorphism* of R-modules or an *R-module homomorphism* or an *R-linear map* if

$$\begin{aligned} f(x+y) &= f(x) + f(y), & x, y \in M, \\ f(ax) &= af(x), & a \in R,\ x \in M. \end{aligned}$$

Example 1.1.6 (i) For any fixed $a \in R$, the map $f : M \to M$, given by $f(x) = ax$, is a homomorphism.

(ii) For any submodule N of M, the inclusion map $i : N \to M$ is a homomorphism.

Definition 1.1.7 Let N be a submodule of M. Consider the quotient abelian group M/N with the scalar multiplication given by $a\,.\,(x+N) = ax + N$, $a \in R$, $x \in M$. Then M/N acquires the structure of an R-module called the *quotient module M/N*. The map $p : M \to M/N$, defined by $p(x) = x + N$, $x \in M$, is a homomorphism of modules called the *projection.*

Definition 1.1.8 A homomorphism of modules which is both one-one (injective) and onto (surjective) is called an *isomorphism.*

Proposition 1.1.9 *Let $f : M \to N$ be a homomorphism of M onto N. Then the kernel of $f = \{x \in M \mid f(x) = 0\}$ is a submodule K of M and the quotient module M/K is isomorphic to N.*

Proof Clearly, K satisfies the conditions for a submodule. The map $\overline{f} : M/K \to N$, given by $\overline{f}(x+K) = f(x)$, $x \in M$, is well defined and is an isomorphism. □

Corollary 1.1.10 *Let M be an R-module, N a submodule of M and K a submodule of N. Then $M/N \simeq \frac{M/K}{N/K}$.*

Proof The composition of the projections $M \to M/K \to \frac{M/K}{N/K}$ is surjective and has kernel N. Hence by Proposition 1.1.9, □

$$M/N \simeq \frac{M/K}{N/K}.$$

We now consider some operations on modules. If N and K are submodules of M, then $N \cap K$ is a submodule of M but $N \cup K$ is not in general a submodule of M.

Definition 1.1.11 The smallest submodule of M containing $N \cup K$ is called the *submodule generated by N and K*. More generally, for any subset S of M, the submodule of M generated by S is the smallest submodule of M containing S.

Proposition 1.1.12 *The submodule S generated by N and K is the submodule*

$$N + K = \{x + y \,|\, x \in N, y \in K\}.$$

Proof Clearly, $N + K$ is a submodule of M, $N \subseteq N + K$ and $K \subseteq N + K$, so that $S \subseteq N + K$. Conversely, for any $x \in N, y \in K$, we have $x, y \in S$, so that $x + y \in S$. Thus $N + K \subseteq S$, and $S = N + K$. □

Corollary 1.1.13 *If $N_1, N_2, \ldots, N_k$ are submodules of M, then the submodule generated by $N_1, N_2, \ldots, N_k$ is equal to*

$$\left\{ \sum_{i=1}^{k} x_i \,|\, x_i \in N_i \right\} = N_1 + N_2 + \cdots + N_k.$$

Let A be a subset of M and I an ideal in R. Then the set

$$\left\{ x = \sum_{i=1}^{k} a_i x_i \,|\, a_i \in I,\, x_i \in A \right\}$$

is a submodule of M and is denoted by IA. In particular, if $I = R$ and $A = \{x\}$, IA is denoted by Rx.

Definition 1.1.14 An R-module M is called *cyclic* if $M = Rx$ for some $x \in M$.

Proposition 1.1.15 *An R-module M is cyclic if and only if $M \simeq R/I$ for some ideal I in R.*

Proof If M is cyclic then $M = Rx$ for some $x \in M$. The natural map $\theta : R \to M$, given by $\theta(a) = ax$, is a surjective homomorphism. If $I = \operatorname{Ker}\theta$, then $R/I \simeq M$. Conversely, if $M \simeq R/I$ then M is cyclic, as R/I is cyclic, being generated by $\overline{1} = 1 + I$. □

Definition 1.1.16 The *annihilator* of an R-module M is defined as

$$\operatorname{Ann}(M) = \{a \in R \mid aM = 0\}.$$

Clearly, $\operatorname{Ann}(M)$ is an ideal of R. If M is cyclic and is generated by x, then $\operatorname{Ann}(M)$ is denoted by $\operatorname{Ann}(x)$.

Definition 1.1.17 An R-module M is called a *faithful* R-module if $\operatorname{Ann}(M) = (0)$.

Definition 1.1.18 M is called a *finitely generated* R-module if

$$M = M_1 + M_2 + \cdots + M_n,$$

where each M_i is a cyclic submodule of M. If $M_i = Rx_i$, then $\{x_1, x_2, \ldots, x_n\}$ is called a *generating set* for M.

Example 1.1.19 The module of polynomials over R of degree at most n is generated by $1, X, X^2, \ldots, X^n$. Clearly, this generating set is not unique, as $1, 1 + X, X^2, \ldots, X^n$ is also a generating set for the same module.

Definition 1.1.20 An R-module M is called a *direct sum* of submodules $M_1, M_2, \ldots, M_n$ if every $x \in M$ can be uniquely expressed as

$$x = x_1 + x_2 + \cdots + x_n, \quad x_i \in M_i \ (1 \leq i \leq n).$$

The direct sum is denoted by $M = M_1 \oplus M_2 \oplus \cdots \oplus M_n$.

Proposition 1.1.21 *An R-module $M = M_1 \oplus M_2 \oplus \cdots \oplus M_n$ if and only if*

(i) $M = M_1 + M_2 + \cdots + M_n$, and
(ii) $M_i \cap (M_1 + M_2 + \cdots + M_{i-1} + M_{i+1} \cdots + M_n) = 0$ for all $i\,(1 \leq i \leq n)$.

Proof Suppose $M = M_1 \oplus \cdots \oplus M_n$. Then clearly (i) is true. To prove (ii), suppose x is in the intersection on the left hand side, so that $x \in M_i$ and $x = y_1 + y_2 + \cdots + y_{i-1} + 0 + y_{i+1} + \cdots + y_n$, $y_j \in M_j$, $j \neq i$. Since $x = 0 + 0 + \cdots + 0 + x + 0 \cdots + 0$ with x in the i-th place, we have by uniqueness, $x = 0$.

Conversely, assume conditions (i) and (ii). By (i), each $x \in M$ can be expressed as

$$x = x_1 + x_2 + \cdots + x_n,\ x_i \in M_i.$$

Suppose $x = y_1 + y_2 + \cdots + y_n$, $y_i \in M_i$. Then

$$0 = (x_1 - y_1) + \cdots + (x_i - y_i) + \cdots + (x_n - y_n),$$

so that $(x_i - y_i) \in M_i$ and $x_i - y_i = -[(x_1 - y_1) + \cdots + (x_{i-1} - y_{i-1}) + 0 + (x_{i+1} - y_{i+1}) + \cdots (x_n - y_n)] \in (M_1 + \cdots + M_{i-1} + M_{i+1} + \cdots + M_n)$. Hence by (ii), $x_i - y_i = 0$, i.e., $x_i = y_i$, $1 \leq i \leq n$. Thus, $M = M_1 \oplus M_2 \oplus \cdots \oplus M_n$. □

If $M = M_1 \oplus M_2$, $x \in M$ then x can be uniquely expressed as $x = x_1 + x_2$, $x_1 \in M_1$, $x_2 \in M_2$.

Definition 1.1.22 The mappings $\pi_1 : M \to M_1$, $\pi_2 : M \to M_2$, defined by $\pi_1(x) = x_1$, $\pi_2(x) = x_2$, are called *projections*.

Remark 1.1.23 The definition of direct sum can be extended to any collection of modules. An R-module M is a direct sum of a collection of submodules $\{M_\alpha\}_{\alpha \in I}$ if each non-zero $x \in M$ can be expressed uniquely as $x = x_{\alpha_1} + x_{\alpha_2} + \cdots + x_{\alpha_n}$ for some non-zero $x_{\alpha_i} \in M_{\alpha_i}$, $\alpha_1, \alpha_2, \ldots, \alpha_n \in I$. We denote this by $M = \oplus \Sigma_{\alpha \in I} M_\alpha$.

Definition 1.1.24 A cyclic R-module $M = Rx$ is called *free* if $\operatorname{Ann}(x) = 0$. An R-module M is called *free* if it can be expreesed as a direct sum $M = \oplus_\alpha M_\alpha$, where each M_α is a free cyclic R-module. If $M_\alpha = Rx_\alpha$, then the collection $\{x_\alpha\}$ is called a *basis* of the free module M.

Example 1.1.25 (i) $R^n = \{(a_1, \ldots, a_n) \mid a_i \in R\}$ is a free R-module with a basis $\{e_1 = (1, 0, \ldots, 0), e_2 = (0, 1, \ldots, 0), \ldots, e_n = (0, 0, \ldots, 0, 1)\}$.
(ii) $\mathbb{Z}/n\mathbb{Z}$, the group of integers modulo n is a not free $\mathbb{Z}$-module as each $x \in \mathbb{Z}/n\mathbb{Z}$, has a non-zero annihilator. Clearly, the basis of a free module is not unique.

Proposition 1.1.26 *Any two bases of a free module have the same cardinality.*

Proof Let M be a free module with basis $\{x_\alpha\}_{\alpha \in I}$. Choose a maximal ideal $\mathfrak{m}$ of R and let $R/\mathfrak{m} = k$. Then $V = M/\mathfrak{m}M$ is annihilated by $\mathfrak{m}$ and hence it is a k-vector space. If $\overline{x}_\alpha = x_\alpha + \mathfrak{m}M$, then it is easy to verify that $\{\overline{x}_\alpha\}$ is a basis of V over k. Since any two bases of a vector space have the same cardinality, the result follows. □

Corollary 1.1.27 *If a free module F has a basis with n elements, then any other basis of F also has n elements.*

Definition 1.1.28 In the above case, n is called the *rank* of F.

Example 1.1.29 (i) A cyclic module $M = Rx$ with $\operatorname{Ann}(x) = 0$ is free, of rank one.

(ii) The R-module R^n is a free R-module of rank n.

(iii) The R-module $R[X]$ is free with basis $\{1, X, X^2, \ldots, X^n, \ldots\}$ and has countable rank.

1.1. EXERCISES

1. If N and K are submodules of M, then show that
$$\frac{(N+K)}{K} \simeq \frac{N}{(N \cap K)}.$$

2. Show that if M has a finitely generated submodule N such that the quotient M/N is finitely generated, then M is finitely generated.

3. Show by an example that an increasing union of finitely generated submodules of M need not be finitely generated.

4. A submodule N of M is called a *direct summand* of M if $M = N \oplus N'$ for some submodule N' of M. If $K \subset N \subset M$ are submodules, then show the following:
   (a) If N is a direct summand of M, then N/K is a direct summand of M/K.
   (b) If K is a direct summand of N and N is a direct summand of M, then K is a direct summand of M.
   (c) If K is a direct summand of M, then K is a direct summand of N. If, further, N/K is a direct summand of M/K, then N is a direct summand of M.

5. If $M = \oplus_\alpha M_\alpha$, then show that M is finitely generated if and only if $M_\alpha = 0$ for all but finitely many α, and each M_α is finitely generated.

6. Let N and K be submodules of M with $I = \text{Ann}\,(N)$ and $J = \text{Ann}\,(K)$. Show that $\text{Ann}\,(N \cap K) \supseteq I + J$ and give an example to show that the inclusion can be a strict inclusion.

7. Let N and K be submodules of M. Define quotient
$$(N : K) = \{a \in R \,|\, aK \subset N\}.$$
   Show that
   (a) $(N : K) = \text{Ann}\,\left(\frac{N+K}{N}\right)$.
   (b) $(N : Rx)x = N \cap Rx$, $x \in M$.

8. Show that any R-module M can be expressed as a quotient of a free R-module F. If M is finitely generated, then show that F can be chosen to be free, of finite rank.

9. Let F be a free module with a basis $\{e_1, e_2, \ldots, e_n\}$. Show that a necessary condition for $a \in F$ with $a = \sum_{i=1}^{n} a_i e_i$ to be a part of another basis of F is that a is *unimodular*, i.e., $\Sigma_{i=1}^{n} b_i a_i = 1$ for some $b_i \in R$. Deduce that a necessary condition for a vector $(a_1, a_2, \ldots, a_n) \in R^n$ to be completable to a non-singular matrix over R is that it is unimodular.

1.2 Projective modules

Definition 1.2.1 A sequence of R-modules and R-homomorphisms of the type

$$M_0 \xrightarrow{f_0} M_1 \xrightarrow{f_1} M_2 \to \cdots \to M_n \xrightarrow{f_n} M_{n+1}$$

is called an *exact sequence* if

$$\text{Im } f_i = \text{Ker } f_{i+1},\ 0 \leq i \leq n-1.$$

Example 1.2.2 (i) The sequence $0 \to 2\mathbb{Z} \xrightarrow{i} \mathbb{Z} \xrightarrow{p} \mathbb{Z}/2\mathbb{Z} \to 0$ is an exact sequence, where i is the inclusion map and p is the projection of $\mathbb{Z}$ onto $\mathbb{Z}/2\mathbb{Z}$.

(ii) For any two R-modules M and N, the sequence

$$0 \to M \xrightarrow{i} M \oplus N \xrightarrow{\pi} N \to 0$$

is exact, where i is the inclusion given by $i(x) = (x, 0)$, and π is the projection given by $\pi(x, y) = y,\ \ x \in M,\ y \in N.$

Definition 1.2.3 An exact sequence of R modules of the form

$$0 \to M' \xrightarrow{f} M \xrightarrow{g} M'' \to 0$$

is said to be a *split* exact sequence if there exists an R-homomorphism $t : M'' \to M$ such that $gt = \text{Id}_{M''}$, the identity map on M''. The map t is said to a *splitting*.

Example 1.2.4 Example (ii) above is a split exact sequence with the splitting map t : $N \to M \oplus N$ given by $t(y) = (0, y)$. Example (i) has no splitting as the following proposition shows.

Proposition 1.2.5 *If* $0 \to M' \xrightarrow{f} M \xrightarrow{g} M'' \to 0$ *is a split exact sequence with* t : $M'' \to M$ *a splitting, then*

$$M = f(M') \oplus t(M'') \simeq M' \oplus M''.$$

Proof Since $t : M'' \to M$ is a splitting, we have $gt = \mathrm{Id}_{M''}$. This implies that t is injective; for if $t(x) = 0$, $x \in M''$, then $x = gt(x) = 0$. If $x \in M$ then

$$x = tg(x) + (x - tg(x)) \in t(M'') + \mathrm{Ker}\, g,$$

as $g(x - tg(x)) = 0$. Moreover, if $y \in t(M'') \cap \mathrm{Ker}\, g$, then $y = t(z)$, $z \in M''$ and $z = gt(z) = g(y) = 0$. Hence $y = t(z) = 0$. Hence

$$M = t(M'') \oplus \mathrm{Ker}\, g = t(M'') \oplus \mathrm{Im}\, f \simeq M'' \oplus M',$$

as f and t are injective maps. □

Corollary 1.2.6 *Let* $0 \to M' \xrightarrow{f} M \xrightarrow{g} M'' \to 0$ *be an exact sequence of* R*-modules which splits. Then there exists an* R*-module homomorphism* $s : M \to M'$ *such that* $sf = \mathrm{Id}_{M'}$.

Proof Let $t : M'' \to M$ be a splitting, so that by Proposition 1.2.5, we have $M = f(M') \oplus t(M'')$. Take π_1 to be the projection of M onto $f(M')$, and let $s = f^{-1}\pi_1$. Then $sf = \mathrm{Id}_{M'}$. □

Let M and N be R-modules. The set S of all R-homomorphisms of M into N has a natural R-module structure for the operations

$$\begin{aligned} (f+g)(x) &= f(x) + g(x), \quad x \in M, \quad f, g \in S. \\ (af)(x) &= af(x), \quad a \in R,\, x \in M, \quad f \in S. \end{aligned}$$

This module is denoted by $\mathrm{Hom}_R(M, N)$. We want to see how this module changes when either of the two modules M or N is fixed and the other module changes.

Let M be a fixed R-module. Any homomorphism $f : N \to N'$ of R-modules induces a homomorphism

$$f_* : \mathrm{Hom}_R(M, N) \to \mathrm{Hom}_R(M, N'),$$

given by $f_*(\alpha) = f\alpha$, $\alpha \in \mathrm{Hom}_R(M, N)$. Then

$$(gf)_* = g_* f_*, \ g \in \mathrm{Hom}_R(N', N'')$$

and $(\mathrm{Id}_N)_* = \mathrm{Id}$, the identity map of $\mathrm{Hom}_R(N, N)$.

Similarly, for any fixed module N and a homomorphism $g : M \to M'$, there exists an R-module homomorphism

$$g^* : \mathrm{Hom}_R(M', N) \to \mathrm{Hom}_R(M, N)$$

given by $g^*(\beta) = \beta g$, $\beta \in \mathrm{Hom}_R(M'N)$.

Then $(gh)^* = h^* g^*$, $h \in \mathrm{Hom}_R(M', M'')$ and $\mathrm{Id}_M^* = \mathrm{Id}$, the identity map of $\mathrm{Hom}_R(M, N)$.

Proposition 1.2.7 *For any R-module M and an exact sequence*

$$0 \to N' \xrightarrow{f} N \xrightarrow{g} N'' \to 0,$$

the induced sequence

$$0 \to \mathrm{Hom}_R(M, N') \xrightarrow{f_*} \mathrm{Hom}_R(M, N) \xrightarrow{g_*} \mathrm{Hom}_R(M, N'')$$

is exact.

Proof Clearly, f_* is injective, for if

$$f_*(\alpha) = 0, \ \alpha \in \mathrm{Hom}_R(M, N'),$$

then $f\alpha = 0$. As f is injective it is easy to see that the map $\alpha = 0$. Moreover, $g_* f_* = (gf)_* = 0$, as $gf = 0$. Thus, $\mathrm{Im}\, f_* \subset \mathrm{Ker}\, g_*$.

Conversely, let $\beta \in \mathrm{Ker}\, g_*$, so that $g_*(\beta) = g\beta = 0$. For any $x \in M$, $g\beta(x) = 0$, and hence $\beta(x) = f(y)$ for some unique $y \in N'$. This defines a map $\alpha : M \to N$ by seting $\alpha(x) = y$, where $\beta(x) = f(y)$. Let $C = \mathrm{Im} f$. Then f induces an isomorphism from N' to C with inverse $f^{-1} : C \to N'$. As $\mathrm{Im}(B) \subseteq \mathrm{Ker} g = \mathrm{Im} f = C$, we have a composite homomorphism $\alpha = f^{-1}\beta : M \to N'$. Then $f\alpha = \beta$. Hence $\beta \in \mathrm{Im}(f_*)$, showing $\mathrm{Im}(f_*) = \mathrm{Ker}\, g_*$ and the sequence is exact. □

Example 1.2.8 It is not true in general that g_* in Proposition 1.2.7 is surjective. Consider the exact sequence

$$0 \to 2\mathbb{Z} \xrightarrow{f} \mathbb{Z} \xrightarrow{g} \mathbb{Z}/2\mathbb{Z} \to 0$$

and take $M = \mathbb{Z}/2\mathbb{Z}$. Then the map

$$g_* : \mathrm{Hom}_{\mathbb{Z}}(\mathbb{Z}/2\mathbb{Z}, \mathbb{Z}) \to \mathrm{Hom}_{\mathbb{Z}}(\mathbb{Z}/2\mathbb{Z}, \mathbb{Z}/2\mathbb{Z})$$

is not surjective as the identity map $\mathrm{Id}_{\mathbb{Z}/2\mathbb{Z}}$ has no pre-image in view of the fact that $\mathrm{Hom}_{\mathbb{Z}}(\mathbb{Z}/2\mathbb{Z}, \mathbb{Z}) = 0$.

Remark 1.2.9 By a similar argument we can also show that for any R-module N, and an exact sequence of R modules

$$0 \to M' \xrightarrow{f} M \xrightarrow{g} M'' \to 0,$$

the induced sequence

$$0 \to \mathrm{Hom}_R(M'', N) \xrightarrow{g^*} \mathrm{Hom}_R(M, N) \xrightarrow{f^*} \mathrm{Hom}_R(M', N)$$

is exact.

Example 1.2.10 It is not in general true that the map f^* above is surjective. Consider the exact sequence $0 \to \mathbb{Z} \xrightarrow{f} \mathbb{Q} \xrightarrow{g} \mathbb{Q}/\mathbb{Z} \to 0$ and take $N = \mathbb{Z}$. Then $\mathrm{Hom}_{\mathbb{Z}}(\mathbb{Q}, \mathbb{Z}) \xrightarrow{f^*} \mathrm{Hom}_{\mathbb{Z}}(\mathbb{Z}, \mathbb{Z})$ is not surjective as $\mathrm{Hom}_{\mathbb{Z}}(\mathbb{Q}, \mathbb{Z}) = 0$.

Certain types of modules P have the property that for any surjective homomorphism $g : M \to M''$, the induced homomorphism $g_* : \mathrm{Hom}_R(P, M) \to \mathrm{Hom}_R(P, M'')$ is surjective. These are the projective modules defined as follows.

Definition 1.2.11 An R-module M is called *projective* if it is a direct summand of a free R-module.

Proposition 1.2.12 *An R-module P is projective if and only if for any surjective homomorphism $g : M \to M''$, the induced homomorphism*

$$g_* : \mathrm{Hom}_R(P, M) \to \mathrm{Hom}_R(P, M'')$$

is surjective.

Proof Assume first that P is free with basis $\{e_i\}_{i \in I}$.

Given $\alpha \in \mathrm{Hom}_R(P, M'')$, let $\alpha(e_i) = x_i$. Since g is surjective, choose $y_i \in M$ with $g(y_i) = x_i$. Then we have an R-linear map $\beta : P \to M$ defined by $\beta(e_i) = y_i$, $i \in I$, i.e., the map given by $\beta(\Sigma a_i e_i) = \Sigma a_i y_i$. Clearly, $g_*(\beta) = \alpha$. This shows that g_* is surjective.

Now assume that P is projective. Then there exists an R-module Q with $P \oplus Q = F$, free. Let $\pi : F \to P$ be the projection map. Given $\alpha \in \mathrm{Hom}_R(P, M'')$, consider $\alpha\pi \in \mathrm{Hom}_R(F, M'')$. There exists $\beta \in \mathrm{Hom}_R(F, M)$ with $g_*(\beta) = \alpha\pi$. Then $\beta_1 = \beta|_P : P \to M$ has the property that $g_*(\beta_1) = \alpha$. Hence g_* is surjective.

Conversely, assume that P has the property stated in the proposition. Express P as a quotient of a free module F. Hence there is a surjective map $g : F \to P$. Then the induced map $g_* : \mathrm{Hom}_R(P, F) \to \mathrm{Hom}_R(P, P)$ is surjective. In particular, there exists $\beta \in \mathrm{Hom}_R(P, F)$ with $g_*(\beta) = g\beta = \mathrm{Id}_P$. This shows that the exact sequence

$$0 \to \mathrm{Ker}\, g \to F \xrightarrow{g} P \to 0$$

splits. By Proposition 1.2.5, P is isomorphic to a direct summand of F. Hence P is projective.

□

Example 1.2.13 (i) Any free R-module is projective.

(ii) If R is a principal ideal domain, then any projective R-module is free, as any submodule of a free R-module is free. (For reference, see the book *Algebra* by Serge Lang.) In particular, any projective module over $K[X]$, K a field, is free.

The following is an example of a non-free projective module.

Example 1.2.14 Let $R = \mathbb{Z}/6\mathbb{Z}$, $P = \{\bar{0}, \bar{2}, \bar{4}\}$ and $Q = \{\bar{0}, \bar{3}\}$, ideals of R. Then $R = P \oplus Q$, so that P is R-projective. Now, P is not R-free as any free module of rank n over $R = \mathbb{Z}/6\mathbb{Z}$ is isomorphic to R^n and has 6^n elements.

Proposition 1.2.15 *A direct sum $P = \oplus \sum_{\alpha} P_\alpha$ is projective $\Leftrightarrow P_\alpha$ is projective for each α.*

Proof: If P is projective, then it is a direct summand of a free module F. Since P_α is a direct summand of P, it is also a direct summand of F and hence projective.
Conversely, assume that each P_α is projective and consider the diagram

$$\begin{array}{ccccc} & & P & & \\ & & \downarrow{\scriptstyle \phi} & & \\ M & \xrightarrow{g} & M'' & \longrightarrow & 0 \end{array}$$

Let $\phi_\alpha = \phi i_\alpha$, where $i_\alpha : P_\alpha \to P$ is the inclusion. Since P_α is projective, there exists an R-linear map $\psi_\alpha : P_\alpha \to M$ such that $g\psi_\alpha = \phi_\alpha$. Define $\psi : P \to M$ by setting

$$\psi(x) = \Sigma\psi_{\alpha_i}(x_{\alpha_i}) \text{ if } x = \sum_i x_{\alpha_i}.$$

Then ψ is R-linear and $g\psi = \phi$, showing that P is projective. □

Proposition 1.2.16 *An R-module P is projective if and only if every exact sequence $0 \to M' \xrightarrow{f} M \xrightarrow{g} P \to 0$ splits.*

Proof If P is projective, then by Proposition 1.2.12 the identity map $\mathrm{Id}_P : P \to P$ can be lifted to a map $\phi : P \to M$ such that $g\phi = \mathrm{Id}_P$, i.e., the exact sequence $0 \to M' \to M \to P \to 0$ splits.

To prove the converse, express P as the quotient of a free module F with kernel K, so that the sequence $0 \to K \to F \to P \to 0$ is exact. By assumption, this sequence splits, so that P is isomorphic to a direct summand of F and hence projective. □

Corollary 1.2.17 *P is a finitely generated projective module $\Leftrightarrow$ P is a direct summand of a free module of finite rank.*

Definition 1.2.18 An R-module M is said to be *finitely presented* if there exists an exact sequence $0 \to N \to F \to M \to 0$ with F free and F, N finitely generated.

Proposition 1.2.19 (Shanuel's Lemma) *Let $0 \to N \to P \to M \to 0$ and $0 \to N' \to P' \to M \to 0$ be two exact sequences of R-modules with P, P' projective. Then $P \oplus N' \simeq P' \oplus N$.*

Proof: Consider the diagram

$$\begin{array}{ccccccccc} 0 & \longrightarrow & N & \xrightarrow{f} & P & \xrightarrow{g} & M & \longrightarrow & 0 \\ & & & & & & \Big\| \text{Id} & & \\ 0 & \longrightarrow & N' & \xrightarrow{f'} & P' & \xrightarrow{g'} & M & \longrightarrow & 0 \end{array}$$

Since P is projective, there exists an R-linear map $\theta : P \to P'$ such that $g'\theta = g$. Since $g'\theta f = gf = 0$, $\text{Im}(\theta f) \subseteq \text{Ker} g' = \text{Im} f'$, and f' is an isomorphism from N' to $\text{Im} f'$; hence θ induces an R-linear map $\theta' : N \to N'$ defined by $f'^{-1}\theta f$. Define $\psi : P \oplus N' \to P'$ by $\psi(x, y) = \theta(x) - f'(y)$, $x \in P$, $y \in N'$ and $\phi : N \to P \oplus N'$ by $\phi(y) = (f(y), \theta'(y))$. Then the sequence

$$0 \to N \xrightarrow{\phi} P \oplus N' \xrightarrow{\psi} P' \to 0 \tag{1.1}$$

can be seen to be exact. Since P' is projective, by Proposition 1.2.16, the sequence splits. Hence $P \oplus N' \simeq P' \oplus N$ by Proposition 1.2.5. □

Corollary 1.2.20 *Let M be finitely presented and $0 \to K' \to F' \to M \to 0$ be any exact sequence with F' free of finite rank. Then K' is finitely generated.*

Proof Since M is finitely presented, there exists an exact sequence $0 \to K \to F \to M \to 0$ with F free and F, K finitely generated. By Proposition 1.2.19,

$$K \oplus F' \simeq K' \oplus F.$$

Since $K \oplus F'$ is finitely generated, so is $K' \oplus F$. Since K' is its quotient, K' is finitely generated. □

1.2. EXERCISES

1. Verify that the sequence (1.1) in Proposition 1.2.19 is exact.

2. Let $\phi : M \to F$ be a surjective homomorphism of a finitely generated module M onto a free module F. Show that $\text{Ker}\,(\phi)$ is finitely generated.

3. Let n be a positive integer with $n = d_1 d_2$ and $M = \mathbb{Z}/n\mathbb{Z}$. Consider the sequence given by

$$0 \to d_2 M \xrightarrow{\beta} M \xrightarrow{\alpha} d_1 M \to 0,$$

where β is the inclusion map and α is the multiplication by d_1. Show that the sequence is exact. Show further that the sequence splits if and only if d_1 and d_2 are relatively prime.

4. For any R-module M, show that

   (i) $\mathrm{Hom}_R(R, M) \simeq M$,

   (ii) $\mathrm{Hom}_R(\oplus\Sigma_{i=1}^n M_i, N) \simeq \oplus\Sigma_{i=1}^n \mathrm{Hom}_R(M_i, N)$,

   (iii) $\mathrm{Hom}_R(M, \oplus\Sigma_{i=1}^n N_i) \simeq \oplus\Sigma_{i=1}^n \mathrm{Hom}_R(M, N_i)$.

5. Verify that $\mathrm{Hom}_{\mathbb{Z}}(\mathbb{Q}, \mathbb{Z}) = 0$.

6. Show that if P is a projective module, then there exists a free module F with $P \oplus F$ free.

7. An element $e \in R$ is called an *idempotent* if $e^2 = e$. Show that if every projective R-module is free, then the only idempotents of R are 0 and 1.

8. Let M be an R-module. An element $x \in M$ with $\mathrm{Ann}(x) = 0$ is called *unimodular* if the cyclic submodule $Rx (\cong R)$ is a direct summand of M. Show that $x \in M$ is unimodular if and only if there exists an R-homomorphism $\phi : M \to R$ with $\phi(x) = 1$. In particular, if M is free with basis $\{e_i\}_{1 \le i \le n}$, then $x = \Sigma a_i e_i$ is unimodular if and only if $(a_1, \ldots, a_n)$ is a unimodular row.

9. A finitely generated projective R-module P is said to be *stably free* if $P \oplus R^m \simeq R^n$ for some m, n. Show that every stably free R-module is free $\Leftrightarrow$ every unimodular row over R can be completed to a non-singular matrix.

   [D. Quillen and A. A. Suslin proved that if $R = k[X_1, \ldots, X_n]$, k a field, then every finitely generated projective R-module is free. See the book *Serre's Conjecture* by T. Y. Lam; Lecture Notes in Mathematics, No. 635, Springer Verlag (1978), or *Serre's Problem for Projective Modules* by T. Y. Lam; Springer Monographs in Mathematics, Springer (2006).]

10. Let M be a finitely presented R-module and $0 \to N' \to N \to M \to 0$ be an exact sequence with N a finitely generated R-module. Prove that N' is also a finitely generated R-module.

1.3 Tensor products

Let M, N, K be R-modules. We recall that a map $\theta : M \times N \to K$ is *R-bilinear* if $\theta(x, y)$ is R-linear in x for each fixed $y \in N$ and R-linear in y for each fixed $x \in M$.

The study of bilinear maps from $M \times N$ to K can be reduced to the study of linear maps on a suitably defined module (tensor product of M and N) with values in K.

Definition 1.3.1 Let M and N be R-modules. The *tensor product* of M with N is a pair (T, θ), where T is an R-module and $\theta : M \times N \to T$ an R-bilinear map with the property that for any R-module K and R-bilinear map $f : M \times N \to K$, there exists a unique R-linear map $\widetilde{f} : T \to K$ with $\widetilde{f}\theta = f$.

Theorem 1.3.2 *The tensor product of M and N exists and it is unique up to isomorphism.*

Proof We first prove the uniqueness. Suppose (T, θ) and (T', θ') satisfy the conditions for the tensor product. Consider the diagrams

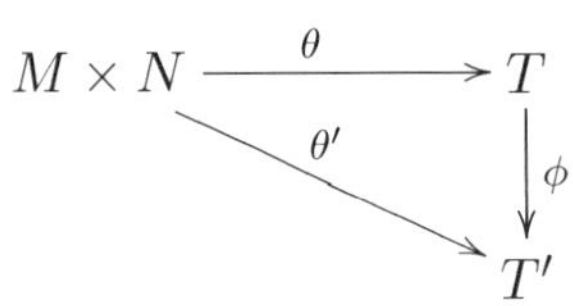

and

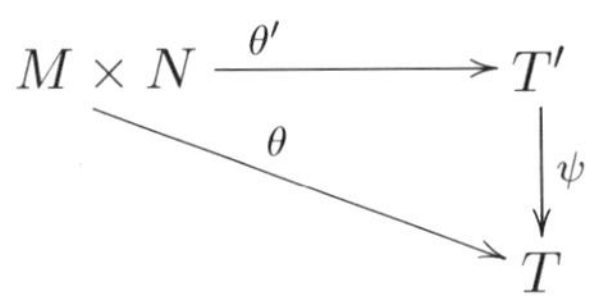

By definition of tensor product, there exist unique R-linear maps $\phi : T \to T'$ and $\psi : T' \to T$ such that $\phi\theta = \theta'$ and $\psi\theta' = \theta$. Consider now the diagrams

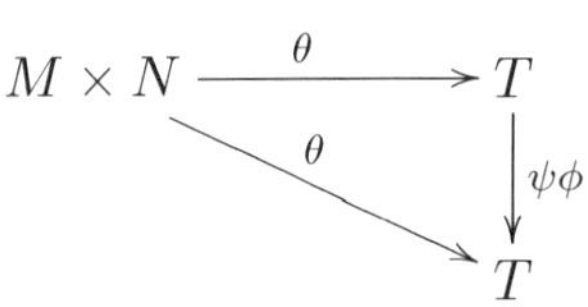

and

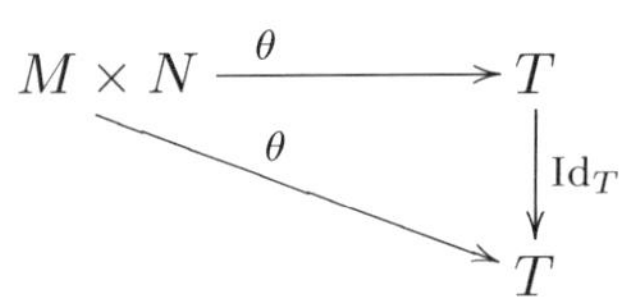

Since $(\psi\phi)\theta = \psi(\phi\theta) = \psi\theta' = \theta$, and $\mathrm{Id}_T\theta = \theta$, we have by uniqueness, $\psi\phi = \mathrm{Id}_T$. Similarly, $\phi\psi = \mathrm{Id}_{T'}$. Hence $T \simeq T'$. We now prove the existence of the tensor product.

Let F be the free R-module on the set $M \times N$, i.e., all formal finite R-linear combinations of the type $\Sigma_i a_i(x_i, y_i)$, $a_i \in R$, $x_i \in M$ and $y_i \in N$, with formal addition and natural scalar multiplication.

Let G be the submodule of F generated by elements of the type

$$(x + x', y) - (x, y) - (x', y), \quad x, x' \in M, \quad y \in N,$$

$$(x, y + y') - (x, y) - (x, y'), \quad x \in M, \quad y, y' \in N,$$

$$(ax, y) - a(x, y), x, y \in M, \quad a \in R,$$

$$(x, ay) - a(x, y), x, y \in M, \quad a \in R.$$

Let $T = F/G$ and $\theta : M \times N \to T$ given by $\theta(x, y) = (x, y) + G$. Then we show that (T, θ) satisfies the conditions of tensor product. Clearly, θ is R-bilinear. Consider now any

R-module K and any R-bilinear map $f : M \times N \to K$. Then f can be extended to an R-linear map $f_1 : F \to K$ by defining

$$f_1 \left[\Sigma_i a_i (x_i, y_i)\right] = \Sigma_i a_i f(x_i, y_i).$$

Then f_1 vanishes on G by the definition of G and the bilinearity of f. It induces therefore an R-linear map $\widetilde{f} : F/G = T \to K$, which satisfies $\widetilde{f}\theta = f$. Hence (T, θ) is the required tensor product. □

Notation. 1.3.3 The tensor product F/G will be denoted by $M \otimes_R N$ or $M \otimes N$ if there is no confusion about the base ring R. The image of the element $(x, y) \in F$ under the natural projection $F \to F/G$ will be denoted by $x \otimes y$.

Proposition 1.3.4

(i) $x \otimes y$ *is* R*-linear in* x *as well as* y.

(ii) $x \otimes y = 0$ *if either* $x = 0$ *or* $y = 0$.

(iii) $M \otimes N$ *is generated by elements of the type*

$$x \otimes y,\ x \in M,\ y \in N.$$

Proof (i) is clear from the definition.

(ii) $(x + x') \otimes y = x \otimes y + x' \otimes y$. In particular, let $x = x' = 0$ and $t = 0 \otimes y$. Then $t = t + t$, so that $t = 0$. Similarly, $x \otimes 0 = 0$.

(iii) If $z \in M \otimes N$, then $z = x + G$, where $x = \Sigma_i a_i (x_i, y_i)$. Hence $z = \Sigma_i a_i (x_i, y_i) + G = \Sigma_i a_i (x_i \otimes y_i)$. □

Corollary 1.3.5 *If* M *and* N *are finitely generated R-modules, then so is* $M \otimes N$.

Proof Let M be generated by $\{x_1, \ldots, x_n\}$ and N by $\{y_1, \ldots, y_m\}$. If $u \otimes v \in M \otimes N$, write $u = \Sigma_i a_i x_i$ and $v = \Sigma_j b_j y_j, a_i, b_j \in R$. Then $u \otimes v = (\Sigma_i a_i x_i) \otimes (\Sigma_j b_j y_j) = \Sigma_{i,j}\, a_i b_j (x_i \otimes y_j)$. Thus, $\{x_i \otimes y_j\}$ generate $M \otimes N$. □

Example 1.3.6 $x \otimes y$ can be zero without either $x = 0$ or $y = 0$. Let $R = \mathbb{Z}$, $M = \mathbb{Z}$ and $N = \mathbb{Z}/2\mathbb{Z}$. Then

$$2 \otimes \overline{1} = 2 \cdot (1 \otimes \overline{1}) = 1 \otimes 2.\overline{1} = 1 \otimes \overline{0} = 0.$$

Proposition 1.3.7 *Let* M, N *and* P *be* R*-modules. Then*

(i) $(M \otimes_R N) \otimes_R P \simeq M \otimes_R (N \otimes_R P)$.

(ii) $M \otimes_R N \simeq N \otimes_R M$.

(iii) $M \otimes_R R \simeq M \simeq R \otimes_R M$.

Proof
(i) The map $(M \otimes_R N) \times P \to M \otimes_R (N \otimes_R P)$ given by

$$(x \otimes y, z) \to x \otimes (y \otimes z).$$

is a well defined R-bilinear map on $(M \otimes_R N) \times P$, which can be extended to an R-linear map $(M \otimes_R N) \otimes_R P \to M \otimes_R (N \otimes_R P)$ given by $(x \otimes y) \otimes z \mapsto x \otimes (y \otimes z)$. This is an isomorphism with inverse given by the map $x \otimes (y \otimes z) \mapsto (x \otimes y) \otimes z$ defined similarly.

(ii) The map $x \otimes y \mapsto y \otimes x$ is well defined and is an isomorphism with the inverse given by $y \otimes x \mapsto x \otimes y$.

(iii) Define R-linear maps $\phi : R \otimes_R M \to M$ and $\psi : M \to R \otimes_R M$ by $\phi(a \otimes x) = ax$ and $\psi(x) = 1 \otimes x$. These are well defined and $\phi\psi = \mathrm{Id}_M$, $\psi\phi = \mathrm{Id}_{R \otimes_R M}$. Hence $M \simeq R \otimes_R M$. Similarly, we can show that $M \simeq M \otimes_R R$. □

Remark 1.3.8 In view of the associativity property (i), the tensor product of a finite number of modules is well defined.

Proposition 1.3.9 *Let $M = \oplus\Sigma_{i=1}^n M_i$ and $N = \oplus\Sigma_{j=1}^m N_j$. Then*

$$M \otimes N \simeq \oplus\Sigma_{i,j}(M_i \otimes N_j).$$

Proof The map $M \times N \to \oplus\Sigma_{i,j}(M_i \otimes N_j)$ given by $(x, y) \mapsto \Sigma_{i,j}(x_i \otimes y_j)$, where $x = \Sigma_i x_i$ and $y = \Sigma_j y_j$, is well defined and extends to an R-linear map $\theta : M \otimes N \to \oplus\Sigma_{i,j}(M_i \otimes N_j)$. The inclusion map $M_i \to M$ and $N_j \to N$, induce R-linear maps $M_i \otimes N_j \to M \otimes N$, for each i and j. These in turn induce an R-linear map $\phi : \oplus\Sigma_{i,j}(M_i \otimes N_j) \to M \otimes N$. Clearly, $\phi\theta = \mathrm{Id}$ and $\theta\phi = \mathrm{Id}$, showing θ is an isomorphism. □

Corollary 1.3.10 *Let M and N be free modules with bases*

$$\{x_1, \ldots, x_n\} \text{ and } \{y_1, \ldots, y_m\}$$

respectively. Then $M \otimes N$ is free with basis $\{x_i \otimes y_j\}$.

Proof By the definition of a free module, $M = \oplus\,\Sigma_i^n Rx_i$ and $N = \oplus\,\Sigma_{j=1}^m Ry_j$, where $Rx_i \simeq R \simeq Ry_j$ for all i and j. By Proposition 1.3.9, $M \otimes N \simeq \oplus\Sigma_{i,j}(Rx_i \otimes Ry_j)$. But $Rx_i \otimes Ry_j \simeq R \otimes_R R \simeq R$ and $Rx_i \otimes Ry_j \simeq R(x_i \otimes y_j)$ under the map $ax_i \otimes by_j \mapsto ab(x_i \otimes y_j)$. Hence

$$M \otimes N \simeq \oplus\Sigma_{i,j}R(x_i \otimes y_j),$$

i.e., it is free with basis $\{x_i \otimes y_j\}$. □

Corollary 1.3.11 *The tensor product of free modules of ranks m and n is free, of rank mn.*

Let $f : R \to S$ be a homomorphism of rings. Any S-module M can be considered as an R-module through the mapping f by defining $ax = f(a)x$, $a \in R$, $x \in M$. Such an R-module is called the module obtained by restricting the scalars through f. In particular, the S-module S can be considered as an R-module through f.

Proposition 1.3.12 *If M is a finitely generated S-module and S is a finitely generated R-module, then M is a finitely generated R-module.*

Proof Let $\{x_1, \ldots, x_n\}$ be a generating set of M over S and $\{s_1, \ldots, s_m\}$ a generating set of S over R. For any $x \in M$, we have $x = \Sigma_i a_i x_i$, $a_i \in S$, and $a_i = \Sigma_j r_{ij} s_j$, $r_{ij} \in R$. Hence $x = \Sigma_i(\Sigma_j r_{ij} s_j)x_i = \Sigma_{i,j} r_{ij}(s_j x_i)$. Thus, the set $\{s_j x_i\}_{i,j}$ is a generating set of M over R. □

Definition 1.3.13 For any R-module M, the module $S \otimes_R M$ is an S-module for the operation defined by $s(s_1 \otimes x) = ss_1 \otimes x$, $s, s_1 \in S$, $x \in M$. (See Exercises.) This S-module is called the *module obtained by extension of scalars*.

Proposition 1.3.14 *If M is a finitely generated R-module, then its scalar extension $S \otimes_R M$ is also a finitely generated S-module.*

Proof If $\{x_i\}$ $(1 \leq i \leq n)$ generate M over R, then $\{1 \otimes x_i\}$ generate $S \otimes_R M$ over S. □

Definition 1.3.15 An *R-algebra* A is a ring A, which is also an R-module satisfying the condition

$$a(xy) = (ax)y = x(ay),\ a \in R,\ x, y \in A.$$

Example 1.3.16 (i) $A = R[X_1, X_2, \ldots, X_n]$, the ring of polynomials in $X_1, X_2, \ldots, X_n$ over R.
(ii) Any extension ring $S \supset R$ is an R-algebra.
(iii) $A = R[[X]]$, the ring of all formal power series in X over R, i.e., $A = \{\Sigma_{i=0}^{\infty} a_i X^i \mid a_i \in R\}$, with the usual addition and multiplication of power series.

Definition 1.3.17 A *homomorphism* $f : A \to B$ of R-algebras is an R-module homomorphism which satisfies the condition $f(xy) = f(x)f(y)$ for $x, y \in A$ and $f(1) = 1$.

Definition 1.3.18 If A and B are R-algebras, the tensor product module $A \otimes_R B$ has a well defined multiplication (see Exercises) defined by

$$(a \otimes b)(a' \otimes b') = aa' \otimes bb',\ a, a' \in A,\ b, b' \in B.$$

This makes $A \otimes_R B$ into an R-algebra, called the *tensor product* of the algebras A and B.

Definition 1.3.19 An R-algebra A is said to be *finitely generated* if there exists a finite set $\{a_1, \ldots, a_n\}$ of elements of A such that the natural map $R[X_1, X_2, \ldots, X_n] \to A$ given by $f(X_1, \ldots, X_n) \mapsto f(a_1, \ldots, a_n)$ is surjective.

Example 1.3.20 (i) $A = \frac{R[X_1,\ldots,X_n]}{I}$, quotient of a polynomial algebra, is finitely generated, generated by $\{a_1, a_2, \ldots, a_n\}$, where

$$a_i = X_i + I \quad (1 \leq i \leq n).$$

(ii) The $\mathbb{Z}$-algebra $\mathbb{Q}$ is not finitely generated. (Verify.)

Proposition 1.3.21 Let A and B be finitely generated R-algebras. Then $A \otimes_R B$ is also a finitely generated R-algebra.

Proof Let A be generated by $\{a_1, \ldots, a_n\}$ and $B = \{b_1, \ldots, b_m\}$. Then there exist surjective algebra homomorphisms

$$R[X_1, \ldots, X_n] \xrightarrow{f} A \text{ and } R[Y_1, \ldots, Y_m] \xrightarrow{g} B.$$

The induced map

$$f \otimes g : R[X_1, \ldots, X_n] \otimes_R R[Y_1, \ldots, Y_m] \to A \otimes_R B$$

is also surjective. But

$$R[X_1, \ldots, X_n] \otimes R[Y_1, \ldots, Y_m] \simeq R[X_1, \ldots, X_n, Y_1, \ldots, Y_m].$$

(Exercise 3, Section 1.3)

Hence $A \otimes_R B$ is finitely generated. □

Remark 1.3.22 (i) If an R-algebra A is finitely generated as an R-module, then it is clearly a finitely generated R-algebra. But the converse is not true as is shown by $A = R[X]$.

(ii) If $f : R \to S$ is a ring homomorphism and A is an R-algebra, then its scalar extension $S \otimes_R A$ is an S-algebra. It will be a finitely generated S-algebra if A is a finitely generated R-algebra.

1.3. EXERCISES

1. Prove rigorously that

   (a) The map $(M \otimes N) \times P \to M \otimes (N \otimes P)$ defined in the proof of Proposition 1.3.7(i) is well defined. (**Hint:** For each $z \in P$, define an R-bilinear map $B_z : M \times N \to M \otimes_R (N \otimes_R P)$ by $(x, y) \mapsto x \otimes (y \otimes z)$ and hence an R-linear map $f_z : M \otimes_R N \to M \otimes_R (N \otimes_R P)$. Now define the R-bilinear map $(M \otimes_R N) \times P \to M \otimes_R (N \otimes_R P)$ by $(u, z) \mapsto f_z(u)$.)

   (b) The maps in Proposition 1.3.7 (ii) and (iii) are all well defined.

   (c) The action $S \times (S \otimes_R M) \to S \otimes_R M$ given by $(s, s_1 \otimes x) \mapsto ss_1 \otimes x$, in Definition 1.3.13 is well defined.

   (d) The action $(A \otimes_R B) \times (A \otimes_R B) \to A \otimes_R B$, given by $(a \otimes b, c \otimes d) \mapsto ac \otimes bd$ in Definition 1.3.18, is well defined.

   (e) The action $S \times (S \otimes_R A) \to S \otimes_R A$ in Remark 1.3.22, is well defined.

2. Let M be an R-module, A an R-algebra and N, P be A-modules. Prove that

$$(M \otimes_R N) \otimes_A P \cong M \otimes_R (N \otimes_A P)$$

   as A-modules.

3. Show that for any commutative ring R,

$$R[X] \otimes_R R[Y] \simeq R[X, Y].$$

4. Show that $\mathbb{Z}/m\mathbb{Z} \otimes_{\mathbb{Z}} \mathbb{Z}/n\mathbb{Z} = 0$ if and only if m and n are relatively prime.

5. If M and N are R-modules and I an ideal of R, then show that

   (a) $R/I \otimes_R M \simeq M/IM$,

   (b) $M/IM \otimes_{R/I} N/IN \simeq \frac{(M \otimes_R N)}{I(M \otimes_R N)}$.

6. Let $f : R \to S$ be a ring homomorphism, M an S-module and M' the R-module M through f. If $S \otimes_R M'$ is the scalar extension of the R-module M', then show that the map $g : M \to S \otimes_R M'$, given by $g(x) = 1 \otimes x$, is an R-linear isomorphism.

7. If A is a finitely generated R-algebra and R is a finitely generated K-algebra, then show that A is a finitely generated K-algebra.

8. Let M, N and K be R-modules. Define a natural map

$$\phi : \mathrm{Hom}_R(M, \mathrm{Hom}_R(N, K)) \to \mathrm{Hom}_R(N \otimes_R M, K)$$

by setting $\phi(f)(x \otimes y) = f(y)(x)$, $x \in N$, $y \in M$. Show that ϕ is an isomorphism.

9. **(Direct product)** For any collection $\{M_\alpha\}$ of R-modules, the product $\Pi_\alpha M_\alpha$ is an R-module for componentwise addition and scalar multiplication. Show by an example that $\Pi_\alpha(M_\alpha) \otimes_R N$ need not be isomorphic to $\Pi_\alpha(M_\alpha \otimes_R N)$.

10. **(Symmetric algebra)** Let M be an R-module. The *tensor algebra* $T(M)$ of M is defined to be the R-algebra

$$T(M) = \oplus\Sigma_{i=0}^{\infty} T_i(M),$$

where $T_i(M) = M \otimes_R M \otimes_R \cdots \otimes_R M$ (i times), $T_0(M) = R$. Define multiplication in $T(M)$ by

$$(x_1 \otimes x_2 \otimes \cdots \otimes x_r)(y_1 \otimes y_2 \otimes \cdots \otimes y_s) = x_1 \otimes x_2 \otimes \cdots \otimes x_r \otimes y_1 \otimes \cdots \otimes y_s, \quad x_i,\ y_j \in M$$

and extended by distributivity.

The *symmetric algebra* $S(M)$ is defined to be the quotient of $T(M)$ by the two-sided ideal generated by elements of the type $\{x \otimes y - y \otimes x\}$, $x,\ y \in M$. Prove that if M is a free R-module with base $\{e_i\}_{i \in I}$, then $S(M)$ is isomorphic to the polynomial algebra in $\{X_i\}_{i \in I}$.

11. Show that for R-modules M and N, $S(M \oplus N) \simeq S(M) \otimes_R S(N)$. Deduce that if P is a projective R-module, then there exists an R-algebra T such that $S(P) \otimes_R T$ is a polynomial algebra.

12. **(Exterior algebra)** Let M be an R-module and $T(M)$ the tensor algebra of M. The quotient of $T(M)$ by the two-sided ideal generated by elements of the type $\{x \otimes x\}$, $x \in M$ is called the *exterior algebra* $\wedge(M)$.

    The canonical image of $x_1 \otimes x_2 \otimes \cdots \otimes x_r$, under the natural projection, is denoted by $x_1 \wedge x_2 \cdots \wedge x_r$ (called *elements of homogeneous components* of degree r). Let $\wedge^r(M)$ denote the r-th homogeneous component of $\wedge(M)$. Prove:

    (a) If M is a free module with basis $\{e_1, \ldots, e_n\}$, then $\wedge(M)$ is free with basis

    $$\{e_{i_1} \wedge e_{i_2} \wedge \cdots \wedge e_{i_r}\}$$

    for $1 \leq i_1 < i_2 < \cdots < i_r \leq n$.

    (b) $\wedge^n(M \oplus N) \simeq \oplus\Sigma_{i+j=n} \wedge^i (M) \otimes_R \wedge^j(N)$.

    (c) If M is a projective R-module, then $\wedge(M)$ is also a projective R-module.

1.4 Flat modules

We now investigate the behaviour of tensor product with respect to exact sequences.

Theorem 1.4.1 *Let $0 \to N' \xrightarrow{f} N \xrightarrow{g} N'' \to 0$ be an exact sequence of R-modules. Then for any R-module M, the sequence*

$$M \otimes N' \xrightarrow{f_*} M \otimes N \xrightarrow{g_*} M \otimes N'' \to 0$$

is exact, where f_* and g_* are defined by $f_*(x \otimes y) = x \otimes f(y)$ and $g_*(x \otimes z) = x \otimes g(z)$.

Proof Since g is surjective, g_* is also surjective. Moreover, $g_* f_* = 0$ as $gf = 0$ so that $\operatorname{Im} f_* \subset \operatorname{Ker} g_*$. To prove $\operatorname{Im} f_* = \operatorname{Ker} g_*$, it is enough to show that the natural map

$$\theta : \frac{M \otimes N}{\operatorname{Im} f_*} \to \frac{M \otimes N}{\operatorname{Ker} g_*}$$

given by

$$x \otimes y + \operatorname{Im} f_* \to x \otimes y + \operatorname{Ker} g_*$$

is an isomorphism. Consider the diagram

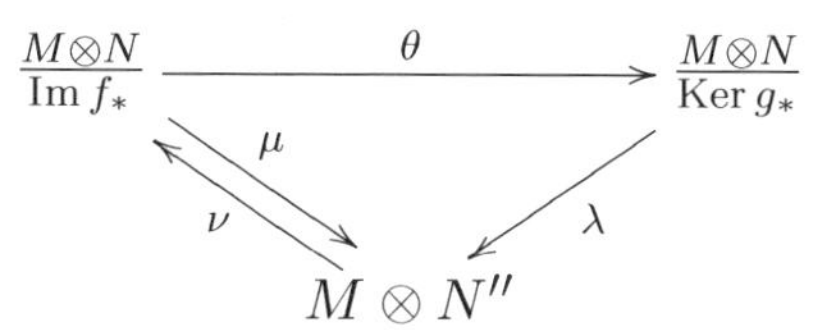

where λ is the isomorphism induced by $g_* : M \otimes N \to M \otimes N''$ and $\mu = \lambda\theta$. We show that θ is an isomorphism. It is sufficient to show that μ is an isomorphism. We construct an inverse ν for μ. Let $B : M \times N \to (M \otimes N)/\operatorname{Im} f_*$ be the R-linear map defined by $B(x, y) = \overline{x \otimes z}$, where $z \in N$ is chosen such that $g(z) = y$. Clearly, B is well defined, for if $g(z_1) = y$ for $z_1 \in N$, then $z - z_1 \in \operatorname{Ker} g = \operatorname{Im} f$, so that $z - z_1 = f(t)$, $t \in N'$ and $x \otimes z - x \otimes z_1 = x \otimes f(t) = f_*(x \otimes t) \in \operatorname{Im} f_*$. Then B induces a R-linear map $\nu : M \otimes N'' \to \frac{M \otimes N}{\operatorname{Im} f_*}$ such that $\nu(x \otimes y) = \overline{x \otimes z}$, $x \in M$, $y \in N''$.
It is easy to verify that μ and ν are inverses of each other. □

Example 1.4.2 It is not in general true that

$$0 \to M \otimes N' \xrightarrow{f_*} M \otimes N$$

is exact. Let $N' = N = \mathbb{Z}$ and $f : \mathbb{Z} \to \mathbb{Z}$ be given by $f(x) = 2x$. Clearly, f is injective. If $M = \mathbb{Z}/2\mathbb{Z}$, then $f_* : \mathbb{Z}/2\mathbb{Z} \otimes_{\mathbb{Z}} \mathbb{Z} \to \mathbb{Z}/2\mathbb{Z} \otimes_{\mathbb{Z}} \mathbb{Z}$ is the zero map, since for any $x \in \mathbb{Z}$, $f_*(\bar{1} \otimes x) = \bar{1} \otimes 2x = 2(\bar{1} \otimes x) = \bar{0} \otimes x = 0$. But $\mathbb{Z}/2\mathbb{Z} \otimes_{\mathbb{Z}} \mathbb{Z} \simeq \mathbb{Z}/2\mathbb{Z} \neq 0$, which shows that f_* is not injective.

Proposition 1.4.3 *For any R-module M, the following conditions are equivalent.*

(i) *For any exact sequence of R-modules*

$$0 \to N' \xrightarrow{f} N \xrightarrow{g} N'' \to 0,$$

the sequence

$$0 \to M \otimes N' \xrightarrow{f_*} M \otimes N \xrightarrow{g_*} M \otimes N'' \to 0$$

is exact.

(ii) *If $N' \xrightarrow{f} N$ is injective, then $f^* : M \otimes N' \to M \otimes N$ is injective.*

(iii) *If $N' \xrightarrow{f} N$ is injective with N and N' finitely generated, then $f^* : M \otimes N' \to M \otimes N$ is injective.*

Proof Theorem 1.4.1 shows the equivalence of (i) and (ii). Clearly, $(ii) \Rightarrow (iii)$ and it is sufficient to show that $(iii) \Rightarrow (ii)$.

Let $f : N' \to N$ be injective, and consider $f_* : M \otimes N' \to M \otimes N$. Let $\Sigma_i(x_i \otimes y_i) \in$ Ker f_* so that $f_*(\Sigma_i x_i \otimes y_i) = \Sigma_i x_i \otimes f(y_i) = 0$. Let N_1' be the submodule of N' generated by the y_i's and let $t = \Sigma_i x_i \otimes y_i \in M \otimes N_1'$. Since $\Sigma_i x_i \otimes f(y_i) = 0$ in $M \otimes N$, there exists a finitely generated submodule N_1 of N containing $f(N_1')$ such that $\Sigma_i x_i \otimes f(y_i) \in M \otimes N_1$ is zero. The restriction of f to N_1' gives an injection $f_1 : N_1' \to N_1$ such that $(f_1)_*(t) = 0$. By (iii), this implies that $t = 0$ in $M \otimes N_1{}'$ and hence $\Sigma x_i \otimes y_i = 0$ in $M \otimes N'$. Hence f^* is injective. □

Definition 1.4.4 An R-module M is said to be *flat* if for every exact sequence of R-modules $0 \to N' \xrightarrow{f} N \xrightarrow{g} N'' \to 0$, the induced sequence $0 \to M \otimes N' \xrightarrow{f_*} M \otimes N \xrightarrow{g_*} M \otimes N'' \to 0$ is exact.

Remark 1.4.5 (i) In view of the above proposition, an R-module M is flat if and only if for every injective map $f : N' \to N$ the map $f_* : M \otimes N' \to M \otimes N$ is injective. Moreover, we may restrict both N and N' to be finitely generated.
(ii) In view of the isomorphism $M \otimes_R N \simeq N \otimes_R M$, an R-module M is flat if and only if for every injective homomorphism $f_1 : N' \to N$, the induced map $(f_1)_* : N' \otimes_R M \to N \otimes_R M$ is injective.

Example 1.4.6 (i) $M = R^n$ is flat. Clearly, R is R-flat as $M \otimes_R R \simeq M$. By Proposition 1.3.9, tensor product commutes with direct sums. Hence $M = R^n$ is flat.

(ii) $\mathbb{Z}/2\mathbb{Z}$ is not flat over $\mathbb{Z}$ by the Example 1.4.2.

Proposition 1.4.7 *Let $f : R \to S$ be a homomorphism of rings.*

(i) If M is R-flat, then the scalar extension $S \otimes_R M$ is S-flat.

(ii) If M is S-flat, and S as an R-module through f is R-flat, then M is R-flat.

Proof (i) Let $0 \to N' \xrightarrow{f} N$ be an exact sequence of S-modules. Since $N' \otimes_S S \otimes_R M \simeq N' \otimes_R M$ and since $0 \to N' \otimes_R M \to N \otimes_R M$ is exact, M being R-flat, it follows that $S \otimes_R M$ is S-flat.

(ii) Let $0 \to N' \to N$ be an exact sequence of R-modules. Since S is R-flat, the sequence $0 \to S \otimes_R N' \to S \otimes_R N$ is exact. Since M is S-flat, the sequence $0 \to M \otimes_S S \otimes N' \to M \otimes_S S \otimes N$ is exact, i.e., $0 \to M \otimes N' \to M \otimes N$ is exact. Hence M is R-flat. □

Theorem 1.4.8 *For an R-module M, the following conditions are equivalent.*
(i) A sequence $0 \to N' \xrightarrow{f} N \xrightarrow{g} N'' \to 0$ of R-modules is exact if and only if the tensored sequence

$$0 \to M \otimes N' \xrightarrow{f_*} M \otimes N \xrightarrow{g^*} M \otimes N'' \to 0 \text{ is exact.}$$

(ii) M is R-flat and for any R-module N, $M \otimes N = 0$ implies $N = 0$.

(iii) M is R-flat and for any R-homomorphism $f : N' \to N$, the induced map $f_* : M \otimes N' \to M \otimes N$ is zero implies that $f = 0$.

Proof $(i) \Rightarrow (ii)$. Clearly, (i) implies that M is flat. If $M \otimes N = 0$, consider the sequence $0 \to N \to 0$. The tensored sequence $0 \to M \otimes N \to 0$ is exact, so that $0 \to N \to 0$ is exact, i.e., $N = 0$.

$(ii) \Rightarrow (iii)$. Let $K = \operatorname{Im} f$, so that $f : N' \to K \to 0$ is exact. Hence, $M \otimes N' \xrightarrow{f_*} M \otimes K \to 0$ is exact. Since $f_* = 0$, $M \otimes K = 0$ and this implies $K = 0$, i.e., $f = 0$.

$(iii) \Rightarrow (i)$. Since M is flat, the exactness of

$$0 \to N' \xrightarrow{f} N \xrightarrow{g} N'' \to 0$$

implies the exactness of

$$0 \to M \otimes N' \xrightarrow{f_*} M \otimes N \xrightarrow{g_*} M \otimes N'' \to 0.$$

Conversely, assume that for any sequence $N' \xrightarrow{f} N \xrightarrow{g} N''$, the tensored sequence $M \otimes N' \xrightarrow{f_*} M \otimes N \xrightarrow{g_*} M \otimes N''$ is exact. Since $g_* f_* = 0$, by (iii), $gf = 0$. Hence $K = \operatorname{Im} f \subset \operatorname{Ker} g = L$. Consider the exact sequence $0 \to K \to L \to L/K \to 0$. Since M is flat, the sequence

$$0 \to M \otimes K \to M \otimes L \to M \otimes L/K \to 0$$

is exact. Hence $(M \otimes L)/K \simeq (M \otimes L)/(M \otimes K) = 0$, since $M \otimes K = \operatorname{Im} f_* = \operatorname{Ker} g_* = M \otimes L$. If p is the projection $p : L \to L/K$, then $p_* : M \otimes L \to M \otimes L/K$ is zero and hence $p = 0$. This implies $K = L$, i.e., the sequence $N' \xrightarrow{f} N \xrightarrow{g} N''$ is exact. □

Definition 1.4.9 An R-module M is called *faithfully flat* if it satisfies any one of the equivalent conditions of the above theorem.

Corollary 1.4.10 *A faithfully flat module M is flat and faithful, i.e.,* $\operatorname{Ann}(M) = 0$.

Proof Let $a \in R$ with $aM = 0$. Consider the mapping $f_a : R \to R$ given by $f_a(b) = ab$. Then $f_a^* : M \otimes R \to M \otimes R$ is zero because $x \otimes b \mapsto x \otimes ab = ax \otimes b = 0$. This implies that $f_a = 0$. Hence $f_a(1) = a = 0$, i.e., $\operatorname{Ann}(M) = 0$. □

Corollary 1.4.11 *An R-module M is faithfully flat if and only if M is flat and for each maximal ideal $\mathfrak{m}$ of R, $\mathfrak{m}M \neq M$.*

Proof $M/\mathfrak{m}M \simeq M \otimes R/\mathfrak{m} \neq 0$ if M is faithfully flat, since $R/\mathfrak{m} \neq 0$.
Conversely, assume $M/\mathfrak{m}M \neq 0$ for all maximal ideals $\mathfrak{m}$ and that M is flat. Let $N \neq 0$ be an R-module. Then N contains a non-zero cyclic submodule $K \simeq R/I$ for a proper ideal I in R. Since I is a proper ideal, $I \subset \mathfrak{m}$ for some maximal ideal $\mathfrak{m}$. Now $M \neq \mathfrak{m}M$ implies $M \neq IM$, i.e., $M \otimes R/I \neq 0$. Hence $M \otimes K \neq 0$. Since M is flat, $M \otimes K$ is a submodule of $M \otimes N$ and so $M \otimes N \neq 0$. This implies that M is faithfully flat. □

Example 1.4.12 A module which is flat and faithful need not be faithfully flat. For instance $\mathbb{Q}$ is flat over $\mathbb{Z}$ (see Exercise 12, Section 1.4) and faithful but $\mathbb{Q}$ is not faithfully $\mathbb{Z}$-flat, as $\mathbb{Q} \otimes_{\mathbb{Z}} \mathbb{Z}/n\mathbb{Z} = 0$ but $\mathbb{Z}/n\mathbb{Z} \neq 0$.

We now study faithfully flat algebras.

Let A be an R-algebra, M and N be R-modules. Then there exists a natural A-homomorphism

$$\phi_M : A \otimes_R \operatorname{Hom}_R(M, N) \to \operatorname{Hom}_A(A \otimes_R M, A \otimes_R N),$$

given by $\phi_M(a \otimes f)\{(b \otimes x)\} = ab \otimes f(x), a, b \in A, x \in M,\ f \in \operatorname{Hom}_R(M, N)$. We want to investigate conditions under which ϕ_M is an isomorphism for all R-modules N. Clearly, ϕ_M is an isomorphism for $M = R$. Using the property that tensor product and Hom commutes with finite direct sums (Exercise 4, Section 1.2), it follows that ϕ_M is an isomorphism when M is a free module of finite rank.

Proposition 1.4.13 *Let A be a flat R-algebra and M and N be R-modules. Let ϕ_M be the natural map defined above. Then*

(i) ϕ_M is injective if M is finitely generated.

(ii) ϕ_M is an isomorphism if M is finitely presented.

Proof (i) If M is finitely generated, then there exists an exact sequence $0 \to K \to F \to M \to 0$, where F is a free module of finite rank. For any fixed R-module N, define

$$T(M) = A \otimes_R \mathrm{Hom}_R(M, N), \text{ and}$$

$$T'(M) = \mathrm{Hom}_A(A \otimes_R M, A \otimes_R N).$$

Since A is R-flat and Hom is left exact (Proposition 1.2.7), we have a commutative diagram

$$\begin{array}{ccccccc} 0 \to & T(M) & \to & T(F) & \to & T(K) \\ & \downarrow \phi_M & & \downarrow \phi_F & & \downarrow \phi_K \\ 0 \to & T'(M) & \to & T'(F) & \to & T'(K) \end{array}$$

with exact rows. Since ϕ_F is an isomorphism, then it follows from the diagram that ϕ_M is injective.

(ii) If M is finitely presented, then by Corollary 1.2.20, K is finitely generated. We can replace K by a free module of finite rank F_1 so that there exists an exact sequence $F_1 \to F \to M \to 0$ with both F and F_1 free, of finite rank. Consider the commutative diagram

$$\begin{array}{ccccccc} 0 \to & T(M) & \to & T(F) & \to & T(F_1) \\ & \phi_M \downarrow & & \phi_F \downarrow & & \phi_{F_1} \downarrow \\ 0 \to & T'(M) & \to & T'(F) & \to & T'(F_1) \end{array}$$

with exact rows. Since ϕ_F and ϕ_{F_1} are isomorphisms, it follows from the above commutative diagram that ϕ_M is an isomorphism. □

Proposition 1.4.14 (i) *Let A be a faithfully flat R-algebra and M an R-module. If $A \otimes_R M$ is finitely generated (finitely presented) over A, then M is finitely generated (finitely presented) over R.*

(ii) *If $A \otimes_R M$ is finitely generated and projective over A, then M is finitely generated and projective over R.*

Proof (i) Suppose $A \otimes_R M$ is finitely generated over A. Choose a generating set of the type

$$\{1 \otimes x_i \mid x_i \in M, 1 \leq i \leq n\}.$$

Let N be the submodule of M generated by

$$\{x_i \mid 1 \leq i \leq n\}.$$

Then the natural map $A \otimes_R N \to A \otimes_R M$, induced by the inclusion $i : N \to M$, is surjective. Since A is faithfully flat, i is surjective, i.e., $N = M$. Hence M is finitely generated.

If $A \otimes_R M$ is finitely presented, it is also finitely generated and so M is finitely generated. Choose an exact sequence

$$0 \to K \xrightarrow{f} F \xrightarrow{g} M \to 0,$$

where F is free, of finite rank. Then the sequence

$$0 \to A \otimes_R K \xrightarrow{f^*} A \otimes_R F \xrightarrow{g_*} A \otimes_R M \to 0$$

is exact. Since $A \otimes_R M$ is finitely presented, $A \otimes_R K$ is finitely generated over A. Hence K is finitely generated over R, showing that M is finitely presented over R.
(ii) Let $A \otimes_R M$ be finitely generated and projective over A. Since it is projective, by Proposition 1.2.16, it is also finitely presented. Hence, by (i), M is finitely presented over R. To show that M is R-projective, it is sufficient to show that for any surjective R-homomorphism $f : N \to L$, the induced homomorphism

$$f_* : \mathrm{Hom}_R(M, N) \to \mathrm{Hom}_R(M, L)$$

is surjective. Since A is faithfully flat over A, it is sufficient to show that

$$\mathrm{Id}_A \otimes_R f_* : A \otimes_R \mathrm{Hom}_R(M, N) \to A \otimes_R \mathrm{Hom}_R(M, L)$$

is surjective.

Consider the diagram

$$\begin{array}{ccc} A \otimes_R \mathrm{Hom}_R(M, N) & \xrightarrow{\mathrm{Id}_A \otimes_R f_*} & A \otimes_R \mathrm{Hom}_R(M, L) \\ \downarrow & & \downarrow \\ \mathrm{Hom}_A(A \otimes_R M, A \otimes_R N) & \longrightarrow & \mathrm{Hom}_A(A \otimes_R M, A \otimes_R L) \end{array}$$

where the horizontal maps are induced by $f : N \to L$ and the vertical maps are isomorphisms by Proposition 1.4.13. Since $A \otimes_R M$ is A-projective, the horizontal map below is surjective. Hence $\mathrm{Id}_A \otimes_R f_*$ is surjective. $\square$

1.4. EXERCISES

1. Show that $M = \Sigma_\alpha M_\alpha$ is flat if and only if each M_α is flat.

2. Show that the tensor product of two flat (resp. faithfully flat) modules is flat (resp. faithfully flat).

3. Let $0 \to M' \to M \to M'' \to 0$ be an exact sequence of R-modules with M'' flat. Show that M is flat if and only if M' is flat.

4. Illustrate by an example that intersection of two flat modules need not be flat.

5. In the proof of Proposition 1.4.3, describe explicitely a finitely generated submodule N_1 of N satisfying the required conditions. Why cannot we chose N_1 to be $f(N_1')$? (**Hint:** For $L_1 \subset L_2$, the induced map $M \otimes L_1 \to M \otimes L_2$ need not be injective.)

6. Show that $M = \oplus\Sigma_\alpha M_\alpha$ is faithfully flat if each M_α is flat and at least one M_α is faithfully flat.

7. Let $\phi : R \to S$ be a homomorphism of rings. Show that S is faithfully R-flat if and only if ϕ is injective and $S/\phi(R)$ is R-flat.

8. Let $\phi : R \to S$ be a homomorphism of rings. Show that if S is faithfully flat over R, then for any R-module M, the induced map $M \to M \otimes_R S$ given by $x \mapsto x \otimes 1$ is injective.

9. Let $\phi : R \to S$ be a homomorphism of rings such that S is faithfully flat over R. Show that for any ideal I of R

$$\phi^{-1}(\phi(I)S) = I.$$

10. Give an example of a faithful and projective module which is not faithfully flat.

11. If K is a field, show that $K(X)$ is not a faithfully flat $K[X]$-module.

12. An R-module M is called *flat for* N if for every submodule N' of N, the natural map $M \otimes_R N' \to M \otimes_R N$ is injective.

    (a) Show that if M is flat for N, M is flat for any quotient module of N.

    (b) Show that if M is flat for each N_α, then M is flat for $\Sigma_\alpha N_\alpha$.

    (c) Deduce that M is flat if and only if M is flat for R, i.e., for each finitely generated ideal I of R, the natural map $M \otimes_R I \to M$ given by $x \otimes a \mapsto ax$ is injective.

(d) Let R be a *Bezout domain*, i.e., an integral domain in which every finitely generated ideal is principal. Show that M is flat if and only if it is torsion-free, i.e., $ax = 0, a \in R,\ x \in M$ implies $a = 0$ or $x = 0$.

(e) Deduce that $\mathbb{Q}$ is flat as a $\mathbb{Z}$-module.

CHAPTER

2

LOCALISATION

The first part of this chapter is concerned with the study of ideals. We start with prime and maximal ideals which correspond geometrically to irreducible varieties and points over an algebraically closed field. We then study ideal operations such as sum, product, extension, contraction and taking radicals.

The second part is concerned with the study of local rings. The ring of meromorphic functions at a point P in an affine variety is a local ring, and this ring reflects the local properties of the variety at P. This ring is generalised to the concept of a ring obtained by localisation with respect to a multiplicatively closed set. The localisation operation, which is defined also for a module, is shown to be well behaved with respect to quotients, tensor products and exact sequences.

Some applications of localisation are given in the study of projective modules as locally free modules.

2.1 Ideals

We recall some preliminary notions about prime and maximal ideals. An ideal P of R is a prime ideal of R if $P \neq R$ and $ab \in P$ implies $a \in P$ or $b \in P$. Clearly, P is a prime ideal if and only if $R/P \neq 0$ and is a domain.

An ideal $\mathfrak{m}$ of R is called a maximal ideal of R if $\mathfrak{m} \neq R$ and for any ideal I with $\mathfrak{m} \subseteq I \subseteq R$, either $\mathfrak{m} = I$ or $R = I$. Clearly, an ideal $\mathfrak{m}$ is a maximal ideal of R if and only if $R/\mathfrak{m}$ is a field.

In particular any maximal ideal is a prime ideal.

Proposition 2.1.1 *Let I be a proper ideal of R. Then there exists a maximal ideal $\mathfrak{m}$ of R with $I \subseteq \mathfrak{m}$.*

Proof Let Σ be the collection of all ideals of R different from R, containing I. Then $\Sigma \neq \emptyset$, as $I \in \Sigma$. Let $\{I_\alpha\}$ be a chain of ideals in Σ. Then, clearly, $I = \underset{\alpha}{\cup} I_\alpha$ is an ideal and $1 \notin I$ as $1 \notin I_\alpha$ for any α. Hence $I \in \Sigma$ and I is an upper bound of the chain $\{I_\alpha\}$. Therefore, by Zorn's Lemma, Σ has a maximal element $\mathfrak{m}$, which is a maximal ideal of R containing I. □

Corollary 2.1.2 *If $a \in R$ is a non-unit, then there exists a maximal ideal $\mathfrak{m}$ of R containing a.*

Proof Take $I = Ra$. □

Proposition 2.1.3 *Let $f : R \to S$ be a homomorphism of rings. If Q is a prime ideal of S, then $P = f^{-1}(Q)$ is a prime ideal of R.*

Proof The given map f induces a natural ring homomorphism $\overline{f} : R/P \to S/Q$, which is injective. If Q is a prime ideal, S/Q is a domain and so is R/P. Hence P is a prime ideal of R. □

Corollary 2.1.4 *If R is a subring of S and Q is a prime ideal of S, then $Q \cap R$ is a prime ideal of R.*

Corollary 2.1.5 *An ideal P of R containing an ideal I is a prime ideal in R if and only if P/I is a prime ideal in R/I.*

Proof Follows from the isomorohism theorem

$$\frac{R/I}{P/I} \cong R/P.$$ □

Example 2.1.6 It is not true that an inverse image of a maximal ideal is a maximal ideal. If $i : \mathbb{Z} \to \mathbb{Q}$ is the inclusion map, the ideal $\{0\}$ is maximal in $\mathbb{Q}$ but not in $\mathbb{Z}$.

Proposition 2.1.7 *Let $I_1, I_2, \ldots, I_n$ be ideals of R and P a prime ideal containing $\cap_{i=1}^{n} I_i$. Then $P \supseteq I_i$ for some i.*

Proof If $I_i \not\subseteq P$ for each i, choose $a_i \in I_i$, $a_i \notin P$. Then $a = \Pi_i a_i \in \cap I_i$, and $a \notin P$, a contradiction. □

Corollary 2.1.8 *If $P = \cap_{i=1}^{n} I_i$, P prime, then $P = I_i$ for some i.*

Proof By Proposition 2.1.7, $P \supset I_i$ for some i. Since $P \subseteq I_j$ for all j, we have $P = I_i$. □

Proposition 2.1.9 *Let I be an ideal with $I \subset \cup_{i=1}^{n} P_i$, P_i prime. Then $I \subset P_i$ for some i.*

Proof The proof is by induction on n. If $n = 1$, there is nothing to prove. Assume the result for $(n-1)$ and let $I \subseteq \cup_{i=1}^{n} P_i$. If I is contained in an $(n-1)$ union of P_i's, then induction applies. Suppose, if possible, that for each i, $I \not\subset \cup_{j \neq i} P_j$. Choose $a_i \in I$, $1 \leq i \leq n$, such that $a_i \notin P_j$ for any $j \neq i$. If for some $i, a_i \notin P_i$, then $I \not\subset \cup_{i=1}^{n} P_i$, a contradiction. Hence assume that $a_i \in P_i$ for each i. Then the element $a = \Sigma_{i=1}^{n} a_1 \cdots a_{i-1} \cdot a_{i+1} \cdots a_n$ is an element of I not in $\cup_{i=1}^{n} P_i$, again a contradiction. □

Remark 2.1.10 Proposition 2.1.9 is known as **Prime Avoidance Theorem**.

Definition 2.1.11 An element $a \in R$ is called *nilpotent* if $a^n = 0$ for some $n \geq 1$.

Proposition 2.1.12 *The set of all nilpotent elements of R is an ideal of R.*

Proof Let I be the set of all nilpotent elements. If $a, b \in I$, then $a^n = 0 = b^m$ for some m, n. By Binomial Theorem,

$$(a+b)^{m+n} = \sum_{i+j=m+n} \binom{m+n}{i} a^i b^j.$$

Each term on the right hand side is zero as either $i \geq n$ or $j \geq m$. Hence $a+b \in I$. Clearly, $-a \in I$ and $ra \in I$ for all $r \in R$. Thus, I is an ideal in R. □

Definition 2.1.13 The ideal of nilpotent elements of R is called the *nil radical* of R.

Example 2.1.14 (i) Any integral domain has nil radical zero.

(ii) $\mathbb{Z}/m\mathbb{Z}$ has nil radical zero $\Leftrightarrow m$ is a product of distinct primes.

(iii) For any R with nil radical $N(R)$, the ring $R/N(R)$ has nil radical zero.

Proposition 2.1.15 *The nil radical $N(R)$ of R is the intersection of all prime ideals of R.*

Proof Let $a \in N(R)$, so that $a^n = 0 \in P$ for any prime ideal P. This implies that $a \in P$.

Conversely, let $a \in R$ be non-nilpotent. We construct a prime ideal P_0 of R with $a \notin P_0$. Consider the set $S = \{a^n \mid n \geq 0\}$. Let Σ be the collection of all ideals of R which do not intersect S. Since a is non-nilpotent, $(0) \in \Sigma$ and hence $\Sigma \neq \emptyset$. By Zorn's Lemma, Σ has a maximal element P_0. Clearly, $a \notin P_0$ and it is sufficient to show that P_0 is a prime ideal.

Let $a' \notin P_0$ and $b \notin P_0$, so that by maximality of P_0,

$$(P_0 + Ra') \cap S \neq \emptyset \text{ and } (P_0 + Rb) \cap S \neq \emptyset.$$

Choose m_1, m_2 such that

$$a^{m_1} = p_1 + \lambda a' \text{ and } a^{m_2} = p_2 + \mu b,$$

where $p_1, p_2 \in P_0$ and $\lambda, \mu \in R$. Then

$$a^{m_1+m_2} = p_3 + \lambda\mu a'b,\ p_3 \in P_0,$$

which shows that $a'b \notin P_0$. □

Definition 2.1.16 The intersection $J(R)$ of all maximal ideals of R is called the *Jacobson radical* of R.

Example 2.1.17 The nil radical $N(R) \subset J(R)$, as every maximal ideal is prime. The inclusion is a strict inclusion for the ring $R = k[[X]]$, the power series ring in X over a field k, as $N(R) = 0$. Now, $J(R) = (X)$, since an element $f \in R$ is a unit $\Leftrightarrow$ f has a non-zero constant term.

Proposition 2.1.18 *An element $a \in J(R)$ if and only if $1 + ab$ is a unit for all $b \in R$.*

Proof Let $a \in J(R)$ and let $1 + ab$ be a non-unit for some $b \in R$. Then there exists a maximal ideal $\mathfrak{m}$ of R with $1 + ab \in \mathfrak{m}$. Since $a \in J(R)$, we have $a \in \mathfrak{m}$, which implies $1 \in \mathfrak{m}$, a contradiction.

Conversely, assume that $1 + ab$ is a unit for all $b \in R$. Suppose $a \notin \mathfrak{m}$ for some maximal ideal $\mathfrak{m}$. Then $\mathfrak{m} + Ra = R$, so that $1 = a_0 - ba$, $a_0 \in \mathfrak{m}$, $b \in R$. This implies that $a_0 = 1 + ba$ is a unit, a contradiction as $\mathfrak{m} \neq R$. □

We now consider various ideal operations and study the relations between them. Given two ideals I and J, their sum is defined to be the ideal $I + J = \{a + b \mid a \in I,\ b \in J\}$. The product IJ of I and J is defined by

$$IJ = \{\Sigma_{i=1}^{n} a_i b_i \mid a_i \in I,\ b_i \in J,\ n \geq 1\}.$$

Clearly, IJ is an ideal contained in $I \cap J$ and is not in general equal to $I \cap J$. However, if $I + J = R$, then $IJ = I \cap J$, for if $1 = a + b$, $a \in I$, $b \in J$, then any $x \in I \cap J$ can be expressed as $x = xa + xb \in IJ$.

Definition 2.1.19 Two ideals I and J are said to be *comaximal* if $I + J = R$.

Example 2.1.20 Let $R = \mathbb{Z}$, $I = (m)$ and $J = (n)$. Then I and J are comaximal if and only if m and n are relatively prime. For comaximal ideals I and J we have $IJ = I \cap J$.

Proposition 2.1.21 (Chinese Remainder Theorem)

Let $I_1, I_2, \ldots, I_n$ be ideals in R and $f : R \to \Pi_{i=1}^{n} R/I_i$ the natural map defined by

$$f(a) = (a + I_1, \ldots, a + I_n).$$

Then f is a ring homomorphism which is surjective if and only if $I_1, I_2, \ldots, I_n$ are mutually pairwise comaximal, i.e., $I_i + I_j = R$, $i \neq j$, $1 \leq i,\ j \leq n$.

Proof Clearly, f is a ring homomorphism. Assume that $I_i + I_j = R$, $i \neq j$. To show that f is surjective, it is enough to show that there exists $a \in R$ with $f(a) = (\overline{0}, \overline{0}, \ldots, \overline{1}, \ldots, \overline{0})$ with $\overline{1}$ at the i-th place. We show this say for $i = 1$. Since $I_1 + I_i = R\,(i > 1)$, write

$1 = u_i + v_i$, $u_i \in I_1$, $v_i \in I_i$. If $x = \Pi_{i=2}^n v_i = \Pi_{i=2}^n (1 - u_i) = 1 + u$, $u \in I_1$, then $x \in I_j$ $(j \geq 2)$ and $f(x) = (\overline{1}, \overline{0}, \ldots, \overline{0})$.

Conversely, assume that f is surjective. To show that $I_i + I_j = R$, $i \neq j$. Consider the element $(\overline{0}, \overline{0}, \ldots, \overline{1}, \ldots, \overline{0})$ with $\overline{1}$ in the i-th place. There exists $a \in R$ with $f(a) = (\overline{0}, \overline{0}, \ldots, \overline{1}, \ldots, \overline{0})$, i.e., $a \in I_j$, $j \neq i$ and $a - 1 \in I_i$. Hence $1 = (1 - a) + a \in I_i + I_j$, i.e., $I_i + I_j = R$. □

Corollary 2.1.22 *If $I_1, I_2, \ldots, I_n$ are mutually pairwise comaximal ideals of R, then*

$$R/\cap I_i \simeq \Pi_{i=1}^n R/I_i.$$

Proof Since $\operatorname{Ker} f = \cap_{i=1}^n I_i$, the result follows from the Proposition 2.1.21. □

Corollary 2.1.23 *Let $m_1, m_2, \ldots, m_n$ be mutually coprime integers and $a_1, a_2, \ldots, a_n$ arbitrary integers. Then there exists an integer a with $a \equiv a_i \pmod{m_i}$, $1 \leq i \leq n$.*

Proof Take $R = \mathbb{Z}$, $I_i = m_i\mathbb{Z}$ $(1 \leq i \leq n)$. Since $I_1, I_2, \ldots, I_n$ are mutually coprime, the map f defined in Proposition 2.1.21 is surjective. This implies that for any $a_1, \ldots, a_n$, there exists a with $a \equiv a_i \pmod{m_i}$, $1 \leq i \leq n$. □

Corollary 2.1.24 *Let $R = k[X], k$ a field, $f_1(X), \ldots, f_n(X) \in R$ $a_1, \ldots, a_n$ distinct elements of k and $\alpha_1, \alpha_2, \ldots, \alpha_n$ positive integers. Then there exists $f(X) \in R$ such that*

$$f(X) \equiv f_i(X) \pmod{(X - a_i)^{\alpha_i}}, \ 1 \leq i \leq n.$$

Proof If $I_i = (X - a_i)^{\alpha_i}$, then $I_1, I_2, \ldots, I_n$ are mutually pairwise comaximal. The result now follows since the map $R \to \Pi_i R/I_i$ is surjective. □

Proposition 2.1.25 *Let $I_1, I_2, \ldots, I_n$ be mutually pairwise comaximal ideals of R. Then*

$$\bigcap_{i=1}^n I_i = \prod_{i=1}^n I_i.$$

Proof The proof is by induction on n. The result is already shown to be true for $n = 2$. Assume the result for $(n - 1)$. Consider $I_1, I_2, \ldots, I_n$, which are mutually pairwise comaximal. If $J = \cap_{i=1}^{n-1} I_i$, then $J = \Pi_{i=1}^{n-1} I_i$ by induction. Now, J and I_n are comaximal, for if $1 = a_i + b_i$, $a_i \in I_i$, $b_i \in I_n, i \neq n$, then

$$a = \Pi_{i=1}^{n-1} a_i = \Pi_{i=1}^{n-1} (1 - b_i) = 1 + b, \ \ b \in I_n,$$

showing $J + I_n = R$. Hence $J \cap I_n = JI_n$, and this implies that $\cap_{i=1}^n I_i = \Pi_{i=1}^n I_i$. □

Definition 2.1.26 For any ideal I of R, the set

$$\{a \in R \mid a^n \in I \text{ for some } n \geq 1\}$$

is an ideal of R called the *radical* of I. The radical of I is denoted by $\sqrt{I}$.

Example 2.1.27 (i) If $I = 0$, then $\sqrt{I} = N(R)$, the nil radical of R.

(ii) $R = k[X, Y]$, k a field, and $I = (X^2, Y)$. Then $\sqrt{I} = (X, Y)$.

Proposition 2.1.28 (i) *For any ideal I of R, the radical $\sqrt{I}$ is the intersection of all prime ideals containing I.*

(ii) *Ideals I and J are comaximal $\Leftrightarrow$ $\sqrt{I}$ and $\sqrt{J}$ are comaximal.*

Proof (i) Consider the projection map $p : R \to R/I$. Clearly,

$$\sqrt{I} = p^{-1}\{N(R/I)\}.$$

Since $N(R/I)$ is the intersection of all prime ideals of R/I (by Proposition 2.1.15), the result follows from Corollary 2.1.5.

(ii) If I and J are comaximal, then $I + J = R$ and since $I \subseteq \sqrt{I}$ and $J \subseteq \sqrt{J}$, we have $\sqrt{I} + \sqrt{J} = R$.

Conversely, assume that $\sqrt{I} + \sqrt{J} = R$. Then $1 = a + b, a \in \sqrt{I}, b \in \sqrt{J}$, so that $a^n \in I$ and $b^m \in J$ for some $m, n \geq 1$. Now,

$$1 = (a + b)^{m+n} = \sum_{i+j=m+n} \binom{m+n}{i} a^i b^j,$$

where the summation is over $i + j = m + n$. As either $i \geq n$ or $j \geq m$, $1 \in I + J$. Hence $R = I + J$, i.e., I and J are comaximal. □

Definition 2.1.29 Let $f : R \to S$ be a homomorphism of rings. If I is an ideal in R, then the ideal generated by $f(I)$ in S is called the *extension* of I under f. If J is an ideal in S, then $f^{-1}(J)$ is an ideal in R, called the *contraction* of J under f.

Definition 2.1.30 An ideal J of S is called an *extended* ideal if for some ideal I of R, its extension $I^c = J$. An ideal I of R is called a *contracted ideal* if for some ideal J of S, its contraction $J^c = I$.

Example 2.1.31 (i) Let R be a subring of S and $i : R \to S$ the inclusion map. The extension of an ideal I of R is the ideal generated by I in S and the contraction of an ideal J of S is $J \cap R$.

(ii) $f : \mathbb{Z} \to \mathbb{Z}[i]$ the inclusion map, $\mathbb{Z}[i]$ the ring of Gaussian integers. The extension I^e of the prime ideal $I = 2\mathbb{Z}$ of $\mathbb{Z}$ is not a prime ideal in $\mathbb{Z}[i]$, as $2 = (1+i)(1-i) \in I^e$, but $1+i$ and $1-i$ do not belong to I^e.

Proposition 2.1.32 *Let $f : R \to S$ be a ring homomorphism. Then the mapping $I \mapsto I^e$ is a bijection between the set C of contracted ideals in R onto the set E of extended ideals in S.*

Proof We first show that for any ideal I in R and any ideal J in S we have

(i) $I^{ece} = I^e$ and
(ii) $J^{cec} = J^c$.

Clearly, $I \subseteq I^{ec}$ and $J \supseteq J^{ce}$. Hence $I^e \subseteq I^{ece}$, and by applying the second relation to $J = I^e$, we have $I^e \supseteq I^{ece}$.This proves (i). The second relation is proved similarly.

These relations show that the map $C \to E$, given by $I \mapsto I^e$, is bijective. □

2.1. EXERCISES

1. Let R be a ring in which $x^n = x$ for all $x \in R$ and for some $n > 1$. Show that every prime ideal of R is maximal.

2. Let $R[X]$ be the polynomial ring and $f(X) \in R[X]$ given by $f(X) = a_0 + a_1X + \cdots + a_nX^n$, where $a_i \in R$. Show that

   (a) f is nilpotent if and only if all the a_i are nilpotent.

   (b) f is a unit if and only if a_0 is a unit and a_i $(i \geq 1)$ are nilpotent.

3. If P is a prime ideal in R, show that its extension $P[X]$ in $R[X]$ is a prime ideal in $R[X]$. Is the same true of maximal ideals? Why?

4. Let $R[[X]]$ be the power series ring in X over R and $f(X) \in R[[X]]$ given by $f(X) = a_0 + a_1X + \cdots + a_nX^n + \cdots$, where $a_i \in R$. Show that

   (a) $f(X)$ is a unit if and only if $a_0 \in R$ is a unit.

   (b) $f(X)$ is nilpotent implies that each a_i is nilpotent $(i \geq 0)$.

5. Let $S = R[[X]]$ be the power series ring in X over R. Show that every maximal ideal of S is of the type $(\mathfrak{m}, X)$, the ideal generated by X and $\mathfrak{m}$, where $\mathfrak{m}$ is a maximal ideal of R.

6. If $a \in R$ is a nilpotent element, show that $1 + a$ is a unit. Deduce that the sum of a nilpotent element and a unit is a unit.

7. For an R-module M and submodule N of M, define $r_M(N) = \{a \in R \mid a^n M \subset N \text{ for some } n \geq 1\}$. Show that $r_M(N)$ is an ideal (called the *radical* of N in M). Show further that $r_M(N) = \sqrt{(N : M)}$.

8. If $I_1, I_2, \ldots, I_n$ are pairwise comaximal ideals and M is an R-module, show that the natural map $M \to \prod_{i=1}^n M/I_iM$, given by

$$x \mapsto (x + I_1M, \ldots, x + I_nM),\ x \in M,$$

is surjective. Show further that $M/IM \simeq \prod_i M/I_iM$, where $I = \cap_i I_i$.

9. Show that an ideal $I = \sqrt{I}$ if and only if I is an intersection of prime ideals.

10. Show that for the polynomial ring $R[X]$, the nil radical is equal to the Jacobson radical.

11. Let R be a ring with nil radical $N(R)$. Show that R has just one prime ideal if and only if $R/N(R)$ is a field.

2.2 Local rings

Definition 2.2.1 A ring R is called a *local ring* if the non-units of R form an ideal.

Example 2.2.2 (i) Any field is a local ring.
(ii) Let k be a field and $a = (a_1, a_2, \ldots, a_n) \in k^n$, $a_i \in k$. Then

$$R = \{f/g \mid f, g \in k[X_1, \ldots, X_n],\ g(a) \neq 0\}$$

is a local ring. Any non-unit of R is of the type f/g, $g(a) \neq 0$ and $f(a) = 0$. These form an ideal of R.
(iii) $R = k[[X]]$, the power series ring in X over a field k, is a local ring as the non-units form an ideal (X).

Proposition 2.2.3 *R is a local ring if and only if it has a unique maximal ideal.*

Proof If R has a unique maximal ideal, it is precisely the set of non-units by Corollary 2.1.2, and hence R is a local ring.

Conversely, if R is a local ring with $\mathfrak{m}$ the ideal of non-units, $\mathfrak{m}$ is clearly a maximal ideal. If $\mathfrak{m}'$ is another maximal ideal, $\mathfrak{m}' \subseteq \mathfrak{m}$, as elements of $\mathfrak{m}'$ are non-units. Hence $\mathfrak{m}' = \mathfrak{m}$. □

Corollary 2.2.4 *If $\mathfrak{m}$ is a maximal ideal of R such that every element of $1 + \mathfrak{m}$ is a unit, then R is a local ring.*

Proof If $a \notin \mathfrak{m}$, then $\mathfrak{m} + Ra = R$,i.e., $1 = a_0 + ba$ for some $a_0 \in \mathfrak{m}$, $b \in R$. This implies that $ba = 1 - a_0$ is a unit. Hence a is a unit and $\mathfrak{m}$ is precisely the set of non-units. Hence R is a local ring. □

We now study modules over local rings and show that for finitely presented modules over a local ring, the concepts of free, projective and flat modules are equivalent.

Proposition 2.2.5 *Let M be a finitely generated R-module, I an ideal in R with $IM = M$. Then there exists some $a \in I$ with*

$$(1 + a)M = 0.$$

Proof Let M be generated by $\{x_1, x_2, \ldots, x_n\}$. Since $IM = M$, we can write $x_i = \Sigma_{i=1}^n a_{ij}x_j$ for some $a_{ij} \in I$. Hence

$$\Sigma_{j=1}^n (\delta_{ij} - a_{ij})x_j = 0,$$

where $\delta_{ij} = 1$ if $i = j$ and 0 if $i \neq j$. This implies that

$$\begin{pmatrix} 1 - a_{11} & -a_{12} & -a_{1n} \\ -a_{21} & 1 - a_{22} & -a_{2n} \\ .. & .. & .. \\ -a_{n1} & .. & 1 - a_{nn} \end{pmatrix} \begin{pmatrix} x_1 \\ .. \\ .. \\ x_n \end{pmatrix} = \begin{pmatrix} 0 \\ .. \\ .. \\ 0 \end{pmatrix}.$$

If Δ is the determinant of the matrix $(\delta_{ij} - a_{ij})$, then by multiplying the matrix by its adjoint on the left, we have $\Delta x_i = 0$, $1 \leq i \leq n$. This implies that $\Delta M = 0$. The expansion for Δ shows that $\Delta = 1 + a$ for some $a \in I$. □

Corollary 2.2.6 *Let M be a finitely generated R-module and $f : M \rightarrow M$ a surjective homomorphism. Then f is an isomorphism.*

Proof Consider M as an $R[X]$ module where $Xx = f(x)$, $x \in M$. If $I = (X)$, then $IM = M$ as f is surjective. Hence there exists $\phi \in I$ with $(1 + \phi)M = 0$. Since $\phi = \psi(X)X$ for some $\psi(X) \in R[X]$, we have $(1 + \psi(X)X)M = 0$. Hence $x + \psi(X)Xx = 0$ for all $x \in M$. Therefore, if $f(x) = 0$, i.e., if $Xx = 0$, then $f = 0$. Thus f is injective. □

Proposition 2.2.7 (Nakayama Lemma) *Let M be a finitely generated R-module and I an ideal contained in $J(R)$, the Jacobson radical of R, such that $IM = M$. Then $M = 0$.*

Proof By Proposition 2.2.5, $(1 + a)M = 0$ for some $a \subset I$. Since $a \in J(R), 1 + a$ is a unit by Proposition 2.1.18, so that $M = 0$. □

Corollary 2.2.8 *Let M be a finitely generated R-module, N a submodule of M and I an ideal contained in $J(R)$. Then $M = N + IM$ implies $N = M$.*

Proof Apply Proposition 2.2.7 to the module M/N. □

Proposition 2.2.9 *Let R be a local ring with unique maximal ideal $\mathfrak{m}$ and $k = R/\mathfrak{m}$. If M is an R-module, $\{x_1, \ldots, x_n\}$ elements of M whose images in $M/\mathfrak{m}M$ is a basis of the k-vector space $M/\mathfrak{m}M$, then $\{x_1, \ldots, x_n\}$ generate M.*

Proof Let N be the submodule of M generated by $\{x_1, \ldots, x_n\}$. Since $\{\overline{x}_1, \overline{x}_2, \ldots, \overline{x}_n\}$ form a k-basis of $M/\mathfrak{m}M$, it follows that $N + \mathfrak{m}M = M$. Hence $N = M$, i.e., M is generated by $\{x_1, x_2, \ldots, x_n\}$ (Corollary 2.2.8). □

Corollary 2.2.10 *With the same assumption as in Proposition 2.2.9, $\{x_1, x_2, \ldots, x_n\}$ is a minimal generating set for M.*

Proof If some proper subset of $\{x_1, \ldots, x_n\}$ generates M, then their images in $M/\mathfrak{m}M$ will generate the k-vector space $M/\mathfrak{m}M$, which is a contradiction as $\{\overline{x}_1, \overline{x}_2, \ldots, \overline{x}_n\}$ is a k-basis of $M/\mathfrak{m}M$. □

Remark 2.2.11 If $\{x_1, \ldots, x_n\}$ is a minimal generating set for a module M over a local ring R with maximal ideal $\mathfrak{m}$ and residue field $k = R/\mathfrak{m}$, then n is the dimension of the k-vector space $M/\mathfrak{m}M$.

Proposition 2.2.12 *Let R be a local ring with maximal ideal $\mathfrak{m}$ and $k = R/\mathfrak{m}$. If M and N are finitely generated R-modules with $M \otimes_R N = 0$, then either $M = 0$ or $N = 0$.*

Proof $M/\mathfrak{m}M$ is annihilated by $\mathfrak{m}$, so that it is a $k(= R/\mathfrak{m})$-vector space. By Exercise 5, Section 1.3, we have

$$\frac{M}{\mathfrak{m}M} \otimes_k \frac{N}{\mathfrak{m}N} \simeq \frac{M \otimes N}{\mathfrak{m}(M \otimes_R N)} = 0.$$

Since k is a field, this implies either $M/\mathfrak{m}M = 0$ or $N/\mathfrak{m}N = 0$, i.e.,

$$M = \mathfrak{m}M \text{ or } N = \mathfrak{m}N.$$

By Nakayama Lemma, we have $M = 0$ or $N = 0$. □

Theorem 2.2.13 *Let R be a local ring. Any finitely generated projective R-module is free.*

Proof Let M be a finitely generated projective R-module. By Proposition 2.2.9 and Corollary 2.2.10, there exists a minimal generating set $\{x_1, \ldots, x_n\}$ for M. Choose a free module F of rank n with basis $\{e_1, \ldots, e_n\}$ and define an R-linear map $f : F \to M$ by $f(e_i) = x_i, 1 \le i \le n$. If $K = \operatorname{Ker} f$, we have an exact sequence

$$0 \to K \to F \xrightarrow{f} M \to 0 \qquad (*).$$

Now, $K \subset \mathfrak{m}F$, for if $y \in K$ with $y = \Sigma_{i=1}^{n} a_i e_i, a_i \in R$, then

$$f(y) = \Sigma_{i=1}^{n} a_i x_i = 0.$$

If some $a_i \notin \mathfrak{m}$, it is a unit; then x_i will be a linear combination of the remaining x_j's, contradicting the minimality of $\{x_1, \ldots, x_n\}$. Hence $K \subset \mathfrak{m}F$.

Since M is projective, the exact sequence (*) splits and $F = K \oplus M'$, where $M \simeq M'$. Then $\mathfrak{m}F = \mathfrak{m}K + \mathfrak{m}M'$. Since $\mathfrak{m}K \subset K \subset \mathfrak{m}F$, we have

$$K = \mathfrak{m}K \oplus (K \cap \mathfrak{m}M').$$

But $K \cap \mathfrak{m}M' \subset K \cap M' = \{0\}$, so that $K = \mathfrak{m}K$.

Alternatively, one may argue as follows:

From (*), we have $F \cong K \oplus M$ and hence by Proposition 1.3.9 and Exercise $5(a)$, Section 1.3,

$$F/\mathfrak{m}F \cong K/\mathfrak{m}K \oplus M/\mathfrak{m}M.$$

Since both $F/\mathfrak{m}F$ and $M/\mathfrak{m}M$ have the same dimension n over k, we have $K = \mathfrak{m}K$.

Since K is finitely generated, $K = 0$ by Nakayama Lemma. Hence $M \simeq F$ and M is free. $\square$

Remark 2.2.14 I. Kaplansky has shown that Theorem 2.2.13 holds even without the hypothesis "finitely generated"; *cf.* Chapter 1 of *Commutative Ring Theory* by H. Matsumura.

Our aim is to prove that finitely presented flat modules over a local ring are free. We need the following proposition.

Proposition 2.2.15 *Consider a commutative diagram of R-modules*

$$\begin{array}{ccccccccc}
 & & M' & \xrightarrow{f} & M & \xrightarrow{g} & M'' & \longrightarrow & 0 \\
 & & \downarrow{\scriptstyle u} & & \downarrow{\scriptstyle v} & & \downarrow{\scriptstyle w} & & \\
0 & \longrightarrow & N' & \xrightarrow{\phi} & N & \xrightarrow{\psi} & N'' & &
\end{array}$$

with exact rows. Then there is a well defined map

$$\delta : \text{Ker } w \to \text{Coker } u = N'/u(M')$$

such that there is an exact sequence

$$\text{Ker}(u) \xrightarrow{f^*} \text{Ker}(v) \xrightarrow{g^*} \text{Ker}(w) \xrightarrow{\delta} \text{Coker}(u) \xrightarrow{\phi^*} \text{Coker}(v) \xrightarrow{\psi^*} \text{Coker}(w)$$

where f^*, g^*, ϕ^* and ψ^* are natural maps induced by the maps f, g, ϕ and ψ respectively. $\square$

Proof Since the diagram is commutative, f(resp. ϕ) maps $\operatorname{Ker} u$ (resp.$\operatorname{Coker} u$) in $\operatorname{Ker} v$ (resp.$\operatorname{Coker} v$). This defines the map f^*(resp. ϕ^*). Similarly, g^* and ψ^* are defined. To define

$$\delta : \operatorname{Ker}(w) \to \operatorname{Coker}(u),$$

let $x \in \operatorname{Ker}(w)$. Choose $y \in M$ with $g(y) = x$. Then $\psi v(y) = wg(y) = w(x) = 0$, so that $v(y) = \phi(z)$ for some unique $z \in N'$. Define $\delta(x) = z + u(M')$. δ is well defined, for if $g(y_1) = x$, then $y - y_1 \in \operatorname{Ker} g = \operatorname{Im} f$, so that $y - y_1 = f(t)$, $t \in M$. Then $v(y - y_1) = vf(t) = \phi u(t)$. If $v(y_1) = \phi(z_1)$, then $\phi(z - z_1) = \phi u(t)$. This implies that $z - z_1 = u(t)$, so that $z + u(M') = z_1 + u(M')$. The verification of the exactness of the sequence is a routine computation and is left as an exercise. □

Theorem 2.2.16 *Let R be a local ring with maximal ideal $\mathfrak{m}$ and M a finitely presented R-module. If the canonical map*

$$u_M : \mathfrak{m} \otimes_R M \to M,$$

given by $u_M(a \otimes x) = ax$, $a \in \mathfrak{m}$, $x \in M$, is injective, then M is free.

Proof Let $k = R/\mathfrak{m}$. Choose a minimal generating set $\{x_1, \ldots, x_n\}$ for M such that $\{1 \otimes x_i\} \in R/\mathfrak{m} \otimes_R M$ is a k-basis of $R/\mathfrak{m} \otimes_R M$. Let F be a free R-module of rank n with basis $\{e_1, \ldots, e_n\}$ and g an R-linear map $g : F \to M$, defined by $g(e_i) = x_i$, $1 \le i \le n$. If $K = \operatorname{Ker}(g)$, we have an exact sequence

$$0 \to K \xrightarrow{f} F \xrightarrow{g} M \to 0.$$

This induces a commutative diagram

$$\begin{array}{ccccccccc}
 & & \mathfrak{m} \otimes K & \longrightarrow & \mathfrak{m} \otimes F & \longrightarrow & \mathfrak{m} \otimes M & \longrightarrow & 0 \\
 & & \downarrow{\scriptstyle u_K} & & \downarrow{\scriptstyle u_F} & & \downarrow{\scriptstyle u_M} & & \\
0 & \longrightarrow & K & \xrightarrow{f} & F & \xrightarrow{g} & M & &
\end{array}$$

with exact rows. By Proposition 2.2.15, we have an exact sequence

$$\operatorname{Ker}(u_M) \xrightarrow{\delta} \operatorname{Coker}(u_K) \xrightarrow{f^*} \operatorname{Coker}(u_F) \xrightarrow{g^*} \operatorname{Coker}(u_M).$$

Now,

$$\operatorname{Coker}(u_M) = M/\mathfrak{m}M \simeq R/\mathfrak{m} \otimes_R M.$$

By the choice of $\{x_1, \ldots, x_n\}$, $\operatorname{Coker}(u_F) \xrightarrow{g^*} \operatorname{Coker}(u_M)$ is an isomorphism given by $1 \otimes e_i \to 1 \otimes x_i$. Hence $\operatorname{Ker} g^* = 0$, i.e., $\operatorname{Im} f^* = 0$. Hence $\operatorname{Im} \delta = \operatorname{Ker} f^* = \operatorname{Coker}(u_K)$.

This implies that δ is surjective. Since $\operatorname{Ker} u_M = 0$ by assumption, we have $\operatorname{Im} \delta = \operatorname{Coker}(u_K) = 0$, i.e., $K = \mathfrak{m}K$. Since M is finitely presented, K is finitely generated. Hence $K = 0$ by Nakayama Lemma, i.e., $M \simeq F$ is free. $\square$

Corollary 2.2.17 *Let R be a local ring and M a finitely presented R-module. Then the following are equivalent.*

(i) *M is free.*

(ii) *M is projective.*

(iii) *M is flat.*

Proof Clearly, $(i) \Rightarrow (ii) \Rightarrow (iii)$. To show that $(iii) \Rightarrow (i)$, assume that M is flat. Then the map $\mathfrak{m} \otimes M \to R \otimes M$, given by $a \otimes x \mapsto a \otimes x$, $a \in \mathfrak{m}, x \in M$, is injective. Hence the map u_M is injective and the result follows from Theorem 2.2.16. $\square$

Remark 2.2.18 Corollary 2.2.17 holds even under the assumption "M finitely generated R-module". (*cf. Commutative Ring Theory* by H. Matsumura.)

2.2. EXERCISES

1. Let I be an ideal of R such that for all finitely generated R-modules M, the condition $IM = M$ implies $M = 0$. Show that $I \subseteq J(R)$ (Converse of Nakayama Lemma).

2. Give an example of a ring R and a non-zero R-module M (necessarily not finitely generated) such that $J(R)M = M$.

3. An element $e \in R$ is called idempotent if $e^2 = e$. Show that the only idempotents of a local ring are 0 and 1.

4. Let $I \subseteq J(R)$ and $f : M \to N$, a homomorphism with N finitely generated. If the induced map $\overline{f} : M/IM \to N/IN$ is surjective, then show that f is surjective.

5. A ring R is called semi-local if it has only finitely many maximal ideals. Show that

   (a) A finite product of semi-local rings is semi-local.

   (b) Quotient of a *semilocal* ring is semi-local.

6. Show that a ring R is semi-local $\Leftrightarrow$ $R/J(R)$ is a finite direct product of fields.

7. Let M'' be a finitely presented R-module and

$$0 \to M' \to M \to M'' \to 0$$

an exact sequence of R-modules with M finitely generated. Show that M' is also finitely generated. (**Hint:** Express M as a quotient of a free module F of finite rank by a sub-module K and define a suitable map from K to M'.)

2.3 Localisation

Local rings are obtained quite often by the process of localisation. We propose to describe algebraically the method of localisation.

Definition 2.3.1 Let R be a ring. A subset $S \subset R$ is called a *multiplicatively closed* set in R if $1 \in S$ and $ab \in S$ whenever $a \in S$ and $b \in S$.

Example 2.3.2 (i) *R is a domain and $S = R - \{0\}$.*

(ii) *P is a prime ideal of R and $S = R - P$.*

(iii) *Let $\{P_i\}$ be a set of prime ideals of R and $S = R - \cup_i P_i$.*

(iv) *R is any ring and S is the set of all non-zero-divisors of R.*

Let M be an R-module. The localisation of M with respect to S is constructed as follows. Consider the set $M \times S$. Define a relation $\sim$ on $M \times S$ by the condition $(x, s) \sim (y, t)$ if and only if $s_1(tx - sy) = 0$ for some $s_1 \in S$. It is easy to verify that $\sim$ is an equivalence relation on $M \times S$. The set of equivalence classes is denoted by M_S. Any element of M_S, being an equivalence class containing (x, s), is denoted by the symbol (x/s), $x \in M$, $s \in S$. In particular, for $M = R$, the set R_S is defined.

Proposition 2.3.3 *The set R_S is a ring for the operations*

$$(a/s) + (b/t) = (at + bs/st)$$

$$(a/s)(b/t) = (ab/st) \quad a, b \in R,\ s, t \in S.$$

The set M_S is an R_S-module for the operations

$$(x/s) + (y/t) = (tx + sy/st) \quad x, y \in M,\ s, t \in S.$$

$$(a/s)(x/t) = (ax/st) \quad a \in R.$$

Proof We first show that the operations are well defined. If $(a/s) = (a'/s')$ and $(b/t) = (b'/t')$, then $s_1(as' - a's) = 0$ and $t_1(bt' - b't) = 0$ for some $s_1, t_1 \in S$. These imply the relations

$$s_1t_1[(at + bs)s't' - (a't' + b's')st] = 0$$

and $s_1t_1(s't'ab - sta'b') = 0$. Thus, addition and multiplication are well defined. The zero element is $(0/s)$, $s \in S$, and the inverse of (a/s) under addition is $(-a/s)$. The distributivity of addition over multiplication and the associativity of multiplication can be easily verified. Hence R_S is a ring with unit element $(1/1)$. Similarly the operations on M_S can be shown to be well defined and also that M_S is an R_S-module. □

The ring R_S has the following universal property.

Proposition 2.3.4 *Let $f : R \to R_S$ be the natural map given by*

$$f(a) = (a/1).$$

For any ring homomorphism $g : R \to R'$ such that $g(s)$ is a unit for every $s \in S$, there exists a unique homomorphism $\bar{g} : R_S \to R'$ such that $\bar{g}f = g$.

Proof Define $\bar{g} : R_S \to R'$ by $\bar{g}(a/s) = g(a)g(s)^{-1}$. We show that $\bar{g}$ is well defined. If $(a/s) = (a'/s')$, then $s_1(as' - a's) = 0$ for some $s_1 \in S$. This implies $g(s_1)[g(a)g(s') - g(a')g(s)] = 0$. Since $g(s_1)$, $g(s)$ and $g(s')$ are units, $g(a)g(s)^{-1} = g(a')g(s')^{-1}$, i.e., $\bar{g}$ is well defined. Clearly, $\bar{g}$ is a ring homomorphism and $\bar{g}f = g$. Now, $\bar{g}$ is also unique, for if $\bar{g}_1$ is another map with $\bar{g}_1 f = g$, then

$$\bar{g}_1(1/s) = [\bar{g}_1(s/1)]^{-1} = [\bar{g}_1 f(s)]^{-1} = g(s)^{-1}$$

and

$$\bar{g}_1(a/s) = \bar{g}_1[(a/1)(1/s)] = g(a)g(s)^{-1} = \bar{g}(a/s).$$ □

Remark 2.3.5

1. If R is a domain and $S = R - \{0\}$, then R_S is nothing but the quotient field of R.

2. Let $R = k[X_1, \ldots, X_n]$, k a field. For a fixed

   $$a = (a_1, \ldots, a_n) \in k^n, \text{ let } P = \{f \in R \,|\, f(a_1, \ldots, a_n) = 0\}.$$

   Then P is a prime ideal of R. If $S = R - P$, then R_S is nothing but the ring defined in Example 2.2.2(ii).

3. More generally, let P be any prime ideal of R and $S = R - P$. An element $(a/s) \in R_S$ is a unit $\Leftrightarrow a \in S$. Hence the set of non-units of $R_S = \{(a/s) \,|\, a \in R - S = P)\}$ and it is an ideal, i.e., R_S is a local ring. The ring R_S for $S = R - P$ is usually denoted by R_P.

4. If $S = \{a^n \,|\, n \geq 0\}$, $a \in R$ non-nilpotent, then R_S is usually denoted by R_a.

Proposition 2.3.6 *The canonical map $f : R_S \otimes_R M \to M_S$, defined by $f[(a/s) \otimes x] \to (ax/s)$, is well defined and is an isomomorphism of R_S-modules.*

Proof We first show that f is well defined. If $(a/s) = (b/t)$, then $s_1(ta - sb) = 0$ for some $s_1 \in S$. This implies that $s_1(ta - sb)x = 0$, i.e., $(ax/s) = (bx/t)$. Clearly, f is surjective as $f((1/s) \otimes x) = (x/s)$. It remains to show that f is injective. Observe that any element of $R_S \otimes M$ is of the form $1/s \otimes y, s \in S,\ y \in M$, because any $z \in R_S \otimes M$ can be written as $z = \Sigma_i (a_i/s_i) \otimes x_i, a_i \in R,\ s_i \in S,\ x_i \in M$. If $s = \Pi_i s_i$ and $t_i = \Pi_{j \neq i} s_j$, then

$$z = \Sigma_i (a_i t_i/s) \otimes x_i = \Sigma_i 1/s \otimes (a_i t_i x_i) = 1/s \otimes y,$$

where $y = \Sigma_i a_i t_i x_i$.

Now, if $z = 1/s \otimes y \in \operatorname{Ker} f$, then $f((1/s) \otimes y) = (y/s) = 0$. This implies that $s_1 y = 0$ for some $s_1 \in S$. Now, $(1/s) \otimes y = (s_1/s_1 s) \otimes y = (1/ss_1) \otimes s_1 y = 0$. Hence f is injective. Clearly, f is an R_S-module homomorphism. Hence it is an isomorphism of R_S-modules. □

Proposition 2.3.7 *Let $f : M \to N$ be a homomorphism of R-modules. Then $f_S : M_S \to N_S$, defined by $f_S(x/s) = (f(x)/s)$, is well defined and is a homomorphism of R_S-modules. Moreover,*

(i) $f_S = \mathrm{Id}$, when $f = \mathrm{Id}$ and

(ii) $(gf)_S = g_S f_S$ for any other R-homomorphism $g : N \to L$.

Proof We first show that f_S is well defined. If $(x/s) = (y/t)$, then $s_1(tx - sy) = 0$ for some $s_1 \in S$. Then $f(s_1(tx - sy)) = s_1 t f(x) - s_1 s f(y) = 0$, i.e., $(f(x)/s) = (f(y)/t)$. Clearly, f_S is an R_S-module homomorphism as $f_S[(a/s)(x/t)] = f_S(ax/st) = (f(ax)/st) = (af(x)/st) = (a/s)(f(x)/t) = (a/s)f_S(x/t)$. The two relations (i) and (ii) are clear from the definition of f_S. □

Proposition 2.3.8 *Let $M' \xrightarrow{f} M \xrightarrow{g} M''$ be an exact sequence of R-modules. Then the induced sequence*

$$M'_S \xrightarrow{f_S} M_S \xrightarrow{g_S} M''_S$$

is exact.

Proof Since $gf = 0$, we have $(gf)_S = g_S f_S = 0$. Hence $\mathrm{Im}(f_S) \subseteq \mathrm{Ker}(g_S)$. Conversely, let $(x/s) \in \mathrm{Ker}(g_S)$, so that $(g(x)/s) = 0$, i.e., $s_1 g(x) = 0$ for some $s_1 \in S$. Then $g(s_1 x) = 0$, i.e., $s_1 x \in \operatorname{Ker} g = \operatorname{Im} f$. Write $s_1 x = f(y)$ for some $y \in M$. Then

$$(x/s) = (s_1 x/s_1 s) = (f(y)/ss_1) = f_S(y/ss_1) \in \operatorname{Im} f_S.$$

Hence $\operatorname{Im} f_S = \operatorname{Ker} g_S$ and the sequence is exact. □

Corollary 2.3.9 *If N is a submodule of M, then N_S is a submodule of M_S and $(M/N)_S \simeq M_S/N_S$.*

Proof The exactness of the sequence $0 \to N \to M \to M/N \to 0$ implies the exactness of $0 \to N_S \to M_S \to (M/N)_S \to 0$. □

Corollary 2.3.10 *R_S is a flat R-module.*

Proof Consider the exact sequence of R-modules $0 \to M' \to M \to M'' \to 0$. Then the sequence $0 \to M'_S \to M_S \to M''_S \to 0$ is exact. By Proposition 2.3.6, $M_S \simeq R_S \otimes_R M$, making the following diagram commutative:

$$\begin{array}{ccccccccc} 0 & \longrightarrow & M'_S & \longrightarrow & M_S & \longrightarrow & M''_S & \longrightarrow & 0 \\ & & \downarrow & & \downarrow & & \downarrow & & \\ 0 & \longrightarrow & R_S \underset{R}{\otimes} M' & \longrightarrow & R_S \underset{R}{\otimes} M & \longrightarrow & R_S \underset{R}{\otimes} M'' & \longrightarrow & 0 \end{array}$$

Hence the sequence $0 \to R_S \otimes_R M' \to R_S \otimes_R M \to R_S \otimes_R M'' \to 0$ is exact. This implies that R_S is R-flat. □

Example 2.3.11 $\mathbb{Q}$ is $\mathbb{Z}$-flat. (Compare with Exercise 10(e), Section 1.4.)

Corollary 2.3.12 *For any two R-modules M and N,*

$$(M \otimes_R N)_S \simeq M_S \otimes_{R_S} N_S.$$

Proof

$$\begin{aligned} (M \otimes_R N)_S &\simeq R_S \otimes_R (M \otimes_R N) \simeq (R_S \otimes_R M) \otimes_R N \\ &\simeq M_S \otimes_R N \simeq (M_S \otimes_{R_S} R_S) \otimes_R N \\ &\simeq M_S \otimes_{R_S} (R_S \otimes_R N) \\ &\simeq M_S \otimes_{R_S} N_S. \end{aligned}$$

□

Corollary 2.3.13 *For any two R-modules M and N and a prime ideal P of R, we have $(M \otimes_R N)_P \simeq M_P \otimes_{R_P} N_P$.*

Proof Take $S = R - P$. □

Proposition 2.3.14 *Let S be a multiplicatively closed set and $f : R \to R_S$ the natural map given by $f(a) = (a/1)$. Then*

(i) Every ideal of R_S is an extended ideal.

(ii) The prime ideals of R_S are in $(1, 1)$ correspondence with the prime ideals of R not intersecting S.

(iii) f preserves the ideal operations of taking finite sums, products, intersection and radical.

Proof (i) Let J be an ideal of R_S and let $I = f^{-1}(J) = \{a \in R \mid (a/1) \in J\}$. Then $I^e \subset J$.

Conversely, if $(a/s) \in J$, then $(a/1) = (s/1)(a/s) \in J$ so that $a \in I$ and $(a/s) \in I^e$. Hence $J = I^e$. This shows that J is the extension of I.

(ii) Any prime ideal of R_S contracts to a prime ideal of R. Conversely, let P be a prime ideal of R and consider its extension $P^e = PR_S$. Then $P^e = R_S$ if and only if $P \cap S \neq \emptyset$. This is so because $t \in P \cap S$ implies $(t/1) \in P^e$ and since $t \in S$, $t/1$ is a unit showing $P^e = R_S$.

Conversely, if $P^e = R_S$, then $(1/1) = (a/s)$ for some $a \in P$, and $s \in S$. This implies that $a \in S$ and hence $P \cap S \neq \emptyset$. Thus, P^e is a proper ideal of R_S if and only if $P \cap S = \emptyset$. It remains to show that for any prime ideal of P of R with $P \cap S = \emptyset$, P^e is a prime ideal of R_S. If a product $(a/s)(b/t) \in P^e$, then for some $s_1 \in S$, $s_1 ab \in P$. Since $P \cap S = \emptyset$, we have $s_1 \notin P$, so that $ab \in P$, i.e., $a \in P$ or $b \in P$. This implies $(a/s) \in P^e$ or $(b/t) \in P^e$.

(iii) The verification that f preserves finite sums, products and intersections of ideal is routine. We now show that f preserves the operation of taking radical. Let I be an ideal of R and I^e its extension under f.

If $b \in (\sqrt{I})^e$, then $b = (a/s)$, $a \in \sqrt{I}$, $s \in S$, so that $b^n = (a^n/s^n) \in I^e$ for some $n \geq 1$, i.e., $b \in \sqrt{I^e}$.

Conversely, assume that $b \in \sqrt{I^e}$ so that $b^n \in I^e$ for some $n \geq 1$. Write $b = (a/s)$ so that $b^n = (a^n/s^n) \in I^e$. This implies that $s_1 a^n \in I$ for some $s_1 \in S$. Hence $s_1^n a^n \in I$, i.e., $s_1 a \in \sqrt{I}$ and $b = (s_1 a/s_1 s) \in (\sqrt{I})^e$. Hence $\sqrt{I^e} = (\sqrt{I})^e$. □

Corollary 2.3.15 *The nil radical of R extends to the nil radical of R_S.*

Proof Consider $I = (0)$ in the relation $\sqrt{I^e} = (\sqrt{I})^e$. □

Corollary 2.3.16 *For any prime ideal P of the ring R, the prime ideals of R_P are in $(1, 1)$ correspondence with prime ideals of R contained in P.*

Proof Take $S = R - P$ in (ii) in Proposition 2.3.14. □

Corollary 2.3.17 *Let S be a multiplicatively closed set. Any ideal P of R maximal with respect to the property $P \cap S = \emptyset$ is a prime ideal.*

Proof Clearly, PR_S is a proper ideal of R_S, as $P \cap S = \emptyset$. Moreover, it is a maximal ideal of R_S as P is maximal with the property $P \cap S = \emptyset$. Hence its contraction under $f : R \to R_S$ is a prime ideal of R but the contraction is P. Hence P is a prime ideal of R. □

Corollary 2.3.18 *Let S be a multiplicatively closed set and I an ideal of R with $I \cap S = \emptyset$. Then there exists a prime ideal $P \supset I$ such that $P \cap S = \emptyset$.*

Proof IR_S is a proper ideal of R_S and hence it is contained in some maximal ideal of R_S. The contraction of this maximal ideal gives the required prime ideal. □

Proposition 2.3.19 (Localisation and Quotients) *Let I be an ideal of R and $p : R \to R/I$ the natural projection. For any multiplicatively closed set S of R not intersecting I, let T be its image under p. Then p defines a ring homomorphism $\overline{p} : R_S \to (R/I)_T$ which is surjective and $R_S/IR_S \simeq (R/I)_T$.*

Proof Since $S \cap I = \emptyset$, we have $0 \notin T$ and T is multiplicatively closed. The mapping $\overline{p} : R_S \to (R/I)_T$, given by $\overline{p}(a/s) = (p(a)/p(s))$, $a \in R$, $s \in S$, is a well defined ring homomorphism and is surjective. If $(a/s) \in \operatorname{Ker} \overline{p}$, then $tp(a) = 0$ for some $t \in T$. If $t = p(s_1)$, then $p(s_1 a) = 0$ and hence $s_1 a \in I$. This implies that $(a/s) \in IR_S$. Hence, $\operatorname{Ker} \overline{p} = IR_S$ and $R_S/IR_S \simeq (R/I)_T$. □

Proposition 2.3.20 (Localisation of Localisation) *Let S and S' be multiplicatively closed sets in R with $S \subset S'$ and $h : R_S \to R_{S'}$ the natural map defined by $h(a/s) = (a/s)$, $s \in S$, $a \in R$. If $f : R \to R_S$ is the natural map given by $a \mapsto (a/1)$, then h maps $f(S')$ onto units of $R_{S'}$ and induces a ring homomorphism $\overline{h} : (R_S)_{f(S')} \to R_{S'}$ which is an isomorphism.*

Proof Clearly, any element of $f(S')$ is of the type $(a/1)$, $a \in S'$ and its image under h is a unit in $R_{S'}$. By Proposition 2.3.4, there is an induced ring homomorphism $\overline{h} : (R_S)_{f(S')} \to R_{S'}$ given by

$$\overline{h}\Big((a/s)/(s'/1)\Big) = (a/ss'),\ s \in S,\ s' \in S',\ a \in R.$$

Clearly, h is surjective. Moreover, $\lambda = (a/s)/(s'/1) \in \operatorname{Ker} \overline{h}$ implies $s'_1 a = 0$ for some $s'_1 \in S'$. This implies $\lambda = 0$ in $(R_S)_{f(S')}$ as $(s'_1/1)(a/s) = 0$. Hence $\overline{h}$ is an isomorphism. □

Corollary 2.3.21 *Let S be a multiplicatively closed set and P a prime ideal of R with $S \cap P = \emptyset$. Then $R_P \simeq (R_S)_P$.*

Proof Take $S' = R - P$. □

A property is called a *local property* if its validity for a ring or a module is equivalent to its validity for each of the localisations under prime (maximal) ideals. We study some local properties.

Proposition 2.3.22 *Let M be an R-module. The following conditions are equivalent:*

(*i*) $M = 0$.

(ii) $M_P = 0$ *for all prime ideals* P.

(iii) $M_{\mathfrak{m}} = 0$ *for all maximal ideals* $\mathfrak{m}$.

Proof Clearly, (i) implies (ii), which in turn implies (iii). Assume now condition (iii) and suppose, if possible, there exists $x \in M, x \neq 0$. If $I = \text{Ann}(x)$, then $I \neq R$, as $1 \notin I$. Choose a maximal ideal $\mathfrak{m}$ with $I \subset \mathfrak{m}$. By (iii), $(x/1) \in M_{\mathfrak{m}}$ is zero, and hence there exists $a \notin \mathfrak{m}$ such that $ax = 0$, a contradiction as $\text{Ann}(x) \subseteq \mathfrak{m}$. □

Corollary 2.3.23 *Let* M *and* N *be an* R*-modules and* $f : M \to N$ *an* R*-homomorphism. The following conditions are equivalent:*

(i) f *is injective (surjective).*

(ii) $f_P : M_P \to N_P$ *is injective (surjective) for all prime ideals* P.

(iii) $f_{\mathfrak{m}} : M_{\mathfrak{m}} \to N_{\mathfrak{m}}$ *is injective (surjective) for all maximal ideals* $\mathfrak{m}$.

Proof Proposition 2.3.8 shows that (i) implies (ii), which clearly implies (iii). Assume now (iii) and consider an R-homomorphism $f : M \to N$. Let $K(\text{resp.}\, L)$ be $\text{Ker}\, f(\text{resp. Coker}\, f)$. Then there is an exact sequence of R-modules

$$0 \to K \xrightarrow{i} M \xrightarrow{f} N \xrightarrow{p} L \to 0.$$

This implies the exactness of

$$0 \to K_{\mathfrak{m}} \xrightarrow{i_{\mathfrak{m}}} M_{\mathfrak{m}} \xrightarrow{f_{\mathfrak{m}}} N_{\mathfrak{m}} \xrightarrow{p_{\mathfrak{m}}} L_{\mathfrak{m}} \to 0$$

for all maximal ideals $\mathfrak{m}$. Since $f_{\mathfrak{m}}$ is injective (surjective), $K_{\mathfrak{m}}$ (resp. $L_{\mathfrak{m}}$) is zero for each $\mathfrak{m}$. This implies K (resp. L) is zero. Hence f is injective (surjective). □

Proposition 2.3.24 *Let* M *be an* R*-module. The following conditions are equivalent:*

(i) M *is* R*-flat.*

(ii) M_P *is* R_P*-flat for every prime ideal* P.

(iii) $M_{\mathfrak{m}}$ *is* $R_{\mathfrak{m}}$*-flat for every maximal ideal* $\mathfrak{m}$.

Proof Since M is R-flat, its scalar extension $M_P \simeq R_P \otimes_R M$ is R_P-flat by Proposition 1.4.7. This shows that (i) implies (ii), which in turn implies (iii). Assume now condition (iii) and consider an injective homomorphism $0 \to N' \xrightarrow{f} N$. It is sufficient to show that $0 \to N' \otimes M \xrightarrow{f \otimes 1} N \otimes M$ is injective. Now, $0 \to N'_{\mathfrak{m}} \xrightarrow{f_{\mathfrak{m}}} N_{\mathfrak{m}}$ is injective, and since $M_{\mathfrak{m}}$ is flat over $R_{\mathfrak{m}}$,

$$0 \to N'_{\mathfrak{m}} \otimes_{R_{\mathfrak{m}}} M_{\mathfrak{m}} \to N_{\mathfrak{m}} \otimes_{R_{\mathfrak{m}}} M_{\mathfrak{m}}$$

is injective. By Corollary 2.3.13, $N_{\mathfrak{m}} \otimes_{R_{\mathfrak{m}}} M_{\mathfrak{m}} \simeq (N \otimes_R M)_{\mathfrak{m}}$ under a natural isomomorphism. It follows that $0 \to (N' \otimes_R M)_{\mathfrak{m}} \to (N \otimes_R M)_{\mathfrak{m}}$ is injective. This implies from Corollary 2.3.23 that $0 \to N' \otimes M \to N \otimes M$ is injective. Hence M is R-flat. $\square$

We show using localisation that for any ring homomorphism $f : R \to S$ with S faithfully flat over R, prime ideals of R are precisely obtained by contracting prime ideals of S and that this property characterises faithfully flat extensions.

Proposition 2.3.25 *Let $f : R \to S$ be a homomorphism of rings and P a prime ideal of R. Then P is the contraction of a prime ideal of S if and only if $P = P^{ec}$.*

Proof If P is the contraction of a prime ideal, then $P = P'^c$, P' a prime ideal in S. Then $P^{ec} = P'^{cec} = P'^c = P$.

Conversely, assume that $P = P^{ec}$ and let $T = f(R - P)$. Then P^e does not intersect T and hence by Corollary 2.3.18, there exists a prime ideal Q of S containing P^e with $Q \cap T = \emptyset$. Since $Q \supseteq P^e$, $Q^c \supseteq P^{ec} = P$. Since $Q \cap T = \emptyset$, we have $Q^c \cap (R - P) = \emptyset$, i.e., $Q^c \subseteq P$ and hence $Q^c = P$. $\square$

Theorem 2.3.26 *Let $f : R \to S$ be a homomorphism of rings. If S is a faithfully flat R-module, then every prime ideal of R is the contraction of a prime ideal of S. Conversely, if $f : R \to S$ is a ring homomorphism such that S is flat over R and every prime ideal of R is the contraction of a prime ideal of S, then S is faithfully flat over R.*

Proof Assume S is faithfully flat over R. Then the map $M \to M \otimes_R S$, given by $x \to x \otimes 1$, is injective. (See Exercises is Section 1.4.) In particular, if $M = R/I$, I any ideal of R, then the map $R/I \to R/I \otimes_R S$ is injective. This implies $I^{ec} = I$ for any ideal I of R. The result now follows from Proposition 2.3.25.

Conversely, assume that S is R-flat and $P^{ec} = P$ for all prime ideals P in R. Then for each maximal ideal $\mathfrak{m}$ of R, $\mathfrak{m}^{ec} = \mathfrak{m}$. In particular, $\mathfrak{m}^e \neq S$. This implies that S is a faithfully flat R-module by Corollary 1.4.11. $\square$

2.3. EXERCISES

1. If M is a finitely generated R-module, then show that $M_S = 0$ if and only if there exists some $s \in S$ with $sM = 0$.

2. Let I be an ideal of R and $S = 1 + I$. Show that the extension of I under $R \to R_S$ is contained in the Jacobson radical of R_S.

3. Let M be an R-module and I an ideal of R such that $M_{\mathfrak{m}} = 0$ for all maximal ideals $\mathfrak{m} \supseteq I$. Show that $M = IM$.

4. Let S be the set of non-zero-divisors of R. The ring R_S is called the total ring of fractions of R. Show

    (a) S is the largest multiplicatively closed set for which $R \to R_S$ is injective.

(b) Every element of R_S is either a unit or a zero-divisor.

(c) If R is such that every non-unit is a zero-divisor, then $R \to R_S$ is bijective.

5. Let $S \subset T$ be multiplicatively closed sets. Show that the natural map $R_S \to R_T$ given by $(a/s) \to (a/s)$ is bijective if and only if every prime ideal of R which intersects T also intersects S.

6. A multiplicatively closed set S is called saturated if $ab \in S$ implies $a \in S$ and $b \in S$. Show that

    (a) S is saturated if and only if $R - S$ is a union of prime ideals.

    (b) Every multiplicatively closed set S is contained in a smallest saturated multiplicatively closed set $\overline{S}$.

    (c) For mulitplicatively closed sets S and T, we have $S \subseteq \overline{T}$ if and only if every R-module M with $M_S = 0$ satisfies $M_T = 0$.

7. Show that for finitely presented R-modules M and N,

$$(\mathrm{Hom}_R(M, N))_S \simeq \mathrm{Hom}_{R_S}(M_S, N_S)$$

for any mulitplicatively closed set S.

8. Show that an R-module (R a domain) is torsion-free if and only if M_P is torsion-free for all prime ideals P.

9. Show that a ring R has no non-zero nilpotent elements if and only if R_P has the same property for every prime ideal P. Is it true that R_P is a domain for every prime ideal P implies that R is a domain? Why?

10. Give an example of an R-module M to show that $\{\mathrm{Ann}_R(M)\}_S$ may not be isomorphic to $\mathrm{Ann}_{R_S}(M_S)$.

2.4 Applications

This section is concerned with the applications of localisation in the study of projective modules as locally free modules.

Proposition 2.4.1 *Let $u : M \to N$ be a homomorphism of R-modules.*

(i) If N is finitely generated and $u_P : M_P \to N_P$ is surjective for some prime ideal P, then there exists some $f \in R - P$ such that $u_f : M_f \to N_f$ is surjective.

(ii) If M is finitely generated, N finitely presented and u_P injective for some prime ideal P, then there exists some $g \in R - P$ such that u_g is bijective.

Proof Let K (resp. L) be the kernel (resp. cokernel) of u. Consider the exact sequence

$$0 \to K \xrightarrow{i} M \xrightarrow{u} N \xrightarrow{P} L \to 0.$$

The sequence $0 \to K_P \to M_P \xrightarrow{u_P} N_P \to L_P \to 0$ is also exact. Assume now that u_P is surjective so that $L_P = 0$. Since N is finitely generated, L is also finitely generated and hence (Exercise 1, Section 2.3) there exists some $f \in R - P$ with $fL = 0$. This implies $L_f = 0$. Hence from the exactness of the sequence $M_f \to N_f \to L_f \to 0$, it follows that $u_f : M_f \to N_f$ is surjective.

Assume now that u_P is bijective. By (i), there exists some $f \in R - P$ with $L_f = 0$. Consider the exact sequence

$$0 \to K_f \to M_f \xrightarrow{u_f} N_f \to 0.$$

Since M is finitely generated and N is finitely presented, by Exercise 10 of Section 1.2, K_f is finitely generated. By Corollary 2.3.21, we have $(K_f)_{PR_f} \simeq K_P = 0$. Hence there exists some $f_1 \in R_f - PR_f$ with $f_1 K_f = 0$. Write $f_1 = a/f^r$, $a \in R - P$. Since $(f/1)$ is invertible in R_f, $(a/1)K_f = 0$. Hence $K_{fa} \simeq (K_f)_{(a/1)} = 0$. Then $g = fa \in R - P$ is the required element. □

Corollary 2.4.2 *Let N be a finitely presented R-module such that N_P is a free R_P-module of rank n. Then there exists some $f \in R - P$ with N_f free of rank n over R_f.*

Proof Choose $\{x_1, \ldots, x_n\}$ in N such that $\{(x_i/1)\}$ is a basis of N_P over R_P. Consider the R-homomorphism $u : R^n \to N$, where R^n is free with basis $\{e_1, \ldots, e_n\}$, and $u(e_i) = x_i$, $1 \leq i \leq n$. Since by assumption u_P is bijective, there exists some $f \in R - P$ with u_f bijective. Hence N_f is free of rank n over R_f. □

Proposition 2.4.3 *Let $(f_i)_{1 \leq i \leq n}$ be elements of R such that $\Sigma_{i=1}^n Rf_i = R$. Then the ring $S = \Pi_{i=1}^n R_{f_i}$ is a faithfully flat R-module.*

Proof By Corollary 2.3.10, each R_{f_i} is R-flat and hence S is flat over R. Let P be a prime ideal of R. Since $\Sigma_{i=1}^n Rf_i = R$, there exists some $f_i \notin P$, say $f_1 \notin P$. Then $P_{f_1} = PR_{f_1}$ is a prime ideal of R_{f_1} and $PS \subset PR_{f_1} \times \Pi_{i=2}^n R_{f_i} \neq S_{f_1}$ since $PR_{f_1} \neq R_{f_1}$. Hence S is faithfully flat over R by Corollary 1.4.11. □

Corollary 2.4.4 *Let $(f_i)_{1 \leq i \leq n}$ be elements of R such that $\Sigma Rf_i = R$. An R-module M is finitely generated (finitely presented) if and only if each M_{f_i} is finitely generated (finitely presented) over R_{f_i}.*

Proof Clearly, M finitely generated (finitely presented) implies the same for each M_{f_i}. Conversely, assume that each M_{f_i} is finitely presented over R_{f_i}, $1 \leq i \leq n$. Then there exists an exact sequence

$$R_{f_i}^m \to R_{f_i}^p \to M_{f_i} \to 0$$

where m and p can be chosen to be independent of i. If $M' = \Pi_{i=1}^{n} M_{f_i}$, then $M' \simeq M \otimes_R S$, where $S = \Pi_i R_{f_i}$. Since S is faithfully flat over R, the result follows from Proposition 1.4.14. The finitely generated case follows similarly. □

Definition 2.4.5 For any commutative ring R, $X = \operatorname{Spec} R$ denotes the set of prime ideals of R.

For an ideal I of R, $V(I)$ denotes the set of prime ideals of R containing I and $D(f) = X - V(Rf)$, $f \in R$.

Proposition 2.4.6 *$X = \operatorname{Spec} R$ is a topological space in which the closed sets are precisely $\{V(I) \mid I$ an ideal of $R\}$ and the collection $\{D(f)\}_{f \in R}$ is a base for open sets. Moreover, X is quasi-compact.*

Proof The axioms for closed sets are satisfied in view of the following properties of $V(I)$, which follow readily from the definition.

(i) $V(R) = \emptyset$, $V(0) = X$.

(ii) $V(I_1) \cup V(I_2) = V(I_1 I_2)$.

(iii) $\cap_\alpha V(I_\alpha) = V(\Sigma_\alpha I_\alpha)$.

Let U be an open set in X so that $U = X - V(I)$ for some ideal I. By (iii), $\cap_{f \in I} V(Rf) = V(I)$, so that $U = X - V(I) = \cup_{f \in I} D(f)$, i.e., $\{D(f)\}$ is a base for open sets.

To prove that X is quasi-compact, consider an open covering of X by basic open sets, viz. $X = \cup_\alpha D(f_\alpha)$. If I is the ideal generated by the $\{f_\alpha\}$, then $V(I) = \cap_\alpha V(Rf_\alpha)$. Hence

$$X - V(I) = \cup_\alpha D(f_\alpha) = X.$$

Hence $V(I) = \emptyset$, i.e., $I = R$. In particular, $1 = \sum_{i=1}^{m} g_{\alpha_i} f_{\alpha_i}$, $g_{\alpha_i} \in R$. This implies $\emptyset = V(1) = V(\Sigma_{i=1}^{m} Rf_{\alpha_i}) = \cap_{i=1}^{m} V(f_{\alpha_i})$, i.e., $X = \cup_{i=1}^{m} D(f_{\alpha_i})$, showing that X is quasi-compact. □

Corollary 2.4.7 $X = \overset{m}{\underset{i=1}{\cup}} D(f_{\alpha_i}) \Leftrightarrow R = \overset{m}{\underset{i=1}{\Sigma}} Rf_{\alpha_i}$.

Proof Follows from the last part of the above proof. □

Theorem 2.4.8 *Let M be an R-module. The following conditions are equivalent:*

(i) M is finitely generated projective over R.

(ii) M is finitely presented and $M_{\mathfrak{m}}$ is a free $R_{\mathfrak{m}}$-module for every maximal ideal $\mathfrak{m}$.

(iii) M is finitely generated and M_P is free R_P-module for every prime ideal P and the map rank: Spec $R \to Z$, *given by $P \mapsto \mathrm{rank}_{R_P} M_P$, is continuous.*

(iv) There exist elements $\{f_i\}_{1 \leq i \leq n}$ in R such that $R = \Sigma_{i=1}^n R f_i$ and each M_{f_i} is a free R_{f_i}-module of finite rank.

(v) For every maximal ideal $\mathfrak{m}$ of R, there exists some $f \in R - \mathfrak{m}$ with M_f free of finite rank over R_f.

Proof $(i) \Rightarrow (ii)$. Since M is finitely generated and projective, M is finitely presented. Moreover, M is finitely generated and projective implies $M_\mathfrak{m} = M \otimes_R R_\mathfrak{m}$ is also finitely generated and $R_\mathfrak{m}$-projective, and since $R_\mathfrak{m}$ is a local ring, $M_\mathfrak{m}$ is $R_\mathfrak{m}$-free (Corollary 2.2.17).

$(ii) \Rightarrow (v)$. Let $\mathfrak{m}$ be a maximal ideal of R. Since $M_\mathfrak{m} = R_\mathfrak{m} \otimes_R M$ is $R_\mathfrak{m}$-free, choose a basis $\{1 \otimes x_i\}$ of $M_\mathfrak{m}$ over $R_\mathfrak{m}$ $(1 \leq i \leq n)$. Consider the map $u : R^n \to M$ given by $u(e_i) = x_i$, $1 \leq i \leq n$, where $\{e_i\}$ is the standard base for R^n. By assumption, $u_\mathfrak{m}$ is an isomorphism. Hence by Corollary 2.4.2, u_f is an isomorphism for some $f \in R - \mathfrak{m}$. This implies (v).

$(v) \Rightarrow (iv)$. Let $Y = \{f \in R \,|\, M_f \text{free of finite rank over} R_f\}$. For any maximal ideal $\mathfrak{m}$ of R, $Y \not\subseteq \mathfrak{m}$ and hence Y generates the whole ring R. In particular, $1 = \Sigma_{i=1}^n a_i f_i$ for some $f_i \in Y$, $a_i \in R$. This implies (iv).

$(iv) \Rightarrow (iii)$. Clearly, M is finitely generated over R by Corollary 2.4.4. Since $\Sigma R f_i = R$, for each prime ideal P, there exists some i $(1 \leq i \leq n)$ such that $f_i \in R - P$. By Corollary 2.3.21, we have $M_P \simeq (M_{f_i})_{PR_{f_i}}$ and hence M_P is free over R_P of the same rank as M_{f_i}. Moreover, rank $M_P = r$ if and only if $P \in \cup_j D(f_j)$, where the union is taken over all j for which M_{f_i} is of rank r. Hence rank: $P \to \mathrm{rank}_{R_P} M_P$ is continuous.

$(iii) \Rightarrow (v)$. Let $\mathfrak{m}$ be a maximal ideal and choose $\{x_1, \ldots, x_n\}$ in M such that $\{(x_i/1)\}(1 \leq i \leq n)$ is a basis of $M_\mathfrak{m}$ over $R_\mathfrak{m}$. Define an R-linear map $u : R^n \to M$ by $u(e_i) = x_i$, $1 \leq i \leq n$, where (e_i) is the standard basis of R^n. The mapping $u_\mathfrak{m} : R^n_\mathfrak{m} \to M_\mathfrak{m}$ is surjective and since M is finitely generated, by Proposition 2.4.1, there exists some $g \in R - \mathfrak{m}$ such that $u_g : R^n_g \to M_g$ is surjective. Since rank $M_\mathfrak{m} = r$ and the rank function is continuous, there exists some $h \in R - \mathfrak{m}$ such that for all $P \in D(h)$, rank $M_P = r$. Let $f = gh$. Then $f \in R - \mathfrak{m}$ and the map $u_f : R^n_f \to M_f$ is surjective, as u_g is so and rank $M_P = r$ for all $P \in D(f)$. Thus, $u_P : R^n_P \to M_P$ is surjective for all $P \in D(f)$. Since R^n_P and M_P are both free modules of the same rank, u_P is an isomorphism for all $P \in D(f)$.

Let $\mathfrak{m}'$ be a maximal ideal of R_f and $\mathfrak{m}''$ its contraction to R, so that $\mathfrak{m}'' \in D(f)$.

Consider the commutative diagram

$$\begin{array}{ccc} (R_f^n)_{\mathfrak{m}'} & \xrightarrow{(u_f)_{\mathfrak{m}'}} & (M_f)_{\mathfrak{m}'} \\ \downarrow & & \downarrow \\ R^n_{\mathfrak{m}''} & \xrightarrow{(u)_{\mathfrak{m}''}} & M_{\mathfrak{m}''} \end{array}$$

where the vertical maps are isomorphisms given by Corollary 2.3.21. Since $\mathfrak{m}'' \in D(f)$, $u_{\mathfrak{m}''}$ is an isomorphism, this implies that the horizontal map above is an isomorphism for each maximal ideal $\mathfrak{m}'$ of R_f. This implies by Corollary 2.3.23 that M_f is isomorphic to R_f^n, i.e., M_f is free of rank n over R_f.

$(iv) \Rightarrow (i)$. For each i, choose a free R_{f_i} module L_i such that M_{f_i} is a direct summand of L_i; we may assume without loss of generality that all the L_i are of the same rank. Then $L = \Pi_i L_i$ is free over $S = \Pi_i R_{f_i}$ and $M' = \Pi_i M_{f_i}$ is a direct summand of L. Hence M' is finitely generated projective over S. Since $M' \simeq M \otimes_R S$ and S is faithfully flat over R, M is finitely generated projective over R (Proposition 1.4.14). □

Corollary 2.4.9 *Any finitely presented flat module is projective.*

Proof Follows from condition (ii) and Corollary 2.2.17. □

If R is a local ring and M a finitely generated projective R-module, then M is free, say of rank n. Then M_P is also R_P-free of rank n, so that the function $P \to$ rank M_P is constant. However, in general, the rank function is only locally constant by condition (iii) of Theorem 2.4.8. It will be a constant if Spec R is connected. The following Proposition gives the condition under which Spec R is connected.

Proposition 2.4.10 $X =$ Spec R *is disconnected* $\Leftrightarrow 1 = e_1 + e_2$, *where* e_1, e_2 *are mutually orthogonal, non-trivial idempotents in* R.

Proof Let X be disconnected so that $X = Y_1 \cup Y_2$, Y_1, Y_2 non-empty closed, $Y_1 \cap Y_2 = \emptyset$. Write $Y_i = V(I_i)$, $i = 1, 2$. Then $V(I_1 + I_2) = Y_1 \cap Y_2 = \emptyset$ implies $I_1 + I_2 = R$. Let $1 = e_1 + e_2$. Since $X = Y_1 \cup Y_2 = V(I_1) \cup V(I_2) = V(I_1 I_2)$, i.e., $I_1 I_2 \subseteq \text{nil}(R)$, we have $e_1 e_2$ is nilpotent, i.e., $e_1^n e_2^n = 0$. Replacing I_1, I_2 by Re_1^n and Re_2^n, we may assume that $I_1 \cap I_2 = 0$. Hence e_1, e_2 are mutually orthogonal idempotents. They are non-trivial as $Y_i \neq \emptyset$, $i = 1, 2$. Conversely, assume that e_1, e_2 are mutually orthogonal non-trivial idempotents with $1 = e_1 + e_2$. Let $Y_i = V(R(1 - e_i))$, $i = 1, 2$. The relation $1 = 1 - e_1 + e_1(1 - e_2)$ shows that $\emptyset - V(1) - Y_1 \cap Y_2$. Also, $Y_1 \cup Y_2 = V(R(1 - e_1)(1 - e_2)) = V(0) = X$. Since e_i are non-trivial, $Y_i \neq \emptyset$ $(i = 1, 2)$ so that S is disconnected. □

Corollary 2.4.11 Spec R *is connected* $\Leftrightarrow$ R *has no idempotents other than* 0 *or* 1. *In particular, if* R *is a domain, then* Spec R *is connected and the rank function is constant.*

Definition 2.4.12 A projective R-module M is said to be of *constant rank* n if for every prime ideal P of R, M_P is free of rank n over R_P.

Corollary 2.4.13 *For an R-module M and $n \geq 1$, the following conditions are equivalent.*

(i) *M is projective of rank n over R.*

(ii) *$M_{\mathfrak{m}}$ is free of rank n over $R_{\mathfrak{m}}$ for every maximal ideal $\mathfrak{m}$.*

(iii) *M_P is free of rank n over R_P for every prime ideal P of R.*

(iv) *For every maximal ideal $\mathfrak{m}$, there exists some $f \in R - \mathfrak{m}$ with M_f free of rank n over R_f.*

Proof Follows from Theorem 2.4.8. □

Proposition 2.4.14 (Patching up of Localisations) *Let R be a commutative ring, $f_1, f_2 \in R$ with $Rf_1 + Rf_2 = R$. Let M_1 and M_2 be R_{f_1} and R_{f_2}-modules respectively such that $(M_1)_{f_2} \simeq (M_2)_{f_1}$. Then there is an R-module M with $M_i \cong M_{f_i}$, $i = 1, 2$. If M_1 and M_2 are finitely generated (finitely presented, finitely generated projective), then M is also finitely generated (finitely presented, finitely generated projective).*

Proof Let $\alpha : (M_1)_{f_2} \to (M_2)_{f_1}$ be the given isomorphism. Define a map $\phi : M_1 \oplus M_2 \to (M_2)_{f_1}$ by $\phi(x, y) = \alpha(x_{f_2}) - y_{f_1}$, where x_{f_2} and y_{f_1} are the images of x and y in $(M_1)_{f_2}$ and $(M_2)_{f_1}$ respectively. If $M = \text{Ker}\, \phi$, then we show that M is the required R-module.

Consider the exact sequence

$$0 \to M \to M_1 \oplus M_2 \xrightarrow{\phi} (M_2)_{f_1}.$$

Localising at f_1, we have an exact sequence

$$0 \to M_{f_1} \to M_1 \oplus (M_2)_{f_1} \xrightarrow{\phi_{f_1}} (M_2)_{f_1}.$$

Now, ϕ_{f_1} restricted to $(M_2)_{f_1}$ is an isomorphism so that $M_{f_1} \simeq M_1$. Similarly, localising at f_2, we have an exact sequence

$$0 \to M_{f_2} \to (M_1)_{f_2} \oplus M_2 \xrightarrow{\phi_{f_2}} (M_2)_{f_1},$$

which gives the isomorphism $M_{f_2} \simeq M_2$.

Suppose now that M_1 and M_2 are finitely generated. Choose $x_1, x_2, \ldots, x_n \in M$ such that if M' is the submodule generated by $\{x_1, \ldots, x_n\}$, then $M'_{f_1} = M_1$ and $M'_{f_2} = M_2$. Then for any prime ideal P, either $f_1 \notin P$ or $f_2 \notin P$ since $Rf_1 + Rf_2 = R$. Hence $M'_P \simeq M_P$. Thus, $M \simeq M'$ and M is finitely generated.

If M_1 and M_2 are finitely presented, then there exists an exact sequence

$$0 \to K \to F \to M \to 0,$$

where K_{f_1} and K_{f_2} are finitely generated and F is free of finite rank. By the above argument, K is finitely generated and hence M is finitely presented.

If M_1 and M_2 are finitely generated and projective, then M_P is free for all prime ideals P and M is finitely presented. Therefore M is finitely generated and projective by Theorem 2.4.8. □

We now study properties of rank one projective modules.

Theorem 2.4.15 *Let M be a finitely generated R-module. The following conditions are equivalent.*

(i) M is projective of rank 1 over R.

(ii) $M \otimes_R N$ is isomorophic to R for some R-module N.

Moreover, if (ii) holds, then N is isomorphic to the dual module $M^* = \mathrm{Hom}_R(M, R)$.

Proof $(i) \Rightarrow (ii)$ Let $u : M \otimes_R M^* \to R$ be the natural map defined by $u(x, f) = f(x)$, $x \in M$, $f \in M^*$. If $\mathfrak{m}$ is any maximal ideal of R, $(M \otimes_R M^*)_\mathfrak{m} \simeq M_\mathfrak{m} \otimes_{R_\mathfrak{m}} (M^*)_\mathfrak{m}$. Since M is finitely presented and $R_\mathfrak{m}$ is a flat R-algebra, it follows from Exercise 7, Section 2.3, that the localisation map $u_\mathfrak{m} : (M \otimes_R M^*)_\mathfrak{m} \to R_\mathfrak{m}$ induces a map $u_\mathfrak{m} : M_\mathfrak{m} \otimes_{R_\mathfrak{m}} (M_\mathfrak{m})^* \to R_\mathfrak{m}$. Since $M_\mathfrak{m}$ is free of rank one over $R_\mathfrak{m}$, $u_\mathfrak{m}$ is an isomorphism and hence u is an isomorphism. This implies (ii).

$(ii) \Rightarrow (i)$. By Theorem 2.4.8, it is enough to show that M is free of rank one under the assumption that R is a local ring. Let $\mathfrak{m}$ be the unique maximal ideal of R and $k = R/\mathfrak{m}$. Given an isomorphism $u : M \otimes_R N \to R$, we have an isomorphism $u : M/\mathfrak{m}M \otimes_k N/\mathfrak{m}N \to k$. Thus, $M/\mathfrak{m}M$ has rank one as a k-vector space. Hence M is R-cyclic. The annihilator of M also annihlates $M \otimes N$ and hence the annihilator of M is zero. Thus, M is isomorphic to R and it is free of rank one over R.

Finally, if $M \otimes N \simeq R$, then $N \simeq M^*$ under the isomorphisms

$$N \simeq R \underset{R}{\otimes} N \simeq (M \underset{R}{\otimes} M^*) \underset{R}{\otimes} N$$

$$\simeq (M \underset{R}{\otimes} N) \underset{R}{\otimes} M^* \simeq R \underset{R}{\otimes} M^* \simeq M^*.$$ □

Corollary 2.4.16 *If M and N are projective R-modules of rank one, $M \otimes_R N$ and M^* are projective R-modules of rank one.*

Proof Clearly, $M \otimes_R N$ and M^* are R-projective if M is R-projective. The relations

(i) $\mathrm{rank}(M \otimes N) = (\mathrm{rank}\, M).(\mathrm{rank}\, N)$ and (*ii*) $\operatorname{rank} M^* = \operatorname{rank} M$

follow by localisation at every prime ideal. □

Definition 2.4.17 Consider the set C of isomorphism classes of projective R-modules of rank one. The operation on C given by $[M] + [N] = [M \otimes_R N]$, where $[M]$ is the isomorphism class containing M, is a well defined operation. Theorem 2.4.15 shows that C is an abelian group for this operation with $[R]$ as the zero element and $[M^*]$ as the inverse of $[M]$. This group is called the *Picard group* and is denoted by $\operatorname{Pic}(R)$.

The Picard group is important in the study of the geometry of hypersurfaces.

2.4. EXERCISES

1. Let P be a finitely generated projective R-module. Show that R and P can be decomposed as $R = \Pi_{i=1}^n R_i$ and $P = \Pi_{i=1}^n P_i$, where P_i is a projective R_i module of rank n_i, and the n_i can be distinct.

2. Let R be a domain with quotient field K and M an R-module. Show that the following conditions are equivalent.

   (i) M is finitely generated and projective over R.

   (ii) M is finitely generated and $M_{\mathfrak{m}}$ is free over $R_{\mathfrak{m}}$ for each maximal ideal $\mathfrak{m}$.

   (iii) M is projective and $\dim_K K \otimes_R M$ is finite.

3. Let M be a projective R-module of rank n. Show that there exists a finitely generated faithfully flat R-algebra A such that the scalar extension $A \otimes_R M$ is free of rank n over A.

4. Let $(f_i)_{1 \leq i \leq n}$ be elements of R such that $\Sigma_{i=1}^n Rf_i = R$ and M an R-module. Suppose $x_i \in M_{f_i}$ are such that x_i and x_j have the same image in $M_{f_i f_j}$ under the natural mappings $M_{f_i} \to M_{f_i f_j}$, and $M_{f_j} \to M_{f_i f_j}$, $i \neq j$. Show that there exists a unique $x \in M$ such that canonical image of x in M_{f_i} is x_i $(1 \leq i \leq n)$.

5. Show that if M and N are finitely generated projective modules of constant rank, then the following modules are also projective of constant rank and the ranks satisfy the conditions as below.

   (i) $\operatorname{rank}(M \oplus N) = \operatorname{rank} M + \operatorname{rank} N$.

   (ii) $\operatorname{rank}(M \otimes N) = (\operatorname{rank} M).(\operatorname{rank} N)$.

   (iii) $\operatorname{rank} \operatorname{Hom}_R(M, N) = (\operatorname{rank} M).(\operatorname{rank} N)$.

   (iv) $\operatorname{rank}(\wedge^k M) = \binom{n}{k}$, where $n = \operatorname{rank} M$.

6. If A is an R-algebra and P is a finitely generated projective R-module of rank n, show that $A \otimes_R P$ is finitely generated A-projective of rank n.

7. Let $f : R \to S$ be a ring homomorphism. Show that the natural map $\mathrm{Pic}(f) : \mathrm{Pic}(R) \to \mathrm{Pic}(S)$, given by

$$[P] \mapsto [S \otimes_R P],$$

is well defined and is a group homomorphism.

8. Let M be a finitely generated R-module. An element $x \in M$ is called *basic* if and only if its image in $\frac{M_P}{PM_P}$ is non-zero for every prime ideal P. Show that for a finitely generated projective R-module M, $x \in M$ is basic if and only if it is unimodular.

CHAPTER

3

NOETHERIAN RINGS

An affine variety obtained as the common zeros of polynomials of an ideal in $k[X_1, X_2, \ldots, X_n], k$ a field, can be obtained as the common zeros of a finite number of polynomials.

This is due to a theorem of Hilbert which says that every ideal in $k[X_1, X_2, \ldots, X_n], k$ a field, is finitely generated. Rings with this property are called Noetherian rings. Every ideal in such a ring can be written as a finite intersection of primary ideals. As one consequence, we get an algebraic analogue of the fact that any affine variety can be expressed as a finite union of irreducible subvarieties. In such a decomposition, there are embedded and non-embedded components. The latter are shown to be unique while the former may not be unique.

Noetherian rings are best studied in a general setup by studying Noetherian modules. These have the property that every submodule is finitely generated or equivalently any strictly increasing chain of submodules is finite. Modules with the corresponding property for decreasing chains lead to the notion of Artinian modules. Artinian rings are shown to be nothing but Noetherian rings in which every prime ideal is maximal. For finitely generated modules over an Artinian ring, we define the conecpt of length which generalises the concept of dimension of a vector space.

3.1 Noetherian modules

Proposition 3.1.1 *Let M be an R-module. The following are equivalent.*

(i) *Any non-empty collection of submodules of M has a maximal element.*

(ii) *For any increasing sequence of submodules of* M,

$$M_1 \subseteq M_2 \subseteq \cdots \subseteq M_n \subseteq \cdots,$$

there exists some integer m *such that* $M_k = M_m$ *for all* $k \geq m$.

(iii) *Every submodule of* M *is finitely generated.*

Proof $(i) \Rightarrow (ii)$. Let any increasing sequence of submodules of M, say $M_1 \subseteq M_2 \subseteq \cdots \subseteq M_n \subseteq \cdots$, be given. Consider the collection $\Sigma = \{M_n\}$. This collection has a maximal element M_m. Then $M_k = M_m$ for all $k \geq m$.

$(ii) \Rightarrow (iii)$. Let N be a submodule of M. Choose any $x_1 \in N$. If $Rx_1 \neq N$, choose $x_2 \in N$, $x_2 \notin Rx_1$. If $Rx_1 + Rx_2$ is not equal to N, continue this process. By assumption (ii), after a finite number of steps we have $Rx_1 + Rx_2 + \cdots + Rx_t = N$. Thus N is finitely generated.

$(iii) \Rightarrow (ii)$. Consider any increasing sequence of submodules of M, say $M_1 \subset M_2 \subseteq \cdots \subseteq M_n \subseteq \cdots$. Then their union $N = \cup_1^\infty M_n$ is a submodule of M, which is finitely generated by (iii). If $\{x_1, x_2, \ldots, x_t\}$ is a generating set for N, choose m such that $N = \{x_1, x_2, \ldots, x_t\} \subset M_m$. Then $M_k = M_m$ for all $k \geq m$.

$(ii) \Rightarrow$ (i). Let Σ be a non-empty collection of submodules of M. Choose any $M_1 \in \Sigma$. If M_1 is not maximal, choose $M_2 \in \Sigma$, $M_1 \subsetneqq M_2$. If M_2 is not maximal, continue this process.

After a finite number of steps, there exists some $M_m \in \Sigma$ which is maximal, for otherwise condition (ii) will be violated. □

Definition 3.1.2 An R-module M is called *Noetherian* if it satisfies any one of the above equivalent conditions.

Definition 3.1.3 A ring R is called a *Noetherian ring* if the R-module R is Noetherian.

Example 3.1.4 (i) Any field R is a Noetherian ring.

(ii) Any finite ring R is Noetherian.

(iii) Any principal ideal ring is Noetherian.

Proposition 3.1.5 *Let* $0 \to M' \to M \to M'' \to 0$ *be an exact sequence of* R*-modules. Then* M *is Noetherian if and only if both* M' *and* M'' *are Noetherian.*

Proof Let i denote the injective map from M' to M. Assume M is Noetherian. Since every submodule of M' is isomorphic to a submodule of M, M' is Noetherian. Now, every submodule L of M'' is isomorphic to $M_1/i(M')$, where M_1 is a submodule of M

containing $i(M')$. Since M is Noetherian, M_1 is finitely generated. Hence L is also finitely generated. This implies that M'' is Noetherian.

Conversely, assume that both M' and M'' are Noetherian. We have $M'' \cong M/i(M')$. We identify M' with $i(M')$. Then, for any submodule N of M, we have $N/(N \cap M') \simeq (N + M')/M'$, which is isomorphic to a submodule of M''. Since M'' is Noetherian, it is finitely generated. Since M' is Noetherian, $N \cap M'$ is finitely generated. These imply that N is finitely generated and hence M is Noetherian. □

Corollary 3.1.6 *If $\{M_i\}_{1 \leq i \leq n}$ are Noetherian, then $\oplus\Sigma_{i=1}^{n} M_i$ is Noetherian.*

Proof Consider the exact sequence

$$0 \to \oplus\Sigma_{i=1}^{n-1} M_i \to \oplus\Sigma_{i=1}^{n} M_i \to M_n \to 0,$$

where the first map is the inclusion and the second map is the projection onto M_n. The result follows by induction on n. □

Corollary 3.1.7 *Any homomorphic image of a Noetherian ring is Noetherian.*

Proof If R is a Noetherian ring and I an ideal in R, R/I is a Noetherian R-module and hence also a Noetherian R/I module. Hence R/I is a Noetherian ring. □

Corollary 3.1.8 *Let R be a Noetherian ring and M a finitely generated R-module. Then M is Noetherian.*

Proof Since M is finitely generated, M is isomorphic to a quotient of the free module R^n. By Corollary 3.1.6, R^n is Noetherian and hence M is Noetherian. □

Proposition 3.1.9 *Let M be a Noetherian R-module and S a multiplicatively closed set in R. Then the R_S-module M_S is Noetherian.*

Proof Let N' be a submodule of M_S and

$$N = \{x \in M | (x/1) \in N'\}.$$

Then N is a submodule of M and $N_s = N'$. Since M is Noetherian, N is finitely generated, say by $\{x_1, x_2, \ldots, x_k\}$. Then the set $\{(x_1/1), \ldots, (x_k/1)\}$ generates N' over R_S. Hence M_S is Noetherian. □

Corollary 3.1.10 If R is a Noetherian ring and S a multiplicatively closed set, the ring R_S is Noetherian.

Theorem 3.1.11 (Hilbert's Basis Theorem) *If R is a Noetherian ring, then the polynomial ring $R[X]$ is Noetherian.*

Proof Let I be a non-zero ideal of $R[X]$. Consider the set J of the highest degree coefficients of elements of I together with 0. Then J is an ideal of R which is finitely generated, say by $\{a_1, \ldots, a_k\}$, as R is Noetherian. Choose $f_i \in I$, with

$$f_i = a_i X^{n_i} + \cdots, \; a_i \in J$$

and let $n = \max(n_i)$. The ideal I' generated by $\{f_1, \ldots, f_k\}$ is contained in I. We show that $I = I' + I \cap K$, where K is the R-submodule of $R[X]$ generated by $\{1, X, X^2, \ldots, X^{n-1}\}$.
Let $f \in I$ so that $f = bX^m + \cdots$, $b \in J$. If $m \leq n-1$ then $f \in I \cap K$. If $m > n-1$, write

$$b = \lambda_1 a_1 + \cdots \lambda_k a_k, \; \lambda_i \in R$$

and consider the polynomial

$$g = f - \lambda_1 f_1 X^{m-n_1} - \cdots - \lambda_k f_k X^{m-n_k}.$$

Then $\deg g \leq m-1$, so that by successive operations of this type we can make the degree of g less than or equal to $n-1$. This implies $f - h \in I \cap K$ for some $h \in I'$. Hence $I = I' + I \cap K$.

Since K is finitely generated and R is Noetherian, $I \cap K$ is finitely generated over R and hence over $R[X]$. Since $I' = \{f_1, \ldots, f_k\}$, I is finitely generated over $R[X]$. Hence $R[X]$ is Noetherian. □

Corollary 3.1.12 *If R is Noetherian, the ring $R[X_1, \ldots, X_n]$ is Noetherian.*

Corollary 3.1.13 *A finitely generated algebra A over a Noetherian ring R is Noetherian.*

Proof If A is a finitely generated R-algebra, then

$$A \simeq \frac{R[X_1, \ldots, X_n]}{I}.$$

The result now follows from Theorem 3.1.11 and Corollary 3.1.7. □

3.1. EXERCISES

1. Let M be an R-module.

   (a) If M_1 and M_2 are submodules of M such that M/M_1 and M/M_2 are Noetherian, then show that $M/M_1 \cap M_2$ is Noetherian.

   (b) If $I = \mathrm{Ann}(M)$ and M is a Noetherian R-module, then show that R/I is a Noetherian ring.

2. Show that if M is Noetherian and N is finitely generated R-module, then $M \otimes_R N$ is Noetherian.

3. Show that if I is an ideal in a Noetherian ring then $(\sqrt{I})^n \subset I$ for some $n \geq 1$. Deduce that the nil radical of a Noetherian ring is nilpotent.

4. Show that every ideal in a Noetherian ring contains a product of prime ideals.

5. Let R be a Noetherian ring and $f(X) = \Sigma_{i=1}^{\infty} a_i X^i$ a power series in X over R such that each a_n is a nilpotent element. Show that f is nilpotent.

6. Show that a ring in which every prime ideal is finitely generated is Noetherian. (**Hint:** Let Σ be the collection of ideals which is not finitely generated. If $\Sigma \neq \emptyset$, choose a maximal element of Σ and show that it is a prime ideal.)

7. Let R be a ring such that (i) for each maximal ideal $\mathfrak{m}$, the ring $R_{\mathfrak{m}}$ is Noetherian and (ii) each non-zero element of R is contained in only finitely many maximal ideals of R. Show that R is Noetherian.

8. If I is a finitely generated ideal of R such that the ring R/I is Noetherian, then show that R need not be Noetherian. Show that under the additional assumption that I is nilpotent, R is Noetherian.

9. Let R be a ring and A a faithfully flat R-algebra. Show that if A is Noetherian, then R is also Noetherian.

10. Let A be a finitely generated R-algebra generated by $\{x_1, x_2, \ldots, x_n\}$ and M a finitely generated A-module. If $x_1, x_2, \ldots, x_n \in \sqrt{\mathrm{Ann}_A M}$, then show that M is a finitely generated R-module.

11. Let A be a finitely generated algebra over a field k generated by $\{x_1, \ldots, x_n\}$ and I the ideal of relations between $\{x_1, \ldots, x_n\}$ with coefficients from k. For any extension field K of k, show that the map $f \mapsto (f(x_1), \ldots, f(x_n))$ is a bijection between the set of k-homomorphisms of A in K onto the set defined by

$$V(I) = \{(a_1, \ldots, a_n) \in K^n \mid f(a_1, \ldots, a_n) = 0,\ f \in I\}.$$

3.2 Primary decomposition

If M is an R-module and $a \in R$, the map $\lambda_a : M \to M$, defined by $\lambda_a(x) = ax$, is R-linear. It is called the homothety defined by a.

Definition 3.2.1 A submodule N of M is called *primary* if $N \neq M$ and for each $a \in R$, the homothety $\lambda_a : M/N \to M/N$ is either injective or nilpotent.

Definition 3.2.2 An ideal I of R is called a *primary ideal* if it is a primary submodule of R.

It is clear from the definition that N is a primary submodule of M if and only if $ax \in N$, $a \in R$, $x \in M$, implies either $x \in N$ or $a^n M \subset N$ for some $n \geq 1$, i.e., $a \in r_M(N) = \sqrt{\text{Ann}(M/N)}$.

Thus the set of all $a \in R$ for which $\lambda_a : M/N \to M/N$ is not injective is an ideal equal to $r_M(N)$. This ideal is a prime ideal, for if $a, b \notin r_M(N)$, then λ_a, λ_b are injective and then $\lambda_{ab} = \lambda_a \lambda_b$ is injective, i.e., $ab \notin r_M(N)$.

Definition 3.2.3 If N is a primary submodule of M and $P = r_M(N)$, then N is called *P-primary*.

If I is an ideal of R, then clearly $r_R(I) = \sqrt{I}$.

If N is a proper submodule of M and $P = r_M(N)$, then N is P-primary if and only if $ax \in N$, $a \in R$, $x \in M$, implies, $x \in N$ or $a \in P$.

Example 3.2.4 (i) $R = \mathbb{Z}$ and $I = (p^n)$, p prime. I is a P-primary ideal where $P = (p)$.

(ii) Let $R = k[X, Y]$, k a field and $I = (X, Y^2)$. Then I is P-primary where $P = (X, Y)$.

(iii) Power of a prime ideal P need not be P-primary.

Let $R = \frac{k[X,Y,Z]}{(XY-Z)^2}$, k a field and $P = (\overline{X}, \overline{Z})$. Then P is a prime ideal in R, but P^2 is not P-primary.

More generally, we have the following proposition.

Proposition 3.2.5 *Let I be an ideal of R with $\sqrt{I} = \mathfrak{m}$, a maximal ideal. Then I is $\mathfrak{m}$-primary.*

Proof Let $ab \in I$, and $b \notin \mathfrak{m}$. Then $\mathfrak{m} + Rb = R$, so that $c + \lambda b = 1$ for some $c \in \mathfrak{m}$, $\lambda \in R$. Since $\sqrt{I} = \mathfrak{m}$, $c^k \in I$ for some $k \geq 1$. Hence $c^k + \lambda' b = 1$ for some $\lambda' \in R$. This implies $a = ac^k + \lambda' ab \in I$, i.e., I is $\mathfrak{m}$-primary. □

Corollary 3.2.6 *For any maximal ideal $\mathfrak{m}$ all its powers $\mathfrak{m}^i$, $i \geq 1$, are $\mathfrak{m}$-primary.*

Example 3.2.7 A primary ideal need not be the power of a prime ideal. Let $R = k[X, Y]$, k a field and $I = (X, Y^2)$. Then $\mathfrak{m} = (X, Y)$ is a maximal ideal with $\sqrt{I} = \mathfrak{m}$, and so I is $\mathfrak{m}$-primary. But $I \neq \mathfrak{m}^i$ for any $i \geq 1$.

Definition 3.2.8 Let N be a submodule of M. A decomposition of the type $N = N_1 \cap N_2 \cap \cdots \cap N_r$, where N_i $(1 \le i \le r)$ are primary submodules of M, is called a *primary decomposition* of N in M.

The primary decomposition is said to be *reduced* if

(i) N cannot be expressed as an intersection of a proper subset of $\{N_1, N_2, \ldots, N_r\}$, and

(ii) N_i are P_i-primary with all the P_i distinct $(1 \le i \le r)$.

From any primary decomposition we can always obtain a reduced primary decomposition in view of the following proposition.

Proposition 3.2.9 *Let $N_1, N_2, \ldots, N_k$ be P-primary submodules of M. Then $N_1 \cap N_2 \cap \cdots \cap N_k$ is P-primary.*

Proof It is sufficient to prove the result for $k = 2$. Let $N = N_1 \cap N_2$ and suppose $ax \in N$, $a \in R$, $x \in M$. If $a \notin P$, then $ax \in N_1 \cap N_2$ and N_1, N_2 are P-primary implies $x \in N_1 \cap N_2 = N$. If $a \in P$ then $a^{n_1}M \subseteq N_1$ and $a^{n_2}M \subseteq N_2$ for some n_1 and n_2. Hence $a^{n_1+n_2}M \subseteq N$, i.e., $a \in r_M(N)$. Hence N is P-primary. □

We now prove the existence of primary decomposition for finitely generated modules over a Noetherian ring.

Definition 3.2.10 A submodule N of M is called *irreducible* if $N \neq M$, and N cannot be expressed as an intersection of two submodules properly containing N.

Proposition 3.2.11 *Let M be a Noetherian module. Then any irreducible submodule N of M is primary.*

Proof Let N be an irreducible submodule and suppose N is not primary. Then there exists some $a \in R$ such that the homothety $\lambda_a : M/N \to M/N$ is neither injective nor nilpotent. Consider the increasing chain of submodules of M/N

$$\text{Ker}\,\lambda_a \subseteq \text{Ker}\,\lambda_a^2 \subseteq \cdots \subseteq \text{Ker}\,\lambda_a^n \subseteq \cdots.$$

Since M/N is Noetherian, there exists some i for which

$$\text{Ker}\,\lambda_a^i = \text{Ker}\,\lambda_a^{i+1} = \cdots.$$

If $\phi = \lambda_a^i$, then $\text{Ker}\,\phi = \text{Ker}\,\phi^2$. This implies that

$$\text{Ker}(\phi) \cap \text{Im}(\phi) = (0);$$

for if $x \in \mathrm{Ker}(\phi) \cap \mathrm{Im}(\phi)$, then $x = \phi(y)$. Now, $0 = \phi(x) = \phi^2(y)$, so that $y \in \mathrm{Ker}\,\phi^2 = \mathrm{Ker}\,\phi$ and $x = \phi(y) = 0$. Since λ_a is neither injective nor nilpotent, $\mathrm{Ker}\,\phi \neq 0$ and $\mathrm{Im}\,\phi \neq 0$.

If $p : M \to M/N$ is the natural projection and $N_1 = p^{-1}(\mathrm{Ker}\,\phi)$, $N_2 = p^{-1}(\mathrm{Im}\,\phi)$, then $N = N_1 \cap N_2$ is a decomposition of N as an intersection of two submodules properly containing N. This contradicts the irreducibility of N. Hence N is primary. □

Theorem 3.2.12 *Let M be a Noetherian R-module. Then any proper submodule N of M admits a reduced primary decomposition $N = \cap_i N_i$, N_i being P_i-primary.*

Proof We first show that N can be expressed as a finite intersection of irreducible submodules. Let Σ be the collection of all submodules of M which cannot be expressed as a finite intersection of irreducible submodules. We claim that $\Sigma = \emptyset$, for if $\Sigma \neq \emptyset$, it has a maximal element N.

Clearly N is not irreducible so that $N = N_1 \cap N_2$, $N \subsetneqq N_1$ and $N \subsetneqq N_2$. The maximality of N implies that N_1 and N_2 can be expressed as a finite intersection of irreducible submodules and N also has such a decomposition. This is a contradiction as $N \in \Sigma$. Hence $\Sigma = \emptyset$, i.e., every submodule of M can be expressed as a finite intersection of irreducible submodules.

Since an irreducible submodule is primary, we have $N = \cap_i N_i$, N_i being primary. By grouping together all P_i-primary submodules, we can assume the P_i are distinct. By deleting the superfluous N_i, i.e., those N_i which satisfy $N_i \supset \cap_{j \neq i} N_j$, we get a reduced primary decomposition for N. □

Remark 3.2.13 The primary decomposition need not be unique. Let $R = k[X, Y]$, k a field and $I = (X^2, XY)$. If $I_1 = (X)$, $I_2 = (Y - \lambda X, X^2)$, $\lambda \in k$, then $I = I_1 \cap I_2$ is a reduced decomposition of I as an intersection of primary ideals. I_2 is not unique as $\lambda \in k$ is arbitrary.

We now investigate uniqueness of the primary decomposition.

Definition 3.2.14 Let M be a R-module. A prime ideal P of R is said to be *associated* to M if P is the annihilator of some non-zero $x \in M$. Clearly, P is associated to M if and only if there exists an injective R-homomorphism $R/P \to M$.

The set $\mathrm{Ass}(M)$ denotes the set of prime ideals associated with M.

Theorem 3.2.15 *Let R be a Noetherian ring and M a finitely generated R-module. Let*

$$(0) = N_1 \cap N_2 \cap \cdots \cap N_r$$

be a reduced primary decomposition of 0 *in M, N_i being P_i-primary* $(1 \leq i \leq r)$. *Then*

$$\mathrm{Ass}(M) = \{P_1, P_2, \ldots, P_r\}.$$

Proof Let $P \in \mathrm{Ass}(M)$, so that $P = \mathrm{Ann}(x)$ for some non-zero $x \in M$. Since $x \neq 0$, by rearranging the N_i, we can find a j such that

$$x \notin N_1 \cup N_2 \cup \cdots \cup N_j$$

and

$$x \in N_{j+1} \cap \cdots \cap N_r,\ 1 \leq j \leq r.$$

For each $a \in P_i$, the homothety $\lambda_a : M/N_i \to M/N_i$ is nilpotent, so that $P_i^{n_i} M \subseteq N_i$ for some $n_i \geq 1$, as P_i is finitely generated. Now,

$$(\Pi_{i=1}^{j} P_i^{n_i})x \subseteq \cap_{i=1}^{r} N_i = (0),$$

so that $\Pi_{i=1}^{j} P_i^{n_i} \subset P$. This implies that $P_l \subseteq P$ for some l. For any $a \in P$, we have $ax = 0$ and this implies that the homothety $\lambda_a : M/N_i \to M/N_i$ is not injective and hence nilpotent, i.e., $a \in P_l$. Hence $P = P_l$.

Conversely, we show that $P_i \in \mathrm{Ass}(M)$ $(1 \leq i \leq r)$. We show for instance that $P_1 \in \mathrm{Ass}(M)$. Since the primary decomposition is reduced, there exists some $x \in N_2 \cap \cdots \cap N_r$, $x \notin N_1$. Since N_1 is P_1-primary, there exists some $n \geq 1$ such that $P_1^n x \subset N_1$ and $P_1^{n-1} x \nsubseteq N_1$. Let $y \in P_1^{n-1} x$, with $y \notin N_1$. Then $P_1 y \subset \cap_{i=1}^{r} N_i = 0$ and hence $P_1 \subseteq \mathrm{Ann}(y)$.

Conversely, if $ay = 0$, $a \in R$, then $\lambda_a : M/N_1 \to M/N_1$ is not injective and hence $a \in P_1$. Thus $P_1 = \mathrm{Ann}(y)$ and $P_1 \in \mathrm{Ass}(M)$. □

Corollary 3.2.16 (First Uniqueness Theorem) *Let M be a finitely generated module over a Noetherian ring R. If $N = \cap_1^r N_i$ is a reduced primary decomposition of N, N_i being P_i-primary, then the P_i are uniquely determined by N.*

Proof Clearly, $(0) = \cap_{i=1}^{r}(N_i/N)$ is a reduced primary decomposition of (0) in M/N, N_i/N being P_i-primary. Hence $\mathrm{Ass}(M/N) = \{P_1, \ldots, P_r\}$ is uniquely determined by N. □

Corollary 3.2.17 *With the same assumption as in Theorem 3.2.15,*

(i) $\mathrm{Ass}(M)$ is a finite set, and

(ii) $M = 0$ if and only if $\mathrm{Ass}(M) = \emptyset$.

Proof Follows from Theorem 3.2.15. □

Definition 3.2.18 Let M be an R-module. An element $a \in R$ is called a *zero-divisor* of M if there exists some non-zero $x \in M$ with $ax = 0$.

Proposition 3.2.19 *Let R be a Noetherian ring and M a finitely generated R-module. Then the set of zero-divisors of M is equal to $\bigcup_{P \in \mathrm{Ass}\, M} P$.*

Proof Let $a \in P$ for some $P \in \text{Ass}(M)$. Since P is the annihilator of some non-zero $x \in M$, we have $ax = 0$, i.e., a is a zero-divisor of M.

Conversely, let a be a zero-divisor of M so that there exists some $x \in M, x \neq 0$, such that $ax = 0$. Let $0 = \cap_{i=1}^{r} N_i$ be a reduced primary decomposition of 0 in M. Then $x \notin N_i$ for some i. If P_i is the primary ideal corresponding to N_i, then $ax = 0 \in N_i, x \notin N_i$ implies that $a \in P_i$. □

Definition 3.2.20 Let $N = \cap_{i=1}^{r} N_i$ be a reduced primary decomposition of N, N_i being P_i-primary. N_i is called a *minimal primary component* if the corresponding P_i is a minimal element of the set $\{P_1, P_2, \ldots, P_r\}$, i.e., $P_i \not\supset P_j$ for any $j \neq i\,(1 \leq j \leq r)$. A component which is not minimal is called an *embedded component.*

Example 3.2.21 Let $R = k[X, Y], k$ a field, $I = (X^2, XY)$. Let $I_1 = (X)$ and $I_2 = (Y - \lambda X, X^2)$, $\lambda \in k$. Then $I = I_1 \cap I_2$ is a reduced primary decomposition of I and I_2 is P_2-primary, where $P_2 = (X, Y)$. Since $P_2 \supsetneq P_1 = I_1$, I_2 is an embedded component. The P_2-primary component is not unique.

We now show that the minimal primary components of N are uniquely determined.

Proposition 3.2.22 *Let M be an R-module and S a multiplicatively closed set in R. Let C be the set of prime ideals of R which do not intersect S. Then the mapping $P \mapsto P_S$ is an injection of $\text{Ass}_R(M) \cap C$ into $\text{Ass}_{R_S}(M_S)$ and the map is a bijection if R is Noetherian.*

Proof Clearly, $P \to P_S$ is a bijection of C onto the set of prime ideals of R_S. If $P \in C \cap \text{Ass}_R(M)$, then P is the annihilator of $x \in M$, $x \neq 0$, and hence $P_S \in \text{Ass}_{R_S}(M_S)$.

Conversely, assume that P is finitely generated and $P_S \in \text{Ann}_{R_S}(M_S)$. We show that $P \in \text{Ann}_R(M)$. Let P_S be the annihilator of $(x/t) \in M_S, x \in M$, $t \in S$, and let P be generated by $\{a_1, a_2, \ldots, a_n\}$. Then $(a_i/1)(x/t) = 0$ shows that for each i, $s_i a_i x = 0$, $s_i \in S$. If $s = \Pi s_i$, then $sax = 0$ for all $a \in P$, so that $P \subset \text{Ann}(sx)$. Conversely, if $b \in \text{Ann}(sx)$ then $bsx = 0$, so that $(b/1) \in P_S$ and hence $b \in P$. Hence $P = \text{Ann}(sx) \in \text{Ass}_R(M)$. □

Theorem 3.2.23 *Let N be a P-primary submodule of M and P' any prime ideal of R. Let $M' = M_{P'}$, $N' = N_{P'}$ and $f : M \to M_{P'}$ the canonical map. Then*

(i) $N' = M'$ if $P \not\subset P'$,

(ii) $N = f^{-1}(N')$ if $P \subseteq P'$.

Proof (i) We have $M'/N' = M_{P'}/N_{P'} \simeq (M/N)_{P'}$. By Proposition 3.2.22, $\text{Ass}_{R_{P'}}(M'/N') \simeq \text{Ass}_R(M/N) \cap C$, where C is the set of prime ideals of R contained in P'. Since N is P-primary, $\text{Ass}(M/N) = \{P\}$ and if $P \not\subset P'$, we have $\text{Ass}_{R_{P'}}(M'/N') = \emptyset$, so that $M' = N'$.

(ii) Since $\text{Ass}(M/N) = \{P\}$ and $P \not\subseteq P'$, the multiplicatively closed set $R - P'$ does not contain any zero-divisor of M/N. Therefore the natural map $(M/N) \to (M/N)_{P'} \simeq M'/N'$ is injective. This implies that $f^{-1}(N') = N$. □

Corollary 3.2.24 (Second Uniqueness Theorem) *Let $N = \cap_{i=1}^{r} N_i$ be a reduced primary decomposition of N, N_i being P_i-primary. If P_i is minimal, then N_i is uniquely determined by N.*

Proof Suppose P_1 is minimal. Then $P_i \not\subset P_1$ for $i > 1$. From Theorem 3.2.23, we have $(N_i)_{P_1} = M_{P_1}$ for $i > 1$ and $f^{-1}[(N_1)_{P_1}] = N_1$, where f is the natural map $M \to M_{P_1}$. But $N_{P_1} = \cap_{i=1}^{r}(N_i)_{P_1} = (N_1)_{P_1}$. Hence $N_1 = f^{-1}(N_{P_1})$ depends only on N and P_1 and not on the decomposition. □

We now consider some applications of primary decomposition.

Proposition 3.2.25 *Let R be a Noetherian ring and M a finitely generated R-module. Then*

$$\bigcap_{P \in \text{Ass}(M)} P = \sqrt{\text{Ann}_R(M)}.$$

Proof Let $M \neq 0$ and let $0 = \cap_{i=1}^{r} N_i$ be a reduced decomposition of 0 in M, N_i being P_i-primary. Then $\text{Ass}(M) = \{P_1, P_2, \ldots, P_r\}$. If $a \in \sqrt{\text{Ann}_R(M)}$, then $a^n M = 0$ for some $n \geq 1$, as M is finitely generated. Hence $\lambda_a : M/N_i \to M/N_i$ is nilpotent for each $i\,(1 \leq i \leq r)$, i.e., $a \in \cap_{i=1}^{r} P_i$.

Conversely, if $a \in \cap_{i=1}^{r} P_i$, $\lambda_a : M/N_i \to M/N_i$ is nilpotent for each i so that $a^{n_i} M \subset N_i \geq 1$. If $n = \max_i n_i$, then $a^n M = 0$ and $a \in \sqrt{\text{Ann}_R(M)}$. □

Corollary 3.2.26 *If R is a Noetherian ring, then the nil radical $N(R) = \bigcap_{P \in \text{Ass}(R)} P$.*

Proof Take $M = R$. □

Definition 3.2.27 For an R-module M, the set

$$\{P \in \text{Spec}\, R \mid M_P \neq 0\}$$

is called the *support* of M and is denoted by $\text{Supp}(M)$.

Clearly, $M = 0$ if and only if $\text{Supp}(M) = \emptyset$.

Proposition 3.2.28 *Let R be a Noetherian ring and M a finitely generated R-module. For any prime ideal P of R, the following conditions are equivalent.*

(i) $P \in \text{Supp}(M)$.

(ii) $P \supseteq P'$ *for some* $P' \in \text{Ass}(M)$.

(iii) $P \supseteq \text{Ann}_R(M)$.

Proof Let $\text{Ass}(M) = \{P_1, P_2, \ldots, P_n\}$ so that

$$\bigcap_{i=1}^{n} P_i = \sqrt{\text{Ann}_R(M)}.$$

$(i) \Rightarrow (ii)$. Let $P \in \text{Supp}(M)$ and suppose $P \not\supseteq P_i$ for any i. Then $P \not\supseteq \cap_i P_i$, i.e., $P \not\supseteq \text{Ann}_R(M)$. Hence $M_P = 0$, a contradiction.

$(ii) \Rightarrow (iii)$. If $P \supseteq P', P' \in \text{Ass}(M)$, we have $P \supseteq \text{Ann}_R(M)$, since $P' \supseteq \sqrt{\text{Ann}_R(M)} \supseteq \text{Ann}_R(M)$.

$(iii) \Rightarrow (i)$. Let $P \supseteq \text{Ann}_R(M)$ and suppose $M_P = 0$. Since M is finitely generated, there exists some $s \notin P$ with $sM = 0$. This is a contradiction as $P \supseteq \text{Ann}_R(M)$. □

Corollary 3.2.29 $\text{Ass}(M) \subseteq \text{Supp}(M)$ *and the minimal elements of* $\text{Ass}(M)$ *and* $\text{Supp}(M)$ *are the same.*

Proposition 3.2.30 *Let* $0 \to M' \to M \to M'' \to 0$ *be an exact sequence of R-modules. Then*

(i) $\text{Supp}(M) = \text{Supp}(M') \cup \text{Supp}(M'')$.

(ii) *For finitely generated R-modules L and K,*

$$\text{Supp}(L \otimes_R K) = \text{Supp}(L) \cap \text{Supp}(K).$$

Proof (i) For each prime ideal P of R, the sequence

$$0 \to M'_P \to M_P \to M''_P \to 0$$

is exact. Now,

$$\begin{aligned} P \in \text{Supp}(M) &\Leftrightarrow M_P \neq 0 \\ &\Leftrightarrow \text{either } M'_P \neq 0 \text{ or } M''_P \neq 0 \\ &\Leftrightarrow P \in \text{Supp}(M') \text{ or } P \in \text{Supp}(M''). \end{aligned}$$

(ii) $(L \otimes_R K)_P \simeq L_P \otimes_{R_P} K_P$. Since R_P is a local ring, $L_P \otimes_{R_P} K_P \neq 0 \Leftrightarrow L_P \neq 0$ and $K_P \neq 0$. This proves (ii). □

3.2. EXERCISES

1. Let $R = k[X, Y, Z]$, k a field, $P_1 = (X, Y)$, $P_2 = (X, Z)$ and $\mathfrak{m} = (X, Y, Z)$. Show that for the ideal $I = P_1 P_2$, the decomposition $I = P_1 \cap P_2 \cap \mathfrak{m}^2$ is a reduced primary decomposition.

2. Show that for an ideal I, $I = \sqrt{I}$ implies that all the primary components of I are minimal.

3. Let $R \to R[X]$ be the inclusion map. Show that a reduced primary decomposition of I in R extends to a reduced primary decomposition of $I[X]$ in $R[X]$.

4. Let $R = k[X_1, X_2, \ldots, X_n], k$ a field, and $I_i = (X_1, X_2, \ldots, X_i)$ $(1 \leq i \leq n)$. Show that the powers of I_i are primary ideals.

5. Let $f : M \to M'$ be a surjective homomorphism. If $N' \subset M'$ is a primary submodule, show that $f^{-1}(N') = N$ is a primary submodule of M and $r_M(N) = r_{M'}(N')$.

6. Let M be an R-module, N a submodule of M. Prove that $\text{Ass}(N) \subseteq \text{Ass}(M) \subseteq \text{Ass}(N) \cup \text{Ass}(M/N)$.

7. Let M be a finitely generated R-module such that for every multiplicatively closed set $S \subset R$ the kernel of $M \to M_S$ is either (0) or M. Show that $\{0\}$ is primary in M.

8. Let P be a minimal prime ideal of R and $K = \text{Ker} f$, where $f : R \to R_P$ is the natural map. Show that

   (a) K is a P-primary ideal and it is the smallest P-primary ideal.

   (b) The intersection of all K, when P runs through all the minimal prime ideals of R, is contained in the nil radical of R.

3.3 Artinian modules

Proposition 3.3.1 *Let M be an R-module. The following conditions are equivalent.*

(i) Any non-empty collection of submodules of M has a minimal element.

(ii) Any decreasing sequence of submodules of M of the type

$$M_0 \supset M_1 \supset M_2 \supset \ldots \supset M_n \supset \ldots$$

is stationary, i.e., $M_k = M_{k+1} = \ldots$ for some $k \geq 0$.

Proof $(i) \Rightarrow (ii)$. Consider a decreasing sequence of submodules $M_0 \supseteq M_1 \supseteq M_2 \supseteq \dots \supseteq M_n \supseteq \dots$. Then the collection $\Sigma = \{M_n\}_{n \geq 0}$ has a minimal element M_i. Then $M_i = M_{i+1} = \dots$.

$(ii) \Rightarrow (i)$. Let Σ be a non-empty collection of submodules of M. Since $\Sigma \neq \emptyset$, choose any $M_0 \in \Sigma$. If M_0 is not minimal, then there exists some $M_1 \in \Sigma$, with $M_1 \subsetneqq M_0$. If M_1 is not minimal, there exists $M_2 \in \Sigma$ with $M_2 \subsetneqq M_1$. This process will stop after a finite number of steps yielding a minimal element of Σ, for otherwise condition (ii) will be violated. □

Definition 3.3.2 An R-module M is called *Artinian* if it satisfies any one of the above two equivalent conditions.

Definition 3.3.3 A ring R is called an *Artinian ring* if it is an Artinian R-module.

Example 3.3.4 (i) Any field k is Artinian.

(ii) Any finite ring is Artinian.

(iii) Let $G \subset \mathbb{Q}/\mathbb{Z}$ consist of all elements whose orders are powers of a fixed prime p. Then it can be shown that any proper subgroup of G is finite. Hence G is an Artinian $\mathbb{Z}$-module. It is not Noetherian over $\mathbb{Z}$.

(iv) $R = \mathbb{Z}$ is not Artinian ring as the sequence of ideals

$$(n) \supset (n^2) \supset (n^3) \supset \dots$$

is not stationary, $n \in \mathbb{Z}, n > 1$.

(v) $R = k[X]$, k a field, is not Artinian as the sequence of ideals

$(X) \supset (X^2) \supset \dots (X^n) \supset \dots$ is not stationary. But $k[X]$ is a Noetherian ring.

(vi) Let $R = k[X_1, X_2, \dots, X_n, \dots]$, the polynomial ring in infinite number of variables $X_1, X_2, \dots X_n, \dots$ over a field k. Then R is neither an Artinian nor a Noetherian ring.

Proposition 3.3.5 *Let $0 \to M' \to M \to M'' \to 0$ be an exact sequence of R-modules. Then M is Artinian if and only if both M' and M'' are Artinian.*

Proof Identifying M' as a submodule of M, any strictly descending chain of submodule of M' (or M'') gives rise to a strictly descending chain of submodules of M. Hence M Artinian implies both M' and M'' are Artinian.

Conversely, assume that both M' and M'' are Artinian and $M' \subseteq M$, and let $M = M_0 \supset M_1 \supseteq M_2 \supseteq \ldots$ be a decreasing sequence of submodules of M. Consider the sequence

$$M/M' \supseteq (M_1 + M')/M' \supseteq (M_2 + M')/M' \supseteq \ldots.$$

Since $M/M' \simeq M''$ is Artinian, there exists some i for which $M_i + M' = M_{i+1} + M' = \ldots$. Since M' is Artinian, the sequence

$$M' \cap M \supseteq M' \cap M_1 \supseteq M' \cap M_2 \supseteq \ldots$$

is stationary. Therefore, there exists some integer which can be assumed to be i (by choosing the maximum) such that $M' \cap M_i = M' \cap M_{i+1} = \ldots$. Since

$$M_{i+1} \subseteq M_i,\ M_{i+1} \cap M' = M_i \cap M' \text{ and } M_{i+1} + M' = M_i + M',$$

we have by an easy computation that $M_i = M_{i+1}$. Hence M is Artinian. □

Corollary 3.3.6 *If $\{M_i\}_{1 \le i \le n}$ are Artinian R-modules, then $\oplus \Sigma_i M_i$ is Artinian.*

Proof Consider the exact sequence

$$0 \to \oplus \sum_{i=1}^{n-1} M_i \to \oplus \sum_{i=1}^{n} M_i \to M_n \to 0,$$

where the first map is the inclusion and the second map is the projection onto the n-th component. The result follows by induction on n. □

Corollary 3.3.7 *Any finitely generated module over an Artinian ring is Artinian.*

Proof Let M be a finitely generated R-module. Write $M \simeq F/K$, where $F \simeq R^n$ is a free R-module of rank n. Since R is an Artinian R-module, so is F by Corollary 3.3.6. This implies that $M \simeq F/K$ is Artinian. □

Corollary 3.3.8 *If I is an ideal in an Artinian ring, then the ring R/I is an Artinian ring.*

Proof Since R is Artinian, R/I is an Artinian R-module. Since it is also an R/I-module, it is an Artinian R/I module, i.e., an Artinian ring. □

Proposition 3.3.9 *Let R be an Artinian ring. Then every prime ideal of R is maximal.*

Proof By passing to the quotient, it is sufficient to show that an Artinian domain is a field. Let R be an Artinian domain, $a \in R, a \neq 0$. The sequence of ideals $(a) \supset (a^2) \supset \ldots \supset (a^n) \supset \ldots$ is stationary, so that $(a^i) = (a^{i+1})$ for some i. If $a^i = ba^{i+1}, b \in R$, then $1 = ba$, as $a^i \neq 0$. Hence R is a field. □

Corollary 3.3.10 *For an Artinian ring, the nil radical is equal to the Jacobson radical.*

Proposition 3.3.11 *An Artinian ring has only finitely many maximal ideals.*

Proof Consider the family Σ of finite intersections of maximal ideals of R. Since $\Sigma \neq \emptyset$ and R is Artinian, Σ has a minimal element, say $I = \mathfrak{m}_1 \cap \mathfrak{m}_2 \cap \ldots \cap \mathfrak{m}_r$. Now, for any maximal ideal $\mathfrak{m}$ of R, $\mathfrak{m} \cap \mathfrak{m}_1 \cap \cdots \mathfrak{m}_r \subseteq I$, and is equal to I by the minimality of I. Hence $I \subseteq \mathfrak{m}$ and this implies that $\mathfrak{m}_i \subseteq \mathfrak{m}$ for some i $(1 \leq i \leq r)$, *i.e.*, $\mathfrak{m}_i = \mathfrak{m}$. Hence the only maximal ideals of R are $\mathfrak{m}_1, \mathfrak{m}_2, \ldots, \mathfrak{m}_r$. □

Proposition 3.3.12 *If R is an Artinian ring, then the nil radical $N(R)$ is nilpotent.*

Proof Let $N(R) = I$, and consider the sequence of ideals

$$I \supseteq I^2 \supseteq \ldots \supseteq I^n \supset \ldots.$$

There exists some i for which $I^i = I^{i+1} = \ldots = J$ (say). If $J = 0$, then I is nilpotent. If $J \neq 0$, then we obtain a contradiction as follows. Let $\Sigma = \{K | K \text{ ideal in } R \text{ with } KJ \neq 0\}$. Now, $\Sigma \neq \emptyset$, as $J^2 = J \neq 0$. Hence Σ has a minimal element K_0. Now, K_0 is principal, for if $a \in K_0$, with $aJ \neq 0$, then $Ra \subseteq K_0$ and hence $Ra = K_0$ by minimality. Since $(aJ)J = aJ^2 = aJ \neq 0$, by minimality $aJ = K_0 = (a)$. This implies that $a = ab$ for some $b \in J$. But $b \in J \subseteq I = N(R)$, so that $b^n = 0$ for some $n \geq 1$. Hence $a = ab = ab^2 = \ldots = ab^n = 0$, a contradiction. □

Corollary 3.3.13 *In an Artinian ring R, the Jacobson radical is nilpotent.*

Proof Follows from Corollary 3.3.10 and Proposition 3.3.12. □

Theorem 3.3.14 (Structure of Artinian Rings) *An Artinian ring is uniquely isomorphic to a finite direct product of Artinian local rings.*

Proof Let $\mathfrak{m}_1, \mathfrak{m}_2, \ldots, \mathfrak{m}_r$ be the distinct maximal ideals of R. Then $J(R) = \cap_{i=1}^r \mathfrak{m}_i = \Pi_{i=1}^r \mathfrak{m}_i$ is nilpotent. Hence $(\Pi_i^r \mathfrak{m}_i)^k = 0$ for some $k \geq 1$. By the Chinese Remainder Theorem, the natural map $R \to \Pi_{i=1}^r R/\mathfrak{m}_i^k$ is an isomorphism. Since $R/\mathfrak{m}_i^k$ has a unique prime ideal $\mathfrak{m}_i/\mathfrak{m}_i^k$, it is a local ring. Since R is Artinian, so is $R/\mathfrak{m}_i^k$. Thus R is a finite product of Artinian local rings.

To prove the uniqueness, assume that $R = \Pi_{i=1}^r R_i$, where each R_i is an Artinian local ring. Since every prime ideal of an Artinian ring is maximal, each R_i has unique prime ideal.

Let $\pi_i : R \to R_i$ be the i-th projection and $I_i = \operatorname{Ker} \pi_i$. Since $R = \Pi_1^r R_i$, by the Chinese Remainder Theorem, the I_i $(1 \leq i \leq r)$ are comaximal and $\cap_{i=1}^r I_i = (0)$. Let P_i be the unique prime ideal of R_i and $\mathfrak{m}_i = \pi_i^{-1}(P_i)$ $(1 \leq i \leq r)$. Then $\mathfrak{m}_i$ is a prime ideal of R and hence maximal. Since P_i is nilpotent, it follows that I_i is $\mathfrak{m}_i$-primary. Hence

$(0) = \cap_1^r I_i$ is a reduced primary decomposition of (0) in R. Since $I_i \ (1 \leq i \leq r)$ are pairwise comaximal, so are $\mathfrak{m}_i \ (1 \leq i \leq r)$. Hence $I_i \ (1 \leq i \leq r)$ are isolated components of the primary decomposition of 0. Hence $I_i \ (1 \leq i \leq r)$ are uniquely determined by R, i.e., $R_i \simeq R/I_i$ are unique $(1 \leq i \leq r)$. □

3.3. EXERCISES

1. Let N_1 and N_2 be submodules of M such that both M/N_1 and M/N_2 are Artinian modules. Show that $M/N_1 \cap N_2$ is an Artinian module.
2. Let M be an Artinian R-module and $f : M \to M$ an injective homomorphism. Show that f is an isomorphism.
3. Let M be an R-module which is both Artinian and Noetherian and $f : M \to M$ an R-homomorphism. Show that for some $n \geq 1$, $M = f^n(M) \oplus \mathrm{Ker}(f^n)$. Deduce that if $f : V \to V$ is a linear transformation of a finite dimensional vector space V, then $V = V_1 \oplus V_2$, where $f|_{V_1}$ is non-singular and $f|_{V_2}$ is nilpotent.
4. Let M be an Artinian R-module. Is the ring $R/\mathrm{Ann}(M)$ Artinian? Why?
5. Show that a finitely generated algebra A over a field k is Artinian if and only if A is finite dimensional over k.
6. A ring R is called a *primary ring* if it has just one prime ideal. Show that
   (a) For any $\mathfrak{m}$-primary ideal I with $\mathfrak{m}$ maximal, R/I is a primary ring.
   (b) A primary ring has no idempotents other than 0 or 1. Give an example of a primary ring which is not Noetherian.
7. Let R be an Artinian local ring with maximal ideal $\mathfrak{m}$. Show that the following conditions are equivalent.
   (a) Every ideal of R is principal.
   (b) $\mathfrak{m}$ is principal.
   (c) $\dim_k(\mathfrak{m}/\mathfrak{m}^2) \leq 1$, where $k = R/\mathfrak{m}$.
8. Show that $R = \Pi_{i=1}^n R_i$ is an Artinian ring if and only if each R_i is an Artinian ring.
9. Let R be an Artinian ring with $N(R) = 0$. Show that R has only finitely many ideals.

3.4 Length of a module

Definition 3.4.1 An R-module $M \neq 0$ is called a *simple* module if it has no submodules except 0 and M.

Example 3.4.2 (i) Any one dimensional vector space over a field k is a simple k-module.

(ii) Any group of prime order is a simple $\mathbb{Z}$-module.

Definition 3.4.3 A *composition series* of an R-module M is a finite decreasing sequence of submodules

$$M = M_0 \supset M_1 \supset M_2 \supset \ldots \supset M_n = \{0\}$$

such that each M_i/M_{i+1} $(0 \leq i \leq n-1)$ is simple.

Example 3.4.4 (i) Let V be a vector space of dimension n with basis $\{e_1, e_2, \ldots, e_n\}$ and V_{n-i} the subspace generated by $\{e_1, e_2, \ldots, e_i\}$. Then $V = V_0 \supset V_1 \supset \ldots \supset V_n = \{0\}$ is a composition series of V.

(ii) Let $G = (a)$ be a cyclic group of order 6 and H the subgroup generated by a^2. Then $G \supset H \supset \{e\}$ is a composition series of G.

(iii) An R-module M may not always have a compositions series. The $\mathbb{Z}$-module $M = \mathbb{Z}$ has no composition series.

Definition 3.4.5 Given a composition series of M, say $M = M_0 \supset M_1 \supset \ldots \supset M_n = \{0\}$, the number n is called the *length* of the composition series.

Example 3.4.6 In the first example above, the length of the composition series is n, the dimension of the vector space.

The following theorem shows that if an R-module M has a composition series, its length does not depend on the composition series.

Theorem 3.4.7 (Jordan Hölder Theorem) *Let M be an R-module having a composition series of length n. Then any other composition series of M also has length n and any decreasing chain of submodules can be extended to a composition series of M.*

Proof Let $l(M)$ denote the least length of a composition series of M. If N is a submodule of M, then we show that $l(N) \leq l(M)$ and for a proper submodule N of M, we show that $l(N) < l(M)$.

Let $M = M_0 \supset M_1 \supset \ldots \supset M_t = \{0\}$ be a composition series of M of least length $t = l(M)$. Then

$$N = N \cap M_0 \supseteq N \cap M_1 \supseteq \ldots \supseteq N \cap M_t = \{0\}$$

is a decreasing sequence of submodules of N. For each i, the natural map

$$(N \cap M_i)/(N \cap M_{i+1}) \to M_i/M_{i+1}$$

is injective. Since M_i/M_{i+1} is simple, either $N \cap M_i = N \cap M_{i+1}$ or $N \cap M_i/N \cap M_{i+1}$ is simple. This gives rise to a composition series of N of length $\leq t$ by omitting the repeated terms. Hence $l(N)$, the least length of any composition series of N, is $\leq t = l(M)$. Suppose $l(N) = l(M)$. Then, for each $i\,(0 \leq i \leq t-1)$, we have $(N \cap M_i)/(N \cap M_{i+1}) \neq 0$ and is isomorphic to M_i/M_{i+1}. Since $M_t = 0$, we have $N \cap M_{t-1} = M_{t-1}$. This implies in turn $N \cap M_{t-2} = M_{t-2}$, etc., and finally, $N \cap M_0 = M_0$, i.e., $N = M$. Hence, for a proper submodule N of M, $l(N) < l(M)$.

Now, consider any composition series of M of length k, say

$$M = M_0 \supset M_1 \supset \ldots \supset M_k = (0).$$

This implies

$$t = l(M) > l(M_1) > \ldots > l(M_k) = (0).$$

Hence, $t \geq k$. Since t is the least length of a composition series of M, we have $t \leq k$. Hence all the composition series of M have the same length $t = l(M)$.

Now, consider any decreasing chain of submodules

$$M = M_0 \supsetneq M_1 \supsetneq \ldots \supsetneq M_m = \{0\}.$$

If $m < l(M)$, then it is not a composition series of M and hence, for some i, M_i/M_{i+1} is not simple, i.e., there exists a submodule L with $M_i \supsetneq L \supsetneq M_{i+1}$. Consider the new decreasing chain by adding L to the chain

$$M = M_0 \supsetneq M_1 \supsetneq \ldots \supset M_i \supsetneq L \supsetneq M_{i+1} \supsetneq \ldots \supsetneq M_m = \{0\}.$$

If $m+1 < l(M)$, then we repeat the process. After a finite number of steps we get a composition series of M. □

Definition 3.4.8 If an R-module M has a composition, then the length of any composition series of M is called the *length* of M and is denoted by $l_R(M)$. If M has no composition series, then $l_R(M)$ is defined to be ∞.

Proposition 3.4.9 *An R-module M has a composition series if and only if it is both a Noetherian and an Artinian module.*

Proof Assume that M has a composition series of length n. Then any strictly increasing or decreasing sequence of submodules of M has length $\leq n$ by Theorem 3.4.7. Hence M is both a Noetherian and an Artinian module.

To prove the converse, assume that M is both Noetherian and Artinian. Since M is Noetherian, it has a proper maximal submodule M_1. Now, M_1 is also Noetherian, and hence it has a proper maximal submodule M_2. Continuing this process, we have a decreasing chain of submodules.

$$M = M_0 \underset{\neq}{\supset} M_1 \underset{\neq}{\supset} M_2 \underset{\neq}{\supset} \dots.$$

Since M is Artinian, $M_k = (0)$ for some k. Then

$$M = M_0 \supset M_1 \supset \dots \supset M_k = (0)$$

is a composition series of M. □

Proposition 3.4.10 *Let* $0 \to M' \xrightarrow{f} M \xrightarrow{g} M'' \to 0$ *be an exact sequence of R-modules. Then* $l(M) = l(M') + l(M'')$.

Proof M is Noetherian (Artinian) if and only if both M' and M'' are Noetherian (Artinian). Hence $l(M)$ is finite if and only if both $l(M')$ and $l(M'')$ are finite. The relation is clearly true if any one of the modules has length ∞. Assume now that all the lengths are finite. Let $M' = M_0 \supset M_1' \supset \dots \supset M_k' = (0)$ be a composition series of M', with $l(M') = k$ and $M'' = M_0'' \supset M_1'' \supset \dots \supset M_t'' = 0$, a composition series of M'' with $l(M'') = t$. Consider the series

$$\begin{aligned} M &= g^{-1}(M'') \supset g^{-1}(M_1'') \supset \dots \supset g^{-1}(M_t'') \\ &= f(M') \supset f(M_1') \supset \dots f(M_k'') \\ &= \{0\} \end{aligned}$$

By Corollary 1.1.10,

$$\frac{g^{-1}(M_i'')}{g^{-1}(M_{i+1}'')} \simeq \frac{M_i''}{M_{i+1}''}$$

is simple for each i. Hence the above series is a composition series of M of length $k + t$, i.e., $l(M) = l(M') + l(M'')$. □

Proposition 3.4.11 *Let R be a Noetherian ring and $M \neq 0$ a finitely generated R-module. Then there exists a decreasing chain*

$$M = M_n \supset M_{n-1} \supset \dots \supset M_0 = 0,$$

such that $M_i/M_{i-1} \simeq R/P_i$, P_i *prime.*

Proof Let Σ be the collection of submodules of M which have the required property. Note that $0 \notin \Sigma$ by construction as the hypothesis asserts that $M \neq 0$. As $M \neq 0$, by Corollary 3.2.17 (ii), $\text{Ass}(M) \neq \emptyset$. Let $P \in \text{Ass}(M)$ and let $x(\neq 0) \in M$ be such that

$P = \mathrm{Ann}(x)$. Then, as we have a decreasing chain $Rx \supset 0$ with $Rx \cong R/P$, it follows that $Rx \in \Sigma$, i.e., $\Sigma \neq \emptyset$.

Let N be a maximal element of Σ. If $N \neq M$, then $M/N \neq 0$. Hence $\mathrm{Ass}(M/N) \neq \emptyset$. Choose $P_1 \in \mathrm{Ass}(M/N)$ so that $R/P_1 \simeq N_1/N$. Then N_1 also has the required property and $N_1 \supsetneqq N$, contradicting the maximality of N. Hence $N = M$. □

Corollary 3.4.12
With the same notation as above, we have $\mathrm{Ass}(M) \subseteq \{P_1, P_2, \ldots, P_n\} \subseteq \mathrm{Supp}(M)$, *and the minimal elements of all the three sets are the same.*

Proof Consider the chain $M = M_n \supseteq M_{n-1} \supseteq \ldots \supseteq M_0 = \{0\}$ with $M_i/M_{i-1} \simeq R/P_i$, P_i prime $(1 \leq i \leq n)$. Then by Exercise 6, Section 3.2,

$$\mathrm{Ass}(M) \subseteq \mathrm{Ass}(M/M_{n-1}) \cup \mathrm{Ass}(M_{n-1}/M_{n-2}) \cup \ldots \cup \mathrm{Ass}(M_1) \subseteq$$

$$\{P_1, P_2, \ldots, P_n\} \text{ as } M_i/M_{i-1} \simeq R/P_i.$$

Since $P_i \in \mathrm{Supp}(R/P_i) = \mathrm{Supp}(M_i/M_{i-1})$ and $\mathrm{Supp}(M_i/M_{i-1}) \subseteq \mathrm{Supp}(M_i) \subseteq \mathrm{Supp}(M)$, we have $\{P_1, \ldots, P_n\} \subseteq \mathrm{Supp}(M)$. Thus the minimal elements of all three sets are the same by Corollary 3.2.29. □

Theorem 3.4.13 *Let M be a finitely generated module over a Noetherian ring R. Then the following are equivalent.*

(i) $l_R(M)$ is finite.

(ii) Every $P \in \mathrm{Ass}(M)$ is a maximal ideal.

(iii) Every $P \in \mathrm{Supp}(M)$ is a maximal ideal.

Proof $(i) \Rightarrow (ii)$. As in Proposition 3.4.11, choose a decreasing chain

$$M = M_n \supset M_{n-1} \supset \ldots \supset M_0 = \{0\}$$

with $M_i/M_{i-1} \simeq R/P_i$, P_i prime $(1 \leq i \leq n)$. Since $l_R(M) < \infty$ and R/P_i is isomorphic to a subquotient of M, we have $l_R(R/P_i) < \infty$ for each i. This implies that R/P_i is an Artinian R-module and hence an Artinian R/P_i-module, i.e., an Artinian ring. Hence, it is a field, i.e., P_i is maximal. Since $\mathrm{Ass}(M) \subseteq \{P_1, \ldots, P_n\}$, every $P \in \mathrm{Ass}(M)$ is maximal.

$(ii) \Rightarrow (iii)$. Every $P \in \mathrm{Supp}(M)$ contains some $P \in \mathrm{Ass}(M)$. $(iii) \Rightarrow (i)$. Choose a decreasing chain

$$M = M_n \supset M_{n-1} \supset \ldots \supset M_0 = \{0\}$$

with $M_i/M_{i-1} \simeq R/P_i$, P_i-prime $(1 \leq i \leq n)$.

Since

$$\{P_1, \ldots, P_n\} \subseteq \text{Supp}(M)$$

and every $P \in \text{Supp}(M)$ is maximal, we have that R/P_i is a field $(1 \leq i \leq n)$. This implies that $l_R(M_i/M_{i-1}) = 1 \quad (1 \leq i \leq n)$. Hence $l_R(M) < \infty$ by Proposition 3.4.10.

□

Corollary 3.4.14 *Let M be a finitely generated R-module over a Noetherian ring R with $l_R(M) < \infty$. Then* $\text{Ass}(M) = \text{Supp}(M)$.

Proof If $P \in \text{Supp}(M)$, then $P \supseteq P'' \in \text{Ass}(M)$ and hence $P = P'' \in \text{Ass}(M)$. Hence $\text{Supp}(M) = \text{Ass}(M)$.

Corollary 3.4.15 *Let M be a finitely generated module over a Noetherian ring R and P a prime ideal of R. Then $M_P \neq 0$ is of finite length over R_P if and only if P is a minimal element of* $\text{Ass}(M)$.

Proof: We have by Proposition 3.2.22,

$$\text{Ass}_{R_P}(M_P) = \{Q_P \mid Q \in \text{Ass}(M),\, Q \subset P\}.$$

Now, $l_{R_P}(M_P) < \infty$ if and only if every element of $\text{Ass}_{R_P}(M_P)$ is maximal. Since R_P is local with unique maximal ideal PR_P, we have $l_{R_P}(M_P) < \infty$ if and only if there exists no $Q \in \text{Ass}(M)$ with $Q \subsetneqq P$. But $M_P \neq 0 \Leftrightarrow P \in \text{Supp}(M) \Leftrightarrow P \supseteq P'$, $P' \in \text{Ass}(M)$. Hence $M_P \neq 0$ and $l_{R_P}(M_P) < \infty \Leftrightarrow P \in \text{Ass}(M)$ and P is minimal. □

Proposition 3.4.16 *Let R be a Noetherian ring. The following conditions are equivalent.*

(i) R is Artinian.

(ii) Every prime ideal of R is maximal.

(iii) All elements of $\text{Ass}(R)$ *are maximal.*

Proof $(i) \Rightarrow (ii)$ by Proposition 3.3.9.

$(ii) \Rightarrow (iii)$ is obvious.

$(iii) \Rightarrow (i)$. By Theorem 3.4.13, we have $l_R(R) < \infty$ and this implies (i).

Proposition 3.4.17 *A ring R is Artinian $\Leftrightarrow$ length of R as an R-module is finite.*

Proof If $l_R(R) < \infty$, clearly R is Artinian.

Conversely, assume that R is Artinian. Then $J(R)$ is nilpotent by Corollary 3.3.13. Since $J(R) = \cap_{i=1}^{t} \mathfrak{m}_i, \mathfrak{m}_i$ maximal, and $J(R)^n = (\cap_i \mathfrak{m}_i)^n = \Pi_i \mathfrak{m}_i^n = 0$, (0) is a product of maximal ideals, say $0 = \mathfrak{m}_1 \mathfrak{m}_2 \ldots \mathfrak{m}_k$, $\mathfrak{m}_t$ not necessarily distinct. Consider the sequence of ideals $R = \mathfrak{m}_0 \supset \mathfrak{m}_1 \supset \mathfrak{m}_1\mathfrak{m}_2 \supset \ldots \supset \mathfrak{m}_1\mathfrak{m}_2\ldots\mathfrak{m}_{k-1} \supset \mathfrak{m}_1\mathfrak{m}_2\ldots \supset \mathfrak{m}_k = 0$. Since R is an Artinian R-module,

$$\frac{\mathfrak{m}_1\ldots\mathfrak{m}_{i-1}}{\mathfrak{m}_1\ldots\mathfrak{m}_i}$$

is also an Artinian R-module ($1 \leq i \leq k$) and since it is also an $R/\mathfrak{m}_i$-module, it is an Artinian $R/\mathfrak{m}_i$-module, i.e., a finite dimensional vector space over $R/\mathfrak{m}_i$. Hence it has a finite length as $R/\mathfrak{m}_i$-module and also as an R-module. This implies that $l_R(R) < \infty$. □

Corollary 3.4.18 *R is an Artinian ring if and only if R is a Noetherian ring and every prime ideal of R is maximal.*

Proof If R is Noetherian and every prime ideal of R is maximal, then R is Artinian by Proposition 3.4.16.

Conversely, assume R is Artinian. Then $l_R(R) < \infty$ and hence R is Noetherian. From Proposition 3.3.9, every prime ideal of R is maximal. □

Corollary 3.4.19 *Let R be an Artinian ring and M a finitely generated R-module. Then $l_R(M) < \infty$.*

Proof Since R is also a Noetherian ring and M is finitely generated, M is both an Artinian and a Noetherian module. Hence

$$l_R(M) < \infty.$$

□

3.4. EXERCISES

1. Let $0 \to M_1 \to M_2 \to \ldots \to M_n \to 0$ be an exact sequence of R-modules of finite length. Show that $\Sigma_{i=1}^{n}(-1)^i l_R(M_i) = 0$.

2. Show that if M is an R-module of finite length over a Noetherian ring R, then

$$l_R(M) = \sum_{P\in \text{Ass}(M)} l_{R_P}(M_P).$$

3. Let $P_1, P_2, \ldots, P_n$ be the prime ideals associated to the primary decomposition of (0) in a Noetherian ring R. Show that if all the P_i are minimal, and $S = R - \cup_{i=1}^{n} P_i$, then R_S is an Artinian ring.

4. Let R be a Noetherian ring and M a finitely generated R-module with decreasing chain as in Proposition 3.4.11. Let $P \in \text{Ass}(M)$ be a minimal element. Show that $l_{R_P}(M_P)$ is equal to the number of i for which $P = P_i$.

5. Let M be a finitely generated module over a Noetherian ring R. Consider a decreasing chain as in Proposition 3.4.11. Illustrate by an example that $\{P_1, P_2, \ldots, P_n\}$ may not be unique.

6. Let I be an ideal in a Noetherian ring R having primary decomposition without embedded components and S the complement of the union of prime ideals associated to I. Show that $\lambda = l_{R_S}(R_S/IR_S)$ is finite. If R is Artinian, show that $\lambda(I) = l_R(R) - l_R(I)$.

CHAPTER 4

INTEGRAL EXTENSIONS

The study of solutions of Diophantine equations in number theory led to the notion of algebraic integers. A parallel development in algebraic geometry arising out of a study algebraic curves led to the notion of algebraic functions of one variable and integral functions. Both these developments are unified under the study of integral extensions.

Integral extensions are well behaved with respect to extension and contraction of prime and maximal ideals. A basic theorem due to Noether on normalisation shows that any finitely generated k-algebra is integral over a suitably chosen polynomial sub-algebra. From this theorem we obtain Hilbert's Nullstellensatz, which says that if I is a proper ideal in the polynomial ring $k[X_1, \ldots, X_n]$, k an algebraically closed field, then there is at least one common solution for the polynomials in I.

Varieties whose local rings are integrally closed are called normal varieties. The normalisation operation, *i.e.*, taking the integral closure is a standard tool in geometry.

4.1 Integral elements

We assume throughout this chapter that R is a commutative ring with unit element and S is an extension ring of R with the same unit element.

Definition 4.1.1 An element $\alpha \in S$ is said to be *integral* over R if it satisfies a monic polynomial relation of the type

$$\alpha^n + a_1\alpha^{n-1} + \cdots + a_n = 0, \;\; a_i \in R \;\; (1 \leq i \leq n).$$

Example 4.1.2 (i) Any $\alpha \in R$ is integral over R.

(ii) Let K be a field extension of k and $\alpha \in K$ algebraic over k. Then α is integral over k.

(iii) Let $R = \mathbb{Z}$ and $S = \mathbb{Z} + i\mathbb{Z}$, the ring of Gaussian integers. Then $\alpha = 1 + i$ is integral over $\mathbb{Z}$.

(iv) Let $R = \mathbb{Z}$, and $S = \mathbb{Q}$. Then $\alpha \in \mathbb{Q} - \mathbb{Z}$ cannot be integral over $\mathbb{Z}$; for if $\alpha = p/q, (p,q) = 1$, satisfies a relation of the type $\alpha^n + a_1\alpha^{n-1} + \cdots + a_n = 0, a_i \in \mathbb{Z}$, then

$$p^n + a_1p^{n-1}q + \cdots + a_nq^n = 0.$$

This implies that q divides p^n and hence a prime factor of q divides p^n. This contradicts the assumption that $(p,q) = 1$.

We have the following equivalent characterisations for integral elements.

Theorem 4.1.3 *The following conditions are equivalent.*
(i) *The element* $\alpha \in S$ *is integral over* R.

(ii) $R[\alpha]$ *is a finitely generated* R*-module.*

(iii) $R[\alpha] \subset R' \subset S$, *where* R' *is a subring of* S *which is a finitely generated as an* R*-module.*

(iv) *There exists a faithful* $R[\alpha]$*-module* M, *which is finitely generated as an* R*-module.*

Proof $(i) \Rightarrow (ii)$. If α satisfies the relation

$$\alpha^n + a_1\alpha^{n-1} + \cdots + a_n = 0, a_i \in R,$$

then $\alpha^n = -a_1\alpha^{n-1} - a_2\alpha^{n-2} - \cdots - a_n$, so that $1, \alpha, \alpha^2, \ldots, \alpha^{n-1}$ generate $R[\alpha]$ as an R-module.

$(ii) \Rightarrow (iii)$. Take $R' = R[\alpha]$.

$(iii) \Rightarrow (iv)$. Let $R[\alpha] \subset R' \subset S$, where R' is a subring of S finitely generated as an R-module. Then $M = R'$ is an $R[\alpha]$-module, which is faithful because $xR' = 0,\ x \in R[\alpha]$, implies $x.1 = x = 0$.

$(iv) \Rightarrow (i)$. Let M be a finitely generated R-module which is faithful as an $R[\alpha]$-module. The result follows from the following lemma by taking $I = R$.

□

Lemma 4.1.4 *Let* M *be a finitely generated* R*-module which is faithful as an* $R[\alpha]$*-module and* I *an ideal of* R *such that* $\alpha M \subseteq IM$. *Then* α *satisfies a relation of the type*

$$\alpha^n + a_1\alpha^{n-1} + \cdots + a_n = 0, a_i \in I.$$

Proof Let M be generated by $\{x_1, x_2, \ldots, x_n\}$ over R. Since $\alpha M \subseteq IM \subseteq \Sigma_{i=1}^n Ix_i$, we have $\alpha x_i = \Sigma_{j=1}^n a_{ij}x_j,\ a_{ij} \in I$. Consider the system of equations

$$\sum_{j=1}^{n} (\delta_{ij}\alpha - a_{ij})x_j = 0, \quad 1 \le i \le n.$$

Let Δ be the determinant of the matrix $(\delta_{ij}\alpha - a_{ij})$. Then $\Delta x_i = 0,\ 1 \le i \le n$, and this implies that $\Delta = 0$ as M is a faithful $R[\alpha]$-module. By expanding Δ, we have

$$\alpha^n + a_1\alpha^{n-1} + \cdots + a_n = 0, \quad a_i \in I, \quad 1 \le i \le n. \qquad \square$$

Corollary 4.1.5 *Let $\alpha_1, \alpha_2, \ldots, \alpha_n \in S$ be integral over R. Then $R[\alpha_1, \alpha_2, \ldots, \alpha_n]$ is a finitely generated R-module.*

Proof The proof is by induction on n. For $n = 1$, the result follows by Theorem 4.1.3. Assume the result for $(n-1)$, so that $R[\alpha_1, \alpha_2, \ldots, \alpha_{n-1}]$ is a finitely generated R-module. Now, α_n is integral over R and hence over $R[\alpha_1, \alpha_2, \ldots, \alpha_{n-1}]$. Thus, $R[\alpha_1, \alpha_2, \ldots, \alpha_n]$ is a finitely generated $R[\alpha_2, \ldots, \alpha_{n-1}]$-module and this implies by transitivity that it is a finitely generated R-module. $\square$

Corollary 4.1.6 *The set of elements of S integral over R is a subring of S containing R.*

Proof If $\alpha, \beta \in S$ are integral over R, then $R[\alpha, \beta]$ is a finitely generated R-module. Since $\alpha \pm \beta$, and $\alpha\beta$ lie in $R[\alpha, \beta]$, the result follows from Theorem 4.1.3. $\square$

Definition 4.1.7 The subring of all elements of S integral over R is called the *integral closure* of R in S.

Example 4.1.8 (i) The integral closure of $\mathbb{Z}$ in $\mathbb{Q}$ is $\mathbb{Z}$.

(ii) The integral closure of $\mathbb{Z}$ in $\mathbb{Q}(i)$ is $\mathbb{Z} + i\mathbb{Z}$.

(iii) The integral closure of $\mathbb{Z}$ in $\mathbb{C}$, the field of complex numbers, is the ring of algebraic integers.

Definition 4.1.9 If the integral closure of R in S coincides with R, then R is said to be *integrally closed* in S.

Definition 4.1.10 If the integral closure of R in S is the whole of S, then S is said to be an *integral extension* of R.

Proposition 4.1.11 *Let $R \subset S \subset T$ be ring extensions. If S is integral over R and T is integral over S, then T is integral over R.*

Proof If $\alpha \in T$, then α satisfies a relation of the type $\alpha^n + a_1\alpha^{n-1} + \cdots + a_n = 0$, $a_i \in S$. Then α is integral over $R' = R[a_1, a_2, \ldots, a_n]$ so that $R'[\alpha]$ is a finitely generated R'-module. Since S is integral over R, R' is a finitely generated R-module by Corollary 4.1.5. Hence $R'[\alpha]$ is a finitely generated R-module. Since $R \subset R[\alpha] \subset R'[\alpha] \subset T$, α is integral over R by Theorem 4.1.3. □

Corollary 4.1.12 *If $\overline{R}$ is the integral closure of R in S, then $\overline{R}$ is integrally closed in S.*

Proof Let $\alpha \in S$ be integral over $\overline{R}$. Since $\overline{R}$ is integral over R, α is integral over R and hence $\alpha \in \overline{R}$. Thus $\overline{R}$ is integrally closed in S. □

Proposition 4.1.13 (i) *Let S be an integral extension of R, J any ideal of S and $I = J \cap R$. Then S/J is integral over R/I.*

(ii) *For any multiplicatively closed set T in R, S_T is integral over R_T.*

Proof (i) Let $\overline{\alpha} = \alpha + J \in S/J$. Clearly, R/I is isomorphic to a subring of S/J. If α satisfies a relation of the type $\alpha^n + a_1\alpha^{n-1} + \cdots + a_n = 0, a_i \in R$, by passing to the quotient modulo J and identifying R/I as a subring of S/J, we have $\overline{\alpha}^n + \overline{a}_1\overline{\alpha}^{n-1} + \cdots + \overline{a}_n = 0, \overline{a}_i \in R/I$. Hence S/J is integral over R/I.

(ii) Let $(\alpha/t) \in S_T, \alpha \in S, t \in T$. Let α satisfy a relation of the type

$$\alpha^n + a_1\alpha^{n-1} + \cdots + a_n = 0, \quad a_i \in R.$$

This implies $(\alpha/t)^n + (a_1/t)(\alpha/t)^{n-1} + \cdots + (a_n/t^n) = 0, \ (a_i/t^i) \in R_T$. Hence S_T is integral over R_T. □

4.1. EXERCISES

1. Show that a unique factorisation domain is integrally closed in its quotient field.
2. Show that the integral closure of $\mathbb{Z}$ in $\mathbb{Q}(\sqrt{2})$ is the ring of elements of the type $a + b\sqrt{2}, a, b \in \mathbb{Z}$.
3. Let $R \subset S$ be an integral extension and $\mathfrak{n}$ a maximal ideal of S and $\mathfrak{m} = R \cap \mathfrak{n}$. Illustrate by an example that $S_{\mathfrak{n}}$ need not be integral over $R_{\mathfrak{m}}$.
4. Let R be a ring and α an invertible element of $S \supset R$. Show that $R[\alpha] \cap R[\alpha^{-1}]$ is an integral extension of R.
5. Let K be a field and $S = K[X, Y]$. Let R be the subalgebra of S generated by the monomials of the type $X^{i+1}Y^i (i \geq 0)$. Show that $\alpha = XY$ is such that $R[\alpha]$ is contained in a finitely generated R-module, but α is not integral over R. (Theorem 4.1.3 is not true if R' is not a subring).

6. An R-algebra A is said to be integral over R if it is integral over R', the image of R under the natural homomorphism given by $a \to a.1_A,\ a \in R$. Show that if $A_i\ (i \leq i \leq n)$ are integral over R, then $\Pi_{i=1}^n A_i$ is integral over R.

7. Let A be an R-algebra which is integral over R and $f : R \to S$ a ring homomorphism. Show that the scalar extension $S \otimes_R A$ is integral over S.

8. Show that if A and B are R-algebras, both integral over R, then $A \otimes_R B$ is integral over R.

9. Let $R \subset S$ be such that $S - R$ is closed under multiplication. Show that R is integrally closed in S.

4.2 Integral extensions

Proposition 4.2.1 *Let $R \subset S$ be domains and S integral over R. Then R is a field if and only if S is a field.*

Proof Assume that R is a field and let $\alpha \in S, \alpha \neq 0$. Then α satisfies a relation of the type $\alpha^n + a_1\alpha^{n-1} + \cdots + a_n = 0,\ a_i \in R$ and n can be chosen to be minimum. This implies that $a_n \neq 0$, as S is a domain. Hence $(-1/a_n)(\alpha^n + a_1\alpha^{n-1} + \cdots + a_{n-1}\alpha) = 1$, so that α has an inverse $\beta = (-1/a_n)(\alpha^{n-1} + a_1\alpha^{n-2} + \cdots + a_{n-1})$. Hence S is a field.

Conversely, assume that S is a field and let $a \in R$ with $a \neq 0$. Then $\alpha = a^{-1} \in S$ is integral over R. It satisfies a relation of the type $\alpha^n + a_1\alpha^{n-1} + \cdots + a_n = 0,\ a_i \in R$. Multiplying by a^n on both sides, we have $1 + a_1 a + \cdots + a_n a^n = 0$, so that $a \in R$ has an inverse $b \in R$ given by $b = -(a_1 + a_2 a + \cdots + a_n a^{n-1})$. □

Corollary 4.2.2 *Let $R \subset S$ with S is integral over R, P' a prime ideal of S and $P = P' \cap R$. Then P is a maximal ideal of R if and only if P' is a maximal ideal of S.*

Proof R/P may be identified with a subring of S/P' and S/P' is integral over R/P. Hence R/P is a field if and only if S/P' is a field. □

Corollary 4.2.3 *Let $R \subset S$ with S is integral over R, P_1, P_2 prime ideals of S with $P_1 \subseteq P_2$ and $P_1 \cap R = P_2 \cap R = P$. Then $P_1 = P_2$.*

Proof S_P is integral over R_P by Proposition 4.1.13 (ii) and $P_1S_P \subseteq P_2S_P$ are both prime ideals of S_P. Since $P_1S_P \cap R_P = P_2S_P \cap R_P = PR_P$, a maximal ideal of R_P, both P_1S_P and P_2S_P are maximal ideals of S_P. Hence $P_1S_P = P_2S_P$. This implies that $P_1 = P_2$. □

Proposition 4.2.4 *Let $R \subset S$ with S is integral over R. For any prime ideal P of R, there exists a prime ideal P' of S such that $P' \cap R = P$.*

Proof Since S is integral over R, S_P is integral over R_P. Consider the commutative diagram.

$$\begin{array}{ccc} R & \xrightarrow{i} & S \\ \downarrow{\scriptstyle p} & & \downarrow{\scriptstyle \overline{p}} \\ R_P & \xrightarrow{\overline{i}} & S_P \end{array}$$

where the horizontal map below is induced by the inclusion map and the vertical maps are natural homomorphisms.

Let $\mathfrak{m}$ be any maximal ideal of S_P. Then $\mathfrak{m} \cap R_P$ is a maximal ideal of R_P. Hence it must be equal to the unique maximal ideal of R_P, *i.e.*, $\mathfrak{m} \cap R_P = PR_P$. Let P' be the contraction of $\mathfrak{m}$ under the map $\overline{p}$. Then P' is a prime ideal of S and $P' \cap R = P$. □

Definition 4.2.5 Let $R \subset S$ be an extension of rings. A prime ideal P' of S is said to *lie above* the prime ideal P of R if $P = P' \cap R$.

Remark 4.2.6 Let $R \subset S$ be an integral extension. Proposition 4.2.4 shows that for any prime ideal P of R there exists a prime ideal P' of S lying above P. Moreover, two such prime ideals, P_1 and P_2, satisfying $P_1 \subseteq P_2$ must coincide by Corollary 4.2.3.

Example 4.2.7 It is possible for two distinct prime ideals P_1' and P_2' of S to lie over the same prime ideal P of R. Consider $R = \mathbb{Z}$ and $S = \mathbb{Z}[i]$, the ring of Gaussian integers. Then $P_1' = (2+i)$ and $P_2' = (2-i)$ are prime ideals of S lying over the same prime ideal $P = (5)$ of R.

Theorem 4.2.8 (Going Up Theorem) *Let S be an integral extension of R. Let $P_1 \subset P_2 \subset \cdots \subset P_n$ be a chain of prime ideals of R and $P_1' \subset P_2' \subset \cdots \subset P_m'$, $(m < n)$ a chain of prime ideals of S with $P_i' \cap R = P_i$ $(1 \leq i \leq m)$. Then there exist prime ideals $P_{m+1}', \ldots, P_n'$ of S such that $P_m' \subset P_{m+1}' \subset \cdots \subset P_n'$ and $P_i' \cap R = P_i$ $(1 \leq i \leq n)$.*

Proof It is sufficient to prove the result for $n = 2$. Let $P_1 \subset P_2$ be prime ideals of R and P_1' a prime ideal of S with $P_1' \cap R = P_1$. Then S/P_1' is an integral extension of R/P_1 and P_2/P_1 is a prime ideal of R/P_1. Hence there exists a prime ideal of S/P_1', say P_2'/P_1', P_2' prime in S, lying above P_2/P_1. Clearly, $P_1' \subset P_2'$ and $P_2' \cap R = P_2$. □

Proposition 4.2.9 *Let S be an integral extension of R and $f : R \to \Omega$ a ring homomorphism of R into an algebraically closed field Ω. Then f can be extended to a ring homomorphism $g : S \to \Omega$.*

Proof Let $P = \text{Ker}(f)$, so that P is a prime ideal of R. By Proposition 4.2.4, there exists a prime ideal P' of S lying above P. Now, f induces an injective ring homomorphism $R/P \to \Omega$, which can be extended to an injection $\overline{f} : K \to \Omega$, where K is the quotient field of R/P. Let L be the quotient field of S/P'. Then L is an algebraic extension of K as

S/P' is an integral extension of R/P. Since Ω is algebraically closed, $\overline{f}$ can be extended to an isomorphism $\overline{g}$ of L into Ω. Let $g = \overline{g}p$, where $p : S \to S/P'$ is the natural projection. Then g is the required ring homomorphism. □

Proposition 4.2.10 *Let S be a ring extension of R which is finitely generated as an R-module. Then for any prime ideal P of R, the set of prime ideals of S lying above P is a finite set.*

Proof By considering the extension S_P over R_P, we assume without loss of generality that R is a local ring with unique maximal ideal P. By considering extension S/PS over R/P, we may also assume that $P = (0)$, *i.e.*, R is a field. Since the extension S is a finitely generated R-module, where R is a field, S is Artinian by Exercise $5, 3.3$. Hence S has only finitely many prime ideals. □

4.2. EXERCISES

1. Let $R \subset S$, where S is integral over R. Show that $a \in R$ is a unit in S if and only if it is a unit in R. Show further that the Jacobson radical $J(R) = R \cap J(S)$.

2. Let $R \subset S$ be domain, where S is integral over R. Show that if f is a ring homomorphism of S into a domain such that the restriction of f to R is injective, then f is injective.

3. Let G be a finite group of automorphisms of a ring S and $R = S^G = \{a \in S \mid \sigma(a) = a \text{ for all } \sigma \in G\}$. Show that S is an integral extension of R.

4. Let G be a finite group of automorphisms of S and $S^G = R$, the subring of invariants, as in Exercise 3. Let T be a multiplicatively closed set in S with $\sigma(T) \subseteq T$ for all $\sigma \in G$. Show that G acts on S_T and $(S_T)^G$ is isomorphic to R_{T_G}, where $T_G = T \cap R$.

5. Let G be a finite group of automorphisms of S and $S^G = R$, the subring of invariants. Show that G acts transitively on the set of prime ideals of S lying above a given prime ideal P of R and deduce that this set is finite.

6. Let $R \subset S$ be a ring extension for which the Going Up Theorem holds. Show that the contraction of a maximal ideal of S is a maximal ideal in R.

7. Let $R \subset S$ be a ring extension. Show that the following conditions are equivalent.

   (i) Going Up Theorem holds for the extension $R \subset S$.

   (ii) For every prime ideal P of R and $T = R - P$, the ideal P' of S maximal with respect to the property $P' \cap T = \emptyset$ lies above P.

4.3 Integrally closed domains

Definition 4.3.1 A domain R is said to be integrally closed if it is *integrally closed* in its quotient field.

Example 4.3.2 (i) The argument in Example $((iv), 4.1.2)$ shows that any unique factorisation domain is integrally closed. In particular, $k[X_1, X_2, \ldots, X_n]$, k a field, is integrally closed.
(ii) The domain $R = k[X,Y]/(X^2 - Y^3)$ is not integrally closed as $\alpha = \overline{X}/\overline{Y}$ is integral over R but does not belong to R.

Proposition 4.3.3 *Let $R \subset S$ be an extension of rings and $T \subset R$ a multiplicatively closed set in R. If $\overline{R}$ is the integral closure of R in S, then $\overline{R}_T$ is the integral closure of R_T in S_T.*

Proof Clearly, $\overline{R}_T$ is integral over R_T by Proposition 4.1.13. Let $(\alpha/t) \in S_T$ be integral over R_T. Then it satisfies a relation of the type

$$(\alpha/t)^n + (a_1/t_1)(\alpha/t)^{n-1} + \cdots + (a_n/t_n) = 0, \quad a_i \in R, \quad t_i, t \in T.$$

If $t' = \Pi_i t_i$, multiplying both sides by $(tt')^n$, we have an integral dependence of $t'\alpha$ over R. Hence $t'\alpha \in \overline{R}$, *i.e.*, $(\alpha/t) = (t'\alpha/t't) \in \overline{R}_T$. □

Corollary 4.3.4 *If R is an integrally closed domain, then R_T is integrally closed for any multiplicatively closed set $T \subset R$. In particular, R_P is integrally closed for every prime ideal P of R.*

Proof Since $\overline{R} = R$, $\overline{R}_T = R_T$. □

Example 4.3.5 *If R is an integrally closed domain and P a prime ideal of R, then R/P need not be integrally closed. Let $R = k[X,Y]$ and $P = (X^2 - Y^3)$. Then R/P is not integrally closed.*

Proposition 4.3.6 *Let R be a domain. The following conditions are equivalent.*
(i) R is integrally closed.

(ii) R_P is integrally closed for every prime ideal P of R.

(iii) $R_{\mathfrak{m}}$ is integrally closed for every maximal ideal $\mathfrak{m}$ of R.

Proof $(i) \Rightarrow (ii)$ by Corollary 4.3.4.

$(ii) \Rightarrow (iii)$ is clear.

To show that $(iii) \Rightarrow (i)$, let $\overline{R}$ be the integral closure of R in its quotient field K and $i : R \to \overline{R}$ the inclusion map. Then $i_{\mathfrak{m}} : R_{\mathfrak{m}} \to \overline{R}_{\mathfrak{m}}$ is surjective for every maximal ideal $\mathfrak{m}$ by assumption (iii). Hence i is surjective. This implies (i). □

Proposition 4.3.7 *Let R be a ring and $f(X)$ and $g(X)$ be monic polynomial over $R_1 \supset R$ such that the product $h(X) = f(X)g(X)$ has coefficients integral over R. Then $f(X)$ and $g(X)$ have coefficients integral over R.*

Proof We observe that (as in the case of a field) for any monic polynomial $f(X) \in R_1[X]$, there exists an extension ring $R' \supset R_1$, in which $f(X)$ is a product of linear factors. Choose a ring $R' \supset R_1$ such that both $f(X)$ and $g(X)$ split into linear factors over R', *i.e.*,

$$f(X) = \Pi_i(X - \alpha_i)$$

and

$$g(X) = \Pi_j(X - \beta_j),$$

$\alpha_i, \beta_j \in R'$. Let S be the integral closure of R in R'. Since $h(X) = f(X)g(X)$ has coefficients integral over R, $h(X) \in S[X]$. Now, $h(\alpha_i) = 0$ and $h(\beta_j) = 0$ are integral relations for the α_i and the β_j over S. Since S is the integral closure of R in R', all the α_i and the β_j belong to S. Since the coefficients of $f(X)$ and $g(X)$ are sums and product of the α_i and the β_j respectively, the coefficients of $f(X)$ and $g(X)$ belong to S, *i.e.*, they are integral over R. □

Corollary 4.3.8 *Let R be a domain with quotient field K and K' an extension field of K. If $\alpha \in K'$ is integral over R, then the minimum polynomial of α over K has all its coefficients integral over R.*

Proof Let $h(\alpha) = 0, h(X) \in R[X]$, be an integral dependence of α over R and let $f(X) \in K[X]$ be the minimum polynomial of α over K. Then $f(X)$ divides $h(X)$ in $K[X]$ and

$$h(X) = f(X)g(X), g(X) \in K[X].$$

By Proposition 4.3.7, $f(X)$ has coefficients integral over R. □

Corollary 4.3.9 *Let R be an integrally closed domain with quotient field K and K' an extension field of K. If $\alpha \in K'$ is integral over R, then the minimum polynomial of α over K has all its coefficients in R.*

Proof Follows from Corollary 4.3.8. □

Proposition 4.3.10 *Let $R \subset S$ be extension of rings and $\overline{R}$ the integral closure of R in S. Then the integral closure of $R[X]$ in $S[X]$ is $\overline{R}[X]$.*

Proof Since $\overline{R}$ is integral over R, $\overline{R}[X]$ is integral over $R[X]$ (Exercise 7, Section 4.1).

Conversely, let $f(X) \in S[X]$ be integral over $R[X]$. It satisfies a relation of the type

$$f^n + g_1 f^{n-1} + \cdots + g_n = 0, \; g_i \in R[X].$$

Choose m to be greater than n, and the degrees of f and g_i, $(1 \leq i \leq n)$. If $h(X) = f(X) - X^m$, then

$$(h + X^m)^n + g_1(h + X^m)^{n-1} + \cdots + g_n = 0,$$

i.e., $h^n + h_1 h^{n-1} + \cdots + h_n = 0$, where

$$h_n = X^{mn} + g_1 X^{m(n-1)} + \cdots + g_n \in R[X].$$

This implies that $-h(h^{n-1} + h_1 h^{n-2} + \cdots + h_{n-1}) = h_n \in R[X]$. Now, h_n has coefficients in R, *i.e.*, h_n is integral over R. Applying Proposition 4.3.7 to the pair of polynomials on the left hand side, we have $h(X) \in \overline{R}[X]$. This implies that $f(X) = X^m + h(X) \in \overline{R}[X]$. □

Corollary 4.3.11 *If R is an integrally closed domain, then $R[X]$ is also integrally closed.*

Proof Let K be the quotient field of R. Let $\alpha \in K(X)$ be integral over $R[X]$ so that it is integral over $K[X]$. Since $K[X]$ is a unique factorisation domain, it is integrally closed so that $\alpha \in K[X]$. Since $R[X]$ is integrally closed in $K[X]$, $\alpha \in R[X]$. □

Corollary 4.3.12 *If R is an integrally closed domain, then $R[X_1, \ldots, X_n]$ is integrally closed.*

Proof Follows from Corollary 4.3.11. □

In Theorem 4.2.8 we studied the going up property for an integral extension, i.e., that an increasing chain of prime ideals in R can be lifted to a corresponding chain in S. The following example shows that a similar property for decreasing chain of prime ideals is no longer valid without additional assumptions on R and S.

Example 4.3.13 Let $R = \mathbb{Z}$ and $S = \mathbb{Z}[X]/I$, where

$$I = (2X, X^2 - X).$$

Then $P' = (2, \overline{X} - 1)$, where $\overline{X} = X + I$ is a prime ideal of S lying above $P = 2\mathbb{Z}$ and there is no prime ideal of S contained in P' lying above (0).

We propose to study the going down property for ring extensions.

Definition 4.3.14 Let R be an integrally closed domain with quotient field K and L a normal field extension of K. Let G be the group of all K automorphisms of L. If S is the integral closure of R in L, S is called a *normal extension* of R with the Galois group $G = G(L/K)$.

Proposition 4.3.15 *Let R be an integrally closed domain with quotient field K, L a normal field extension of K and S a normal extension of R with Galois group $G = G(L/K)$. Then*

(i) G is the group of R-automorphisms of S.

(ii) Two prime ideals P' and Q' of S lie over the same prime ideal of R if and only if there exists some $\sigma \in G$ with $\sigma(P') = Q'$.

Proof (i) For any $\sigma \in G = G(L/K)$, clearly $\sigma(S) = S$ as $\sigma(S)$ is integral over R, so that σ is an R-automorphism of S. Suppose $\sigma |_S = \tau |_S$ for $\sigma, \tau \in G$. Since L is the quotient field of S we have $\sigma |_L = \tau |_L$, *i.e.*, $\sigma = \tau$. Moreover, any R-automorphism of S can be extended to a K-automorphism of L. This proves (i).

(ii) If there exists $\sigma \in G$ with $\sigma(P') = Q'$, then clearly P' and Q' lie over the same prime ideal of R.

Conversely, assume that P' and Q' lie over the same prime ideal P of R. We prove the result first when G is a finite group. If $Q' \neq \sigma(P')$ for any $\sigma \in G$, then by Proposition 2.1.9 there exists $a \in Q'$ with $a \notin \sigma(P')$ for all $\sigma \in G$.

Let $b = \Pi_{\sigma \in G}\sigma(a)$. Then $\sigma(b) = b$, for any $\sigma \in G$, and hence b is purely inseparable over K, i.e., $b^m \in K$ for some $m \geq 1$. Then $Q' \cap R = P$ contains b^m since $a \in Q'$ and R is integrally closed in K. But $P' \cap R = P$ does not contain b^m since P' is prime and $\sigma(a) \notin P'$ for all $\sigma \in G$. This is a contradiction.

To complete the proof, we have to consider the case when G is infinite. Consider the set Σ of all pairs $(L_\alpha, \sigma_\alpha)$, where L_α is a normal extension of K and $\sigma_\alpha \in G(L_\alpha/K)$ are such that $L_\alpha \subset L$ and $\sigma_\alpha(P' \cap L_\alpha) = Q' \cap L_\alpha$. Introduce an order $\leq$ in Σ by defining $(L_\alpha, \sigma_\alpha) \leq (L_\beta, \sigma_\beta)$ if and only if $L_\alpha \subset L_\beta$ and $\sigma_\alpha = \sigma_\beta |_{L_\alpha}$. Σ is nonempty as $(K, 1_K) \in \Sigma$. By Zorn's Lemma, Σ has a maximal element (L^*, σ^*). It is sufficient to show that $L^* = L$. Then $\sigma^*(P') = Q', \sigma^* \in G(L/K)$.

Suppose, if possible, that $L^* \neq L$; choose $a \in L$ with $a \notin L^*$. Since the number of conjugates of a is finite, there exists a smallest normal extension L_1 containing $L^*(a)$, and finite over L^*. Then σ^* can be extended to an automorphism of the algebraic closure of L and since L is normal, σ^* can be considered as an element of $G = G(L/K)$. Then $\sigma^*(P' \cap L_1)$ and $Q' \cap L_1$ are prime ideals of $S \cap L_1$ lying over $Q' \cap L^*$. Since $G(L_1/L^*)$ is finite, by the first part of the proof there exists $\tau \in G(L_1/L^*)$ with $\tau\sigma^*(P' \cap L_1) = Q' \cap L_1$. Hence $(L_1, \tau\sigma^*) \in \Sigma$, contradicting the maximality of (L^*, σ^*). □

Theorem 4.3.16 (Going Down Theorem) *Let R be an integrally closed domain and S a domain which is an integral extension of R. Let $P_1 \supset P_2 \supset \cdots \supset P_n$ be a descending chain of prime ideals of R and P_1' a prime ideal of S lying above P_1. Then there exists a chain $P_1' \supset P_2' \supset \cdots \supset P_n'$ of prime ideals of S with $P_i = P_i' \cap R$ $(1 \leq i \leq n)$.*

Proof Assume first that S is a normal extension of R. By Going Up Theorem, there exists a chain $P_n'' \subset P_{n-1}'' \subset \cdots \subset P_1''$ of prime ideals of S with $P_i'' \cap R = P_i$, $1 \leq i \leq n$. By Proposition 4.3.15, there exists some $\sigma \in G(S/R)$ with $\sigma P_1'' = P_1'$. If $P_i' = \sigma P_i''$ $(2 \leq i \leq n)$, then $P_1' \supset P_2' \supset \cdots \supset P_n'$ is the required chain.

In the general case, let S^* be a normal extension of R containing S and P_1^*, a prime ideal of S^* lying above P_1'. Then by the first part of the proof, there exists a chain $P_1^* \supset P_2^* \supset \cdots \supset P_n^*$ of prime ideals lying above $P_1 \supset P_2 \supset \cdots \supset P_n$. If $P_i' = P_i^* \cap S$ $(2 \le i \le n)$, then $P_1' \supset P_2' \supset \cdots \supset P_n'$ is the required chain. $\square$

4.3. EXERCISES

1. Let R be the ring of Gaussian integers with even imaginary part. Show that R is not integrally closed. What is its integral closure?

2. Let R be ring of all formal power series in X with coefficients in a field K containing no term in X. Show that R is not integrally closed. What is its integral closure?

3. A domain R is called a *g.c.d. domain* if any two elements of R have a greatest common divisor. Show that any g.c.d. domain is integrally closed.

4. Let $R = k[X^4, Y^4]$ and $S = k[X^4, X^3Y, XY^3, Y^4] \supset R$, where k is a field. Let $P = (Y^4)$ be the prime ideal of R. Show that the ideal (X^4, X^3Y, XY^3, Y^4) of S is a prime ideal associated to PS in S but does not lie above P.

5. Let $R \subset S$ be an extension and $\overline{R}$ the integral closure of R in S. Let I be an ideal of R. Show that if I^e is the extension of I in $\overline{R}$, then the integral closure of I in S is $\sqrt{I^e}$.

6. Let R be a Noetherian ring, K its total quotient ring and $\alpha \in K$. Show that α is integral over R in each of the following cases:

   (i) There is a non-zero-divisor $\beta \in R$ with $\beta\alpha^n \in R$ for each $n \ge 1$.

   (ii) There is an ideal I of R containing a non-zero-divisor such that $\alpha I \subset I$.

7. Let $R \subset S$ be an integral extension, R and S domains and R integrally closed. If P be a prime ideal of R, show that the prime ideals of S lying above P are minimal elements of the set of prime ideals of S containing PS.

4.4 Finiteness of integral closure

Proposition 4.4.1 *Let R be an integrally closed domain with quotient field K and L a finite separable extension field of K. Let S be the integral closure of R in L. Then there exists a basis $\{e_1, e_2, \ldots, e_n\}$ of L over K, such that*

$$S \subset \Sigma_{i=1}^n Re_i.$$

Proof We first show that there exists a basis of L over K which is contained in S. For any $\alpha \in L$, consider the minimal polynomial relation $\alpha^n + a_1\alpha^{n-1} + \cdots + a_n = 0, a_i \in K$, and choose $b \in R, b \neq 0$ such that $ba_i \in R, (1 \leq i \leq n)$. Then

$$(b\alpha)^n + a_1 b(b\alpha)^{n-1} + \cdots + a_n b^n = 0,$$

so that $b\alpha$ is integral over R, *i.e.*, $b\alpha \in S$ for some $b \in R, b \neq 0$. Therefore, given any basis of L over K, by suitably multiplying by a non-zero element of R, we can assume that the basis lies in S. Let such a basis be $\{e'_1, e'_2, \ldots, e'_n\}$.

Consider the bilinear form $L \times L \to K$ given by $(x, y) \to \text{Tr}(xy)$. It is non-degenerate as L/K is separable. Hence there exists a dual basis $\{e_1, e_2, \ldots, e_n\}$ of L/K such that $\text{Tr}(e'_i e_j) = \delta_{ij}$. If $\alpha \in S$ and $\alpha = \Sigma_i a_i\, e_i$, then $\alpha e'_j \in S$ for all $j \;\; (1 \leq j \leq n)$, so that $\text{Tr}(\alpha e'_j) = \text{Tr}(\underset{i}{\Sigma} a_i e_i e'_j) = a_j$. Now, by Corollary 4.3.9, the trace of any element of S belongs to R. Hence $a_j \in R \;\; (1 \leq j \leq n)$. Thus $S \subseteq \Sigma_j R e_j$. □

Corollary 4.4.2 *Let R be a Noetherian integrally closed domain with quotient field K and L a finite separable extension of K. Then the integral closure S of R in L is Noetherian. Also, S is a finite R-module.*

Proof Choose a basis $\{e_1, \ldots, e_n\}$ of L over K such that $S \subset \Sigma_i R e_i$. Since R is a Noetherian ring, S is a Noetherian R-module and hence it is also a Noetherian S-module, i.e., it is a Noetherian ring. □

Corollary 4.4.3 *Let R be a principal ideal domain with quotient field K and L a finite separable extension of K. If S is the integral closure of R in L, then there exists a basis $\{f_1, f_2, \ldots, f_n\}$ of L over K such that $S = \Sigma_{i=1}^n R f_i$.*

Proof By Proposition 4.4.1, we can choose a basis $\{e_1, e_2, \ldots, e_n\}$ of L/K such that $S \subset \Sigma_i R e_i$. Since R is a principal ideal domain and S is a submodule of a free R module, S is also R-free with basis $\{f_1, f_2, \ldots, f_m\} \;\; (m \leq n)$. Now, $\{f_1, f_2, \ldots, f_m\}$ being a basis of S over R is also a basis of L over K. Hence $m = n$. This implies $S = \Sigma_{i=1}^n R f_i$, where $\{f_1, f_2, \ldots, f_n\}$ is a basis of L/K. □

Remark 4.4.4 When $R = \mathbb{Z}$ (and $K = \mathbb{Q}$) and L a number field (i.e., a finite extension of $\mathbb{Q}$), then S is called *the ring of algebraic integers of the number field L*.

Theorem 4.4.5 (Noether's Normalisation Theorem) *Let k be a field and R a finitely generated k-algebra. Then there exist $y_1, y_2, \ldots, y_r \in R$ algebraically independent over k such that R is integral over $k[y_1, y_2, \ldots, y_r]$.*

Proof Let R be generated by $\{x_1, x_2, \ldots, x_n\}$ over k. The proof is by induction on n. Arrange the generating set such that $x_1, x_2, \ldots, x_r$ are algebraically independent over k

and $x_{r+1}, x_{r+2}, \ldots, x_n$ are algebraic over $k[x_1, x_2, \ldots, x_r]$. If $r = n$, there is nothing to prove. Assume that $r < n$, so that x_n is algebraic over $k[x_1, x_2, \ldots, x_{n-1}]$. There exists a nontrivial polynomial $f(X_1, X_2, \ldots, X_{n-1}, X_n)$ satisfied by $x_1, x_2, \ldots, x_n$. We now consider two cases.

Case (i). Let k be an infinite field. If $F(X_1, X_2, \ldots, X_n)$ is the homogeneous component of the highest degree d in the polynomial $f(X_1, \ldots, X_n)$, then there exist $a_1, a_2, \ldots, a_{n-1} \in k$ with

$$F(a_1, a_2, \ldots, a_{n-1}, 1) \neq 0$$

as k is infinite. Making the substitutions

$$x_i' = x_i - a_i x_n, \quad 1 \leq i \leq n-1,$$

the relation $f(x_1, x_2, \ldots, x_n) = 0$ gives a relation of the type

$$F(a_1, a_2, \ldots, a_{n-1}, 1)x_n^d + \text{(terms of degree less than } d \text{ in } x_n) = 0.$$

This shows that x_n is integral over $R' = k[x_1', x_2', \ldots, x_{n-1}']$. By induction, there exist $y_1, \ldots, y_r$ algebraically independent over k such that R' is integral over $k[y_1, \ldots, y_r]$. Since each x_i, $1 \leq i \leq n$, is integral over R', R is integral over R' and hence over $k[y_1, \ldots, y_r]$.

Case (ii). Let k be any field.

Consider the polynomial $f(X_1, X_2, \ldots, X_n)$ such that

$$f(x_1, x_2, \ldots, x_n) = 0.$$

Choose $t >$ total deg f. Attach weights

$$m_1 = t^{n-1}, m_2 = t^{n-2}, \ldots, m_n = 1$$

to the variables $X_1, X_2, \ldots, X_n$ respectively and define the weight $w(M)$ of a monomial $X_1^{\alpha_1} \cdots X_n^{\alpha_n}$ occurring in f, as

$$w(M) = \Sigma_1^n m_i \alpha_i.$$

If $M' = X_1^{\alpha_1'} \cdots X_n^{\alpha_n'}$ is another monomial occurring in f, then it is easy to verify that

$$w(M) > w(M') \Leftrightarrow (\alpha_1, \alpha_2, \ldots, \alpha_n) > (\alpha_1', \alpha_2', \ldots, \alpha_n')$$

in the lexicographic ordering defined by $(\alpha_1, \ldots, \alpha_n) > (\alpha_1', \ldots, \alpha_n')$, when there exists some m with $\alpha_i = \alpha_i'$ $(1 \leq i \leq m)$ and $\alpha_{m+1} > \alpha_{m+1}'$. Thus the distinct monomials occurring in f have different weights.

Choose $M = \lambda X_1^{\alpha_1} \cdots X_n^{\alpha_n}$, $\lambda \neq 0$, $\lambda \in k$ occurring in f with maximum weight. Make the substitutions

$$x_i' = x_i + x_n^{m_i} \text{ for } 1 \leq i \leq n-1$$

in the relation $f(x_1, x_2, \ldots, x_n) = 0$. By the choice of the monomial M, we have a relation of the type $\lambda x_n^{w(M)} +$ (terms of lower degree in x_n) $= 0$, with coefficients in $R' = k[x_1', \ldots, x_{n-1}']$. Hence x_n is integral over $R' = k[x_1', \ldots, x_{n-1}']$. The proof is completed by induction on n as in case (i). □

Corollary 4.4.6 *Let R be a finitely generated k-algebra, k a field, and $\mathfrak{m}$ a maximal ideal R. Then $R/\mathfrak{m}$ is a finite extension of k.*

Proof Since $R' = R/\mathfrak{m}$ is a finitely generated k-algebra, by Theorem 4.4.5, it is integral over $k[y_1, \ldots, y_r]$, where $y_1, \ldots y_r$ are algebraically independent over k. Since R' is a field, so is $k[y_1, \ldots, y_r]$. Hence $r = 0$ and R' is a finite extension of k. □

Corollary 4.4.7 *Let R be a finitely generated k-algebra, k a field and I an ideal of R. Then $\sqrt{I}$ is the intersection of all maximal ideals of R containing I.*

Proof We may assume that $I = 0$. Clearly, $\sqrt{0}$ is contained in every maximal ideal of R.

Conversely, suppose $f \in R - \sqrt{0}$. Then $R_f \neq 0$. Let $\mathfrak{n}$ be a maximal ideal of R_f and $\mathfrak{m} = \mathfrak{n} \cap R$. Now, $R_f/\mathfrak{n}$ is a finitely generated k-algebra and it is a field. Hence it is algebraic over k. But $R_f/\mathfrak{n} \simeq (R/\mathfrak{m})_f$. Hence $R/\mathfrak{m}$ is integral over k. This implies that $R/\mathfrak{m}$ is a field, i.e., $\mathfrak{m}$ is a maximal ideal of R. Since $f \notin \mathfrak{m}$, the intersection of all maximal ideals of R is contained in $\sqrt{0}$. □

Corollary 4.4.8 (Weak Nullstellensatz) *Any maximal ideal $\mathfrak{m}$ of $R = k[X_1, \ldots, X_n]$, k an algebraically closed field, is of the type*

$$\mathfrak{m} = (X_1 - a_1, \ldots, X_n - a_n), \quad a_i \in k \ (1 \leq i \leq n).$$

Proof Since k is algebraically closed by Corollary 4.4.6, $R/\mathfrak{m} \simeq k$. Let a_i be the image of $X_i + \mathfrak{m}$ under this isomorphism. Then $(X_1 - a_1, \ldots, X_n - a_n) \subset \mathfrak{m}$, and since it is maximal, it is equal to $\mathfrak{m}$. □

Corollary 4.4.9 (Hilbert's Nullstellensatz)

Let $R = k[X_1, \ldots, X_n]$, k an algebraically closed field, and I an ideal of R. Let

$$V(I) = \{(a_1, \ldots, a_n) \in k^n \mid f(a_1, \ldots, a_n) = 0 \text{ for all } f \in I\}.$$

Then $J = \{h \in R \mid h(a_1, \ldots, a_n) = 0$ for all $(a_1, \ldots, a_n) \in V(I)\}$ is an ideal equal to $\sqrt{I}$.

Proof Clearly, J is an ideal containing I and hence $J \supseteq \sqrt{I}$. To show that $J \subseteq \sqrt{I}$, consider $g \in J$. Now, for every maximal $\mathfrak{m} \supseteq I$, we have $\mathfrak{m} = (X_1 - a_1, \dots, X_n - a_n) \supseteq I$. This implies

$$a = (a_1, \dots, a_n) \in V(I)$$

and hence $g(a) = 0$, i.e., $g \in \mathfrak{m}$. Hence $g \in \bigcap_{\mathfrak{m} \supseteq I} \mathfrak{m} = \sqrt{I}$. □

We shall now show that the integral closure of an affine domain in a finite extension of its quotient field is an affine domain, in particular, the normalisation of an affine domain is an affine domain.

Theorem 4.4.10 *Let R be a domain which is finitely generated algebra over a field k, K the quotient field of R and L a finite field extension of K. If S is the integral closure of R in L, then S is a module of finite type over R.*

Proof By Theorem 4.4.5, there exists $y_1, \dots, y_r$ algebraically independent over k such that R is integral over $R' = k[y_1, y_2, \dots, y_r]$. Since the integral closure of R' in L is also S, we can assume without loss of generality that R itself is a polynomial algebra, i.e., $R = k[X_1, X_2, \dots, X_n]$. Since L is a finite extension of K, there exists a finite normal extension L' of K such that $L \subseteq L'$. If S' is the integral closure of R in L' and if we show that S' is a finite type R-module, then it will follow that S is also a finite type R-module, as $S \subseteq S'$ and R is Noetherian. Hence we assume without loss of generality that L is a normal extension of K. Then there exists an extension F/K with $K \subset F \subset L$ such that L is separable over F and F is purely inseparable over K. If S_1 is the integral closure of R in F and S is the integral closure of S_1 in L, then S is the integral closure of R in L. It is sufficient to show that S_1 is a finite type R-module; for, by Proposition 4.4.1, S is a finite type S_1-module as L/F is separable. Hence we assume without loss of generality that L/K is purely inseparable, say $L = K(\alpha_1, \alpha_2, \dots, \alpha_m)$, where $\alpha_i^q \in K = k(X_1, X_2, \dots, X_n)$ $(1 \le i \le m)$ for some q, which is a power of the characteristic of k. Write α_i^q as rational functions in $X_1, X_2, \dots, X_n$ $(1 \le i \le m)$ over k, and let $\lambda_1, \lambda_2, \dots, \lambda_r$ be the coefficients which occur in both the numerator and denominator of the α_i^q $(1 \le i \le m)$. Let $k' = k(\lambda_1^{1/q}, \dots, \lambda_r^{1/q})$ and $L' = k'(X_1^{1/q}, X_2^{1/q}, \dots, X_n^{1/q})$. Then $L \subset L'$ and if S' is the integral closure of R in L', then $S \subset S'$ and it is sufficient to show that S' is a finite type R-module. Since k'/k is algebraic, $k'[X_1, \dots, X_n] = S_1$ is integral over $k[X_1, \dots, X_n] = R$, and S_1 is a finitely generated R-module. Now, $k'[X_1^{1/q}, \dots, X_n^{1/q}]$ is integrally closed in L' and $k'[X_1^{1/q}, \dots, X_n^{1/q}]$ is integral over $k'[X_1, \dots, X_n] = S_1$ and S_1 is integral over R. Hence the integral closure of R in L' is $k'[X_1^{1/q}, \dots, X_n^{1/q}]$, i.e., $S' = k'[X_1^{1/q}, \dots, X_n^{1/q}]$. Now, S' is a finitely generated S_1-module, and since S_1 is a finitely generated R-module, S' is a finitely generated R-module and the proof is complete. □

Corollary 4.4.11 *Under the same assumptions as in Theorem 4.4.10, S is a finitely generated k-algebra.*

Proof Since R is a finitely generated k-algebra and S is a module of finite type over R, S is a finitely generated k-algebra. □

4.4. EXERCISES

1. Let $R \subset S$ be domains such that S is a finitely generated R-algebra. Show that there exists $f \in R, f \neq 0$, and elements $\{y_1, \ldots, y_r\} \in S$ algebraically independent over R such that S_f is integral over S'_f, where $S' = R[y_1, y_2, \ldots, y_r]$.

2. Let $R \subset S$ be domains, S a finitely generated R-algebra. Show that there exists $a \in R$, $a \neq 0$, such that any homomorphism $f : R \to \Omega$ with Ω algebraically closed and $f(a) \neq 0$ can be extended to a homomorphism $g : S \to \Omega$.

3. Let k be a field, $S = k[X, Y]/(X^2, XY)$. Let x and y be the cosets of X and Y. Show that x and y^n $(n \geq 0)$ form a basis of S over k and S is not integral over $R = K[x]$. Show further that there is no element of S which is algebraically independent over R.

4. Let $R \subset S$ be domains, S a finitely generated R-algebra. If R has Jacobson radical zero, show that the Jacobson radical of S is also zero.

5. A ring R is called a *Jacobson ring* if every prime ideal of R is an intersection of maximal ideals. Show that R is a Jacobson ring if and only if for every ideal I of R, R/I has both nil and Jacobson radical equal.

6. Show that a finitely generated algebra over a Jacobson ring is a Jacobson ring.

7. Prove the following version of Noether's Normalisation Theorem. Let $R = k[X_1, \ldots, X_n]$, k a field, $h \in R - k$. Then there exist $t_1, \ldots, t_{n-1} \in R$ algebraically independent over k, such that R is integral over $k[t_1, t_2, \ldots, t_{n-1}, h]$.

   Hint: Write $h = \Sigma a_{i_1 \ldots i_n} X_1^{i_1} \cdots X_n^{i_n}$. In the relation

   $$-h + \Sigma a_{i_1 \ldots i_n} X_1^{i_1} \cdots X_n^{i_n} = 0,$$

   make the substitutions $t_i = X_i + X_n^{m_i}$ $(1 \leq i \leq n-1)$ for suitable m_i as in Theorem 4.4.5 and argue as in case (ii) by obtaining an integral relation for X_n over $k[t_1, \ldots, t_{n-1}, h]$.

CHAPTER

5

DEDEKIND DOMAINS

While studying the ring of integers in an algebraic number field, we come across rings which do not have the unique factorisation property for elements. The uniqueness of factorisation can be restored in some cases by replacing elements by ideals, i.e., every proper ideal in such a ring can be expressed uniquely as a product of powers of prime ideals. Rings with this property are called Dedekind domains. They can be characterised equivalently as domains which are Noetherian, integrally closed and having the property that every non-zero prime ideal is maximal. This characterisation is useful in geometry as co-ordinate rings of non-singular curves are of this type. We show that the integral closure of a Dedekind domain in a finite extension is a Dedekind domain.

The ring $\mathcal{O}_P$ of meromorphic functions defined at a non-singular point P of a curve is a special kind of ring which is a local ring as well as a Dedekind domain. Every element $h \in \mathcal{O}_P$ can be assigned an integer $\mathrm{Ord}_P h$ which is nothing but the order of contact of h with the curve at P. The map $h \to \mathrm{Ord}_P h$ is generalised to the concept of a valuation and the ring $\mathcal{O}_P$ is an example of a valuation ring, specifically a discrete valuation ring. The generalised concept of valuation is used to develop the concept of order of contact of two algebraic varieties along generalised branches.

5.1 Valuation rings

Let $R \subset S$ be an extension of rings and Ω an algebraically closed field. The following extension problem arises quite often in geometry. Given a ring homomorphism $h : R \to \Omega$, does it have an extension $\overline{h}$ as a ring homomorphism $\overline{h} : S \to \Omega$? We have seen in Proposition 4.2.9 that such an extension is possible if S is integral over R. If R is a subring of a field K, we look for a subring S of K containing R for which h has a maximal extension. Such maximal extensions turn out to be valuation rings.

Proposition 5.1.1 *Let $R \subset K$, K a field, and $h : R \to \Omega$ a non-trivial ring homomorphism of R into an algebraically closed field Ω. If $\alpha \in K$, $\alpha \neq 0$, then h can be extended to a ring homomorphism $\overline{h}$ of either $R[\alpha] \to \Omega$ or $R[\alpha^{-1}] \to \Omega$.*

Proof We may assume without loss of generality that R is a local ring and $h(R)$ is a subfield of Ω, for if $P = \text{Ker } h$, P is a prime ideal of R and h can always be extended to $\overline{h} : R_P \to \Omega$ by defining $\overline{h}(a/b) = h(a)/h(b)$, $h(b) \neq 0$. Since R_P is local and $\overline{h}(R_P)$ is isomorphic to R_P/PR_p, a subfield of Ω, we can replace (R, h) by $(R_P, \overline{h})$.

Assume therefore that R is a local ring with maximal ideal $\mathfrak{m}$ and $h(R) = F$, a subfield of Ω. Now, h can be extended to a homomorphism $h : R[X] \to F[X]$ by defining

$$h(\sum_i a_i X^i) = \sum_i \overline{a}_i X^i,$$

where $\overline{a}_i = h(a_i)$. Let $I = \{f \in R[X] \mid f(\alpha) = 0\}$ and $J = h(I)$. Then J is an ideal in $F[X]$ and hence $J = (\theta(X))$ for some $\theta(X) \in F[X]$. If $\theta(X)$ is a non-unit, it has a root $\beta \in \Omega$ and $\overline{h} : R[\alpha] \to \Omega$ can be defined by setting $\overline{h}(\alpha) = \beta$, and $\overline{h} \mid_R = h$. Therefore, if h has no extension $\overline{h} : R[\alpha] \to \Omega$, then $\theta(X)$ is a unit and thus we have a relation of the type

$$\sum_{i=0}^{r} a_i \, \alpha^i = 0, \text{ with } \overline{a}_0 = 1, \overline{a}_i = 0, \; i > 1. \tag{5.1}$$

Choose r minimum with this property. Similarly, if h has no extension $\overline{h} : R[\alpha^{-1}] \to \Omega$, α^{-1} satisfies a relation similar to (5.1), i.e.,

$$\sum_{i=0}^{s} b_i \, \alpha^{-i} = 0, \overline{b}_0 = 1, \overline{b}_i = 0, \; i > 1. \tag{5.2}$$

(s minimum). We may assume without loss of generality that $r \geq s$. Since $\overline{b}_0 = 1$, $b_0 - 1 \in \text{Ker } h \subset \mathfrak{m}$ and $b_0 \notin \mathfrak{m}$, i.e., b_0 is a unit. So the second relation may be rewritten, after multiplying by α^s, as

$$\alpha^s + c_1 \alpha^{s-1} + \cdots + c_s = 0, \overline{c}_i = 0, 1 \leq i \leq s. \tag{5.3}$$

By subtracting $a_r \alpha^{r-s}$ multiple of this from (5.1), we obtain a relation of type (5.1) of degree less than r satisfied by α, contradicting the minimality of r. Hence h has an extension $\overline{h} : R[\alpha] \to \Omega$ or $\overline{h} : R[\alpha^{-1}] \to \Omega$. □

Definition 5.1.2 A subring R of a field K is called a *valuation ring* of K if for every $\alpha \in K$, $\alpha \neq 0$, either $\alpha \in R$ or $\alpha^{-1} \in R$.

Example 5.1.3 (i) Any field K is a valuation ring of K.

(ii) Let K be a field and

$$R = \left\{ \frac{f(X)}{g(X)} \mid f(X),\, g(X) \in K(X), \quad \deg f \leq \deg g \right\}.$$

Then R is a valuation ring of K.

(iii) Let K be a field and

$$R = \left\{ \sum_{i=0}^{\infty} a_i X^i \mid a_i \in K \right\}$$

the ring of formal power series in X over K. If

$$L = \left\{ \sum_{i=N}^{\infty} a_i X^i \mid a_i \in K,\, N \in \mathbb{Z} \right\}$$

is the field of all Laurent series in X over K, then R is a valuation ring of L.

(iv) Let p be a fixed prime. Let $R \subset \mathbb{Q}$, the field of rationals, be defined by

$$R = \left\{ p^r \frac{m}{n} \mid r \geq 0,\, (m,p) = 1,\, (n,p) = 1 \right\}.$$

Then R is a valuation ring of $\mathbb{Q}$.

Proposition 5.1.4 *Let R be a subring of a field K and $h : R \to \Omega$ a non-trivial ring homomorphism of R into an algebraically closed field Ω. Then h has a maximal extension $\overline{h} : V \to \Omega$, where V is a valuation ring of K.*

Proof Consider the set Σ of all pairs (R_α, h_α), R_α ring, $R \subset R_\alpha \subset K$ and $h_\alpha : R_\alpha \to \Omega$, a ring homomorphism which is an extension of h. Define $(R_\alpha, h_\alpha) \leq (R_\beta, h_\beta)$ if and only if $R_\alpha \subseteq R_\beta$ and $h_\alpha = h_\beta|_{R_\alpha}$. By Zorn's lemma, Σ has a maximal element $(V, \overline{h})$. Then clearly V is a valuation ring of K, for if $\alpha \in K$, $\alpha \neq 0$, then by Proposition 5.1.1, $\overline{h}$ has an extension $\overline{h}_1 : V[\alpha] \to \Omega$ or $\overline{h}_1 : V[\alpha^{-1}] \to \Omega$. By maximality, $V[\alpha] = V$ or $V[\alpha^{-1}] = V$, i.e., $\alpha \in V$ or $\alpha^{-1} \in V$. □

Proposition 5.1.5 *Let V be a valuation ring of K. Then*

(i) *K is the quotient field of V.*

(ii) *Any subring of K containing V is a valuation ring of K.*

(iii) *V is a local ring.*

(iv) *V is integrally closed.*

Proof (i) Clearly, $1 \in K$ implies $1 = 1^{-1} \in V$. For $\alpha \in K$, $\alpha \neq 0$, either $\alpha \in V$ or $\alpha^{-1} \in V$, so that α can be expressed as $\alpha = \alpha/1$ or $\alpha = 1/\alpha^{-1}$, i.e., K is the quotient field of V.

(ii) If $V \subset V' \subset K$, and $\alpha \in K, \alpha \neq 0$, then $\alpha \in V$ or $\alpha^{-1} \in V$ and hence $\alpha \in V'$ or $\alpha^{-1} \in V'$.

(iii) Let $\mathfrak{m}$ be the set of non-units of V. If $a, b \in \mathfrak{m}$, $a \neq 0, b \neq 0$, then either $a/b \in V$ or $b/a \in V$. If $a/b \in V$, then $a + b = (1 + a/b)b \in \mathfrak{m}$, and if $b/a \in V$, then $a + b = (1 + b/a)a \in \mathfrak{m}$. Moreover, for any $r \in R$ we have $ra \in \mathfrak{m}$. Thus $\mathfrak{m}$ is an ideal and V is a local ring.

(iv) Let $\alpha \in K$ satisfy an integral relation of the type

$$\alpha^n + a_1\alpha^{n-1} + \cdots + a_n = 0,\ a_i \in V.$$

If $\alpha \notin V$, then $\alpha^{-1} \in V$. This implies

$$\alpha = -a_1 - a_2\alpha^{-1} - \cdots - a_n\alpha^{-(n-1)} \in V,$$

a contradiction. Hence $\alpha \in V$, i.e., V is integrally closed in K. □

Proposition 5.1.6 *The ideals of a valuation ring are totally ordered by inclusion. Conversely, if the ideals of a domain V with quotient field K are totally ordered by inclusion, then V is a valuation ring of K.*

Proof Assume that V is a valuation ring of K and I, J ideals of V with $I \not\subset J$. Choose $a \in I$ with $a \notin J$. Then, for any $b \in J$, we must have $b/a \in V$, for otherwise $a/b \in V$ and then $a = b(a/b) \in J$, which is impossible. Now, $b/a \in V$ implies $b = a(b/a) \in I$. Hence $J \subseteq I$.

Conversely, assume that the ideals of V are totally ordered by inclusion and consider $\alpha \in K$, $\alpha \neq 0$. Write $\alpha = a/b,\ a, b \in V$. Then either $(a) \subseteq (b)$ or $(b) \subseteq (a)$, so that $a/b \in V$ or $b/a \in V$, i.e., $\alpha \in V$ or $\alpha^{-1} \in V$. Hence V is a valuation ring of K. □

Corollary 5.1.7 *If V is a valuation ring of K and P is a prime ideal of V, then V_P and V/P are valuation rings.*

Proof If the ideals of V are totally ordered for inclusion, so are the ideals of V_P and V/P. □

Corollary 5.1.8 *Any Noetherian valuation ring is a principal ideal domain.*

Proof Let V be a Noetherian valuation ring and I an ideal of V. If $I = (a_1, \ldots, a_n)$, then there is a largest ideal among $(a_1), (a_2), \ldots, (a_n)$ which must be I. Hence I is principal. □

Corollary 5.1.9 *Let V be a Noetherian valuation ring. Then there exists an irreducible element $p \in V$ such that every ideal I of V is of the type $I = (p^m)$, $m \geq 1$ and $\cap_{m=1}^{\infty}(p^m) = 0$.*

Proof Since V is a principal ideal domain, the maximal ideal $\mathfrak{m}$ of V is principal, i.e., $\mathfrak{m} = (p)$, p irreducible. Clearly, $\cap_{m=1}^{\infty}(p^m) = (0)$, as V is a UFD. Now, every non-unit $a \in V$ is contained in $\mathfrak{m}$ and hence divisible by p, so that by successive division by p, we have $a = up^m$, $m \geq 1$, u a unit. Hence $(a) = (p^m)$. □

Theorem 5.1.10 *Let R be a subring of a field K. Then the integral closure $\overline{R}$ of R in K is the intersection of all valuation rings V of K containing R.*

Proof Let V be a valuation ring of K containing R. Since V is integrally closed, $V \supseteq \overline{R}$.

Conversely, let $a \notin \overline{R}$. We construct a valuation ring V of K such that $V \supseteq R$ and $a \notin V$. Since $a \notin \overline{R}$, we have $a \notin R[a^{-1}] = R'$. Hence a^{-1} is a non-unit in R' and therefore contained in a maximal ideal $\mathfrak{m}'$ of R'. Let Ω be an algebraic closure of $k = R'/\mathfrak{m}'$ and $h : R' \to \Omega$ the composition of the natural projection and inclusion $k \to \Omega$. Now, h can be extended to a maximal homomorphism $\overline{h} : V \to \Omega$, where V is a valuation ring of K containing R' by Proposition 5.1.4. Then $\overline{h}(a^{-1}) = h(a^{-1}) = 0$, as $a^{-1} \in \mathfrak{m}'$. This implies $a \notin V$, for if $a \in V$, then $1 = \overline{h}(1) = \overline{h}(aa^{-1}) = \overline{h}(a)\overline{h}(a^{-1}) = 0$. □

Corollary 5.1.11 *A domain R with quotient field K is integrally closed if and only if it is an intersection of a family $\{V_\alpha\}$ of valuation rings of K.*

Proof If R is integrally closed, then $R = \overline{R} = \bigcap_{V \supseteq R} V$, the intersection of all valuation rings of K containing R.

Conversely, if $R = \bigcap_\alpha V_\alpha$, V_α the valuation ring in K, then R is integrally closed as each V_α is integrally closed. □

We now give another characterisation of valuation rings.

Definition 5.1.12 Let R and S be local rings with unique maximal ideals $\mathfrak{m}_R$ and $\mathfrak{m}_S$ respectively. S is said to *dominate* R if R is a subring of S and $\mathfrak{m}_R = R \cap \mathfrak{m}_S$.

Theorem 5.1.13 *Let K be a field and $\Sigma = \{(R, \mathfrak{m}_R) \mid R \subset K\}$, where R is a local subring of K with maximal ideal $\mathfrak{m}_R$. Define a relation $\leq$ in Σ by $(R, \mathfrak{m}_R) \leq (S, \mathfrak{m}_S)$ if $(S, \mathfrak{m}_S)$ dominates $(R, \mathfrak{m}_R)$. Then $(V, \mathfrak{m}_V) \in \Sigma$ is maximal if and only if V is a valuation ring of K.*

Proof Assume $(V, \mathfrak{m}_V)$ is a maximal element of Σ. Let $k = V/\mathfrak{m}_V$, Ω an algebraic closure of k and $h : V \to \Omega$ the composite of natural projection $V \to V/\mathfrak{m}_V = k$ and the inclusion $k \to \Omega$. By Proposition 5.1.4, it is sufficient to show that (V, h) is a maximal extension. Let (R', h') be an extension of (V, h), where we may assume R' to be a local ring with $\operatorname{Ker} h' = \mathfrak{m}_{R'}$, the unique maximal ideal of R'. Since h' is an extension of

$h, \mathfrak{m}_{R'} \cap V = \mathfrak{m}_V$, so that $(R', \mathfrak{m}_{R'})$ dominates $(V, \mathfrak{m}_V)$. Since $(V, \mathfrak{m}_V)$ is maximal in Σ, $(V, \mathfrak{m}_V) = (R', \mathfrak{m}_{R'})$ and the proof is complete.

Conversely, assume that $(V, \mathfrak{m}_V) \in \Sigma$, where V is a valuation ring of K, and suppose $(V, \mathfrak{m}_V) \leq (R, \mathfrak{m}_R) \in \Sigma$. If $\alpha \in R$, then either $\alpha \in V$ or $\alpha^{-1} \in V$. If $\alpha \notin V$, then $\alpha^{-1} \in V \subseteq R$, i.e., $\alpha^{-1} \in \mathfrak{m}_V = \mathfrak{m}_R \cap V$, i.e., $\alpha^{-1} \in \mathfrak{m}_R$, a contradiction. Hence for every $\alpha \in R$, we have $\alpha \in V$, i.e., $V = R$. Hence $(V, \mathfrak{m}_V)$ is a maximal element of Σ. □

Corollary 5.1.14 *Every subring of K which is a local ring is dominated by at least one valuation ring of K.*

Theorem 5.1.15 *Let R be a principal ideal domain with quotient field K. Then the valuation rings of K containing R and different from K are of the type R_{pR} for some irreducible element $p \in R$.*

Proof If $p \in R$ is an irreducible element, then clearly R_{pR} is a valuation ring of K containing R (Example 5.1.3(iv)).

Conversely, assume that V is a valuation ring of K containing R and different from K. Since $V \neq K$, $\mathfrak{m}_V$ contains a non-zero element $\alpha \in K$. Let $\alpha = a/b$, $a, b \in R$. Then $\mathfrak{m}_V \cap R$ contains the non-zero element a and since it is a prime ideal of R, $\mathfrak{m}_V \cap R = (p)$ for some irreducible element $p \in R$. Then $(R_{pR}, pR_{pR}) \leq (V, \mathfrak{m}_V)$ and by maximality, $R_{pR} = V$. □

Corollary 5.1.16 *The valuation rings of $\mathbb{Q}$ different from $\mathbb{Q}$ are of the type $\mathbb{Z}_{p\mathbb{Z}}$, p prime.*

Corollary 5.1.17 *Let $K(X)$ be the field of rational functions in X over a field K and V a valuation ring of $K(X)$ containing K different from $K(X)$. Then either $V = K[X]_{(f(X))}$ for some $f(X) \in K[X]$ irreducible or $V = K[X^{-1}]_{(X^{-1})}$.*

Proof If $X \in V$, then $K[X] \subseteq V$ and by Theorem 5.1.15, $V = K[X]_{((f(X)))}$ for some irreducible $f(X) \in K[X]$. If $X \notin V$, then $X^{-1} \in V$, so that $K[X^{-1}] \subseteq V$. Then by Theorem 5.1.15, $V = K[X^{-1}]_P$ for some prime ideal P of $K[X^{-1}]$. Since X^{-1} is not invertible in V, $X^{-1} \in P$, i.e., $(X^{-1}) \subset P$. But (X^{-1}) is a maximal ideal, so that $(X^{-1}) = P$. Hence $V = K[X^{-1}]_{(X^{-1})}$. □

We now investigate relations between valuation rings and valuations.

Definition 5.1.18 A group Γ is called an *ordered group* if it has a total ordering $\leq$ which is compatible with the group structure, i.e., $\alpha \leq \beta$ implies $\gamma\alpha \leq \gamma\beta, \alpha\gamma \leq \beta\gamma$ for all $\alpha, \beta, \gamma \in \Gamma$ and $\beta^{-1} \leq \alpha^{-1}$.

Definition 5.1.19 Let Γ be an ordered group. A *valuation* $\| \ \|$ on a field K with values in Γ is a mapping $\| \ \| : K^* \to \Gamma$ satisfying the conditions:

(i) $\| ab \| = \| a \| \| b \|$.

(ii) $\| a + b \| \leq \max(\| a \|, \| b \|)$.

It is sometimes convenient to extend Γ to a larger set $\Gamma \cup \{0\}$ by introducing an element 0 with the properties, $0 \cdot 0 = 0$, $0 \cdot \alpha = \alpha \cdot 0 = 0$, for $\alpha \in \Gamma$ and $0 < \alpha$ for all $\alpha \in \Gamma$. By defining $\| 0 \| = 0$, $\| \ \|$ is extended to all the elements of K and the two conditions (i) and (ii) on $\| \ \|$ are still satisfied.

Example 5.1.20 Every valuation ring defines a valuation. Let V be a valuation ring of K. Define an order $\leq$ in K^* by stipulating $a \leq b$ if and only if $ab^{-1} \in V$ for $a, b \in K^*$. Then $\leq$ satisfies all the conditions of total ordering except the anti-symmetry, because $a \leq b$ and $b \leq a$ imply $ab^{-1} \in V$ and $ba^{-1} \in V$, i.e., ab^{-1} is a unit of V.

Let G denote the multiplicative group of units of V and $\Gamma = K^*/G$. Then the ordering on K^* induces an ordering on Γ which makes it a totally ordered group. The mapping $\| \ \| : K^* \to \Gamma$ given by $\| a \| = aG$ is a valuation on K.

Conversely, every valuation on K defines a valuation ring of K. Let $\| \ \| : K \to \Gamma$ be a valuation on K. Then

$$V = \{a \in K \mid \| a \| \leq 1\}$$

is a valuation ring of K.

There is an alternative notation for valuation which is particularly convenient in geometry. Here the value group is written additively and the order is reversed.

Let Δ be a totally ordered abelian group written additively.

Definition 5.1.21 A *valuation* v on K with values in Δ is a mapping $v : K^* \to \Delta$ satisfying:

(i) $v(ab) = v(a) + v(b)$, (ii) $v(a + b) \geq \min\{v(a), v(b)\}$.

It is sometimes convenient to adjoin an element ∞ to Δ and extend the operations by setting $\alpha + \infty = \infty + \alpha = \infty$, $\alpha \in \Delta$, $\infty + \infty = \infty$ and $\alpha < \infty$, for all $\alpha \in \Delta$. Then v can be extended to a mapping $v : K \to \Delta$ by defining $v(0) = \infty$ and the two conditions on the valuation are still satisfied.

Example 5.1.22 (i) Let $K = k(X)$ be the field of rational functions in X over a field k and $p(X)$ an irreducible polynomial in $k[X]$. Any non-zero element of K can be uniquely expressed as

$$\theta(X) = p(X)^r \cdot \frac{f(X)}{g(X)},$$

$r \in \mathbb{Z}$ and $p(X)$ does not divide $f(X)$ or $g(X)$. Then the map $v : K^* \to \mathbb{Z}$, given by $v(\theta(X)) = r$, defines a valuation on $k(X)$, called the $p(X)$-adic valuation.

(ii) More generally, let R be a PID with quotient field K and $p \in R$ an irreducible element. If $\alpha \in K^*$, write $\alpha = p^r \cdot b/c$ with $(p, b) = 1$, $(p, c) = 1$, $r \in \mathbb{Z}$. Define $v : K^* \to \mathbb{Z}$ by $v(\alpha) = r$. Then v defines a valuation on K, called the p-adic valuation on K. In particular, if $R = \mathbb{Z}$, the p-adic valuation on $\mathbb{Q}$ is defined.

(iii) Let $K = k(X)$ be the field of rational functions in X over k. For any non-zero element $f(X)/g(X) \in k(X)$, define

$$v\left(\frac{f(X)}{g(X)}\right) = \deg\, g(X) - \deg\, f(X).$$

Then v defines a valuation on K.

(iv) Let K be the field of Laurent series in X over k, and for $\alpha \in K, \alpha = \Sigma_{i=N}^{\infty} a_i X^i$, $a_i \in k$, $N \in \mathbb{Z}$. For $\alpha \in K^*$, define $v(\alpha) = N$, the order of α. Then v defines a valuation of K.

Definition 5.1.23 Each of the valuations in the above examples has the value group $\mathbb{Z}$. Such valuations are called *discrete valuations* and the corresponding valuation rings are called *discrete valuation rings*.

The valuation ring corresponding to the valuation v is clearly given by

$$V = \{a \in K \mid v(a) \geq 0\}$$

and the maximal ideal

$$\mathfrak{m} = \{a \in K \mid v(a) > 0\}.$$

5.1. EXERCISES

1. Show that an intersection of a totally ordered collection of valuation rings of K is a valuation ring of K.
2. Prove that in a valuation ring any radical ideal is prime.
3. Let R be a domain with quotient field $K(\neq R)$. Prove that the following conditions are equivalent.

   (i) R is a valuation ring in which every non-zero prime ideal is maximal.

   (ii) There are no rings properly between R and K.
4. Let K be a field, and R an integrally closed subdomain of K with K as its quotient field. Let $\{R_\alpha\}$ be a family of valuation rings of K with $\cap_\alpha R_\alpha = R$. Show that the integral closure of R in L, an extension field of K, is the intersection of all valuation rings of L which dominate one of the R_α.

5. Let R be a local ring with maximal ideal $\mathfrak{m} = (p)$.

   If $\cap_{n=1}^{\infty}(p^n) = (0)$, then show that the only ideals of R are (0) and (p^m) for $m \geq 1$. Show further that either p is nilpotent or R is a valuation ring.

6. Let R be a local domain with quotient field K. Show that R is a valuation ring of K if and only if any ring R' with $R \subsetneqq R' \subset K$ contains the inverse of some non-unit of R.

7. If v is a valuation on K with values in Δ, show that

   (i) $v\left(\Sigma_{i=1}^{n} a_i\right) \geq \min\limits_{1 \leq i \leq n}\{v(a_i)\}$,

   (ii) Equality holds in (i) if there exists some j such that $v(a_j) < \min\limits_{i \neq j}\{v(a_i)\}$.

8. Let $v : K \to \Delta$ be a valuation on K. For $\alpha \in \Delta$ with $\alpha \geq 0$, define $V_\alpha = \{a \in K \mid v(a) > \alpha\}$ and $V'_\alpha = \{a \in K \mid v(a) \geq \alpha\}$. Show that V_α and V'_α are ideals of the valuation ring V and every non-zero ideal of V contains some V'_α.

9. Show that a field which is algebraic over a finite field has no non-trivial valuation.

10. Let $K = k(X)$, the rational function field in X over k and $\alpha \in K$. Show that $\Sigma_v v(\alpha) = 0$, where the summation is taken over all the $p(X)$-adic valuations, each $p(X)$-adic valuation counted $\deg p(X)$ times, and the valuation is as given in Example $((iii)$, 5.1.22).

5.2 Discrete valuation rings

Definition 5.2.1 Let K be a field. A *discrete valuation* on K is a valuation $v : K^* \to \mathbb{Z}$ which is surjective. The corresponding valuation ring is called a *discrete valuation ring*.

Example 5.2.2 All the examples given in 5.1.22 are discrete valuations. The corresponding valuation rings are discrete valuation rings.

Definition 5.2.3 Let v be a discrete valuation on K. An element with $t \in K$ is called a *uniformising parameter* for v if $v(t) = 1$.

Example 5.2.4 In Examples (i), (ii), (iii) and (iv) given in 5.1.22, the uniformising parameters are respectively $(i)\, p(X)$, $(ii)\, p$, $(iii)\, 1/X$ and $(iv)\, X$.

Theorem 5.2.5 *Let v be a discrete valuation on K with discrete valuation ring R and the corresponding maximal ideal $\mathfrak{m}$. Then*

(i) *$\mathfrak{m}$ is principal and $t \in R$ is a uniformising parameter for v if and only if $\mathfrak{m}$ is generated by t.*

(ii) If t is a uniformising parameter, every $a \in K^*$ can be uniquely expressed as $a = ut^n$; u a unit in R, $n \in \mathbb{Z}$ and $K = R_t$.

(iii) Every non-zero ideal of R is uniquely of the type $\mathfrak{m}^n$ $(n \geq 1)$. In particular, R is a principal ideal domain.

Proof (i) We first show that $\mathfrak{m}$ is principal. Since v is surjective, there exists some $t \in R$ with $v(t) = 1$. Clearly, $(t) \subset \mathfrak{m}$ and let $a \in \mathfrak{m}$. Then $v(a) \geq 1$ and $v(at^{-1}) = v(a) - v(t) \geq 0$. Hence $at^{-1} \in R$, so that $a \in (t)$ showing $\mathfrak{m} = (t)$. Thus any uniformising parameter generates $\mathfrak{m}$.

Conversely, assume that $\mathfrak{m} = (t')$. Then $v(t') \geq 1$. Since $t \in (t') = \mathfrak{m}$, we have $t = at'$, $a \in R$, so that $1 = v(t) \geq v(t') \geq 1$, i.e., $v(t') = 1$.

(ii) If $n = v(a)$, then $v(at^{-n}) = 0$ so that $at^{-n} = u$, u unit. This shows that $a = ut^n$ and n is unique as $n = v(a)$. Clearly, this also implies that $K = R_t$.

(iii) Let I be a non-zero ideal of R. Choose $a \in I$ with $v(a) = n$, n, least non-negative integer. If t is a uniformising parameter, then $v(at^{-n}) = 0$, so that $at^{-n} = u$, u unit, i.e., $a = ut^n$. Hence $(t^n) \subset I$. If $b \in I$ with $v(b) = k \geq n$, then $v(bt^{-k}) = 0$, i.e., $b = u't^k$, u' unit, and $b \in (t^n)$. Hence $I = (t^n) = \mathfrak{m}^n$ and n is unique. □

Definition 5.2.6 Let I be a non-zero ideal of a discrete valuation ring R. If $I = \mathfrak{m}^n$, then we define $v(I) = n$.

Proposition 5.2.7 *Let R be a discrete valuation ring with maximal ideal $\mathfrak{m}$ and I a non-zero ideal of R. Then*

(i) $v(I) = l_R(R/I)$.

(ii) If $R/\mathfrak{m}$ *contains a field k, then* $\dim_k(R/I) = [R/\mathfrak{m} : k]\ v(I)$.

Proof (i) Let $\mathfrak{m} = (t)$ and $I = (t^n)$ with $n = v(I)$. Consider the sequence of ideals of R

$$R \supset \mathfrak{m} \supset \mathfrak{m}^2 \supset \cdots \supset \mathfrak{m}^n = I.$$

Then $l_R(R/I) = \sum_{i=0}^{n-1} l_R\left(\frac{\mathfrak{m}^i}{\mathfrak{m}^{i+1}}\right)$ by repeated use of Proposition 3.4.10. Now,

$$l_R\left(\frac{m^i}{m^{i+1}}\right) = \dim_{R/\mathfrak{m}}\left(\frac{\mathfrak{m}^i}{\mathfrak{m}^{i+1}}\right) = 1,$$

as $\mathfrak{m}^i \neq \mathfrak{m}^{i+1}$, for if $\mathfrak{m}^i = \mathfrak{m}^{i+1}$, then $\mathfrak{m} = 0$ by Nakayama Lemma. Hence $l_R(R/I) = n = v(I)$.

(ii) If $R/\mathfrak{m}$ contains a field k, then

$$\dim_k\left(\frac{\mathfrak{m}^i}{\mathfrak{m}^{i+1}}\right) = [R/\mathfrak{m} : k]\ \dim_{R/\mathfrak{m}}\left(\frac{\mathfrak{m}^i}{\mathfrak{m}^{i+1}}\right).$$

Hence

$$\begin{aligned}\dim_k(R/I) &= \sum_{i=0}^{n-1} \dim_k\left(\frac{\mathfrak{m}^i}{\mathfrak{m}^{i+1}}\right)\\ &= [R/\mathfrak{m} : k] \sum_{i=0}^{n-1} \dim_{R/\mathfrak{m}}\left(\frac{\mathfrak{m}^i}{\mathfrak{m}^{i+1}}\right).\end{aligned}$$

But

$$\dim_{R/\mathfrak{m}}\left(\frac{\mathfrak{m}^i}{\mathfrak{m}^{i+1}}\right) = l_{R/\mathfrak{m}}\left(\frac{\mathfrak{m}^i}{\mathfrak{m}^{i+1}}\right) = l_R\left(\frac{\mathfrak{m}^i}{\mathfrak{m}^{i+1}}\right).$$

Hence $\dim_k(R/I) = [R/\mathfrak{m} : k]\ l_R(R/I) = [R/\mathfrak{m} : k]\ v(I)$. □

Theorem 5.2.8 *Let R be a Noetherian local domain with unique maximal ideal $\mathfrak{m} \neq 0$ and K the quotient field of R. The following conditions are equivalent.*

(i) *R is a discrete valuation ring.*

(ii) *R is a principal ideal domain.*

(iii) *$\mathfrak{m}$ is principal.*

(iv) *R is integrally closed and every non-zero prime ideal of R is maximal.*

(v) *Every non-zero ideal of R is a power of $\mathfrak{m}$.*

Proof $(i) \Rightarrow (ii)$ follows from Theorem 5.2.5.
$(ii) \Rightarrow (iv)$ is clear from definition.

$(iv) \Rightarrow (iii)$. Choose $t \in \mathfrak{m}, t \neq 0$. Since $\mathfrak{m}$ is the only non-zero prime ideal, (t) is $\mathfrak{m}$-primary. Choose n such that $\mathfrak{m}^n \subset (t)$ and $\mathfrak{m}^{n-1} \not\subset (t)$. Choose $a \in \mathfrak{m}^{n-1}, a \notin (t)$, and let $\alpha = t/a \in K$. Then $\alpha^{-1} \notin R$ and since R is integrally closed, α^{-1} is not integral over R. This implies by Theorem 4.1.3 that $\alpha^{-1}\mathfrak{m} \not\subset \mathfrak{m}$. But by construction $\alpha^{-1}\mathfrak{m} \subset R$ so that $\alpha^{-1}\mathfrak{m} = R$. Hence $\mathfrak{m} = R\alpha$ is principal.

$(iii) \Rightarrow (ii)$. Let $\mathfrak{m} = (t)$. We claim that $\cap_{n=0}^{\infty}\mathfrak{m}^n = (0)$. If $a \in \mathfrak{m}^n$ for all $n \geq 0$, then $a = b_n t^n, b_n \in R$. Consider the increasing chain of ideals $(b_0) \subset (b_1) \subset \ldots \subset (b_n) \subset \ldots$. Since R is Noetherian, there exists some n such that $(b_n) = (b_{n+1})$, i.e., $b_{n+1} = cb_n, c \in R$. This implies that $b_n(1 - ct) = 0$, so that $b_n = 0$ as t is a non-unit, i.e., $a = 0$. Let I be any non-zero ideal of R. Then there exists an integer n such that $I \subset \mathfrak{m}^n$ and $I \not\subset \mathfrak{m}^{n+1}$. Choose $a \in I, a \notin \mathfrak{m}^{n+1}$. Then $a = ut^n$, $u \notin \mathfrak{m}$ and hence u is a unit. Hence $t^n \in I$, i.e., $I = (t^n)$.

$(ii) \Rightarrow (i)$. Since R is a principal ideal domain, $\mathfrak{m}$ is principal, i.e., $\mathfrak{m} = (t)$ for some $t \in R$. Clearly, $\cap_{n=0}^{\infty}(t^n) = 0$, as shown in the previous implication. Hence every $a \in R$ can be expressed uniquely as $a = ut^n$, u a unit in R. Consequently, if $\alpha \in K$, $\alpha \neq 0$, α can be written uniquely as $\alpha = ut^m$, $m \in \mathbb{Z}$, u a unit in R. If we define $v(\alpha) = m$, v is a discrete valuation on K whose valuation ring is R.

Thus the conditions (i) to (iv) are equivalent. Clearly, $(i) \Rightarrow (v)$ by Theorem 5.2.5. We show that $(v) \Rightarrow (iii)$. By Nakayama Lemma $\mathfrak{m} \neq \mathfrak{m}^2$. Choose $t \in \mathfrak{m} - \mathfrak{m}^2$. Then $(t) = \mathfrak{m}^n$ for some $n \geq 1$ and clearly $n = 1$ by the choice of t.

Hence all the conditions (i) to (v) are equivalent. □

Corollary 5.2.9 *If R is a discrete valuation ring and $t \in \mathfrak{m} - \mathfrak{m}^2$, then t is a uniformising parameter.*

5.2. EXERCISES

1. Show that a valuation ring other than a field is Noetherian if and only if it is a discrete valuation ring.

2. Let R be a domain and P a prime ideal in $R[X]$, which contracts to (0) in R. Show that $R[X]_P$ is a discrete valuation ring.

3. Let R be a domain with quotient field K such that R contains an element $p \neq 0$ with (p) prime ideal and $R[p^{-1}] = K$. Show that R is a discrete valuation ring.

4. Given an example of a local domain whose maximal ideal is principal and which is not a valuation ring.

5. Let R be a local domain with quotient field K and maximal ideal $\mathfrak{m}$. Let v be a valuation on K, whose valuation ring dominates R. Suppose $\mathfrak{m}$ is finitely generated or v is a discrete valuation. Show that there exists $a \in \mathfrak{m}$ with $v(a) = \operatorname{Inf}_{b \in \mathfrak{m}} v(b)$.

6. Let R be a valuation ring which is not a field. Show that R is a discrete valuation ring if and only if every prime ideal of R is principal.

7. Let R be a valuation domain with quotient field K and L a subfield of K. Let $R' = L \cap R$. Show that R' is a valuation domain with quotient field L, and that if R is a discrete valuation ring, so is R'.

5.3 Dedekind domain

Definition 5.3.1 A *Dedekind domain* is a Noetherian integrally closed domain in which every non-zero prime ideal is maximal.

Example 5.3.2 (i) Any principal ideal domain is a Dedekind domain.

(ii) Let K be an algebraic number field, i.e., a finite extension field of the rational number field $\mathbb{Q}$, and let R be the ring of integers in K, i.e., the integral closure of $\mathbb{Z}$ in K. Then R is a Dedekind domain, as shown in Proposition 5.3.3.

In particular, the ring $\mathbb{Z}[\sqrt{-5}] = \{a + b\sqrt{-5} \mid a,\, b \in \mathbb{Z}\}$ is a Dedekind domain.

(iii) The coordinate ring of a non-singular curve is a Dedekind domain.

Proposition 5.3.3 *The ring of integers in an algebraic number field is a Dedekind domain.*

Proof Let K be an extension field of $\mathbb{Q}$ of degree n and let R be the integral closure of $\mathbb{Z}$ in K. By Proposition 4.4.1, there exists a basis $\{e_1, e_2, \ldots, e_n\}$ of $K/\mathbb{Q}$ such that $R \subseteq \Sigma_{i=1}^{n}\mathbb{Z}e_i$. Thus R is a finitely generated $\mathbb{Z}$-module and hence Noetherian. R is integrally closed by definition. Let P be a non-zero prime ideal of R. Now, $P \cap \mathbb{Z} \neq 0$ as (0) is the unique prime ideal of R lying above (0) (Corollary 4.2.3, Proposition 4.2.1). Hence $P \cap \mathbb{Z}$ is a maximal ideal of $\mathbb{Z}$. This implies that P is a maximal ideal of R, showing that R is a Dedekind domain. □

Theorem 5.3.4 *Let R be a Noetherian domain in which every non-zero prime ideal is maximal. The following conditions are equivalent.*

(i) *R is a Dedekind domain.*

(ii) *R_P is a* DVR *(discrete valuation ring) for every non-zero prime ideal P of R.*

(iii) *Every primary ideal of R is a power of a prime ideal.*

Proof R is integrally closed if and only if R_P is integrally closed for every prime ideal P of R. Hence condition (i) implies (ii) by Theorem 5.2.8.

Conversely, condition (ii) implies that R is integrally closed and then R is a Dedekind domain.

Now, we show that $(ii) \Rightarrow (iii)$. Let Q be a P-primary ideal of R and we may assume $P \neq 0$. Then R_P is a DVR. Hence, by Theorem 5.2.8, $QR_P = P^n R_P$ for some $n \geq 1$. Since P is maximal, P^n is P-primary and hence

$$Q = QR_P \cap R = P^n R_P \cap R = P^n.$$

To show $(iii) \Rightarrow (ii)$, let Q' be a non-zero ideal of R_P and let $Q = Q' \cap R$. Since every non-zero prime ideal of R_P is also maximal, Q' is PR_P primary. Hence Q is P-primary and by (iii), $Q = P^n \ (n \geq 1)$. This implies that $Q' = P^n R_P$ and hence R_P is a DVR by Theorem 5.2.8. □

Corollary 5.3.5 *Let R be a Noetherian domain in which every non-zero prime ideal is maximal and let K be the quotient field of R. The following conditions are equivalent.*

(i) *R is a Dedekind domain.*

(ii) *R is the intersection of a collection of* DVRs *of K.*

(iii) *R is the intersection of all the* DVRs *of K containing R.*

Proof $(i) \Rightarrow (ii)$ follows from the relation $R = \bigcap_P R_P$.

$(ii) \Rightarrow (iii)$ is clear.

$(iii) \Rightarrow (i)$. Since each DVR is integrally closed, R is integrally closed and hence R is a Dedekind domain. □

Theorem 5.3.6 *Let R be a Dedekind domain with quotient field K and I a non-zero ideal of R. Then I can be uniquely expressed as*

$$I = P_1^{n_1} P_2^{n_2} \cdots P_r^{n_r} = P_1^{n_1} \cap P_2^{n_2} \cap \cdots \cap P_r^{n_r},$$

where $P_1, P_2, \ldots, P_r$ are prime ideals of R containing I. The positive integers n_i are given by $n_i = v_{P_i}(IR_{P_i})$, where v_{P_i} is the discrete valuation of K corresponding to the valuation ring R_{P_i} $(1 \leq i \leq r)$.

Proof Since every primary ideal of R is a power of a prime ideal, I has a primary decomposition of the type $I = P_1^{n_1} \cap \cdots \cap P_r^{n_r}$, where $P_1, P_2 \ldots, P_r$ are non-zero prime ideals of R and hence maximal. This implies that $P_1, \ldots, P_r$ are pairwise comaximal so that

$$I = P_1^{n_1} \cap \cdots \cap P_1^{n_r} = P_1^{n_1} P_2^{n_2} \cdots P_r^{n_r}.$$

Since $IR_{P_i} = P_i^{n_i} R_{P_i}$, we have $v_{P_i}(IR_{P_i}) = n_i \ (1 \leq i \leq r)$. The uniqueness part follows from the above relation and the uniqueness of primary decomposition. □

A Dedekind domain R can also be characterised by the property stated in Theorem 5.3.6. (See O. Zariski and P. Samuel: *Commutative Algebra-Vol. I*).

We now give another characterisation of a Dedekind domain which is useful in number theory. Let R be a domain with quotient field K.

Definition 5.3.7 A *fractionary* (or *fractional*) ideal of R is an R-submodule M of K such that $aM \subseteq R$ for some $a \in R$ and $a \neq 0$.

Example 5.3.8 (i) Let I be an ideal of R and $\alpha \in K$, $\alpha \neq 0$. Then $M = \alpha I$ is a fractionary ideal of R. In particular, if $\alpha = 1$, then any ideal of R is also a fractionary ideal. These are called *integral ideals*. A fractionary ideal of the type $R\alpha$, $\alpha \in K$, is called a *principal fractionary ideal*.

(ii) If M is a finitely generated R-submodule of K, then clearly M is a fractionary ideal.

Conversely, every fractionary ideal of R is a finitely generated R-module if R is Noetherian. For if M is a fractionary ideal, aM is an integral ideal for some $a \in R$, $a \neq 0$. By choosing a generating set $\{a_1, \ldots, a_n\}$ for aM, we have $\{a^{-1}a_1, \ldots, a^{-1}a_n\}$ as a generating set of M over R.

Definition 5.3.9 A fractionary ideal M is called *invertible* if there exists another fractionary ideal N such that $MN = R$.

Example 5.3.10 The principal fractionary ideal $M = R\alpha, \alpha \in K, \alpha \neq 0$, is invertible with inverse $N = R\alpha^{-1}$.

Remark 5.3.11 (i) If M is invertible, its inverse is unique and is equal to $(R : M) = \{\alpha \in K \mid \alpha M \subseteq R\}$ because $MN = R$ implies

$$N \subseteq (R : M) = (R : M)MN \subseteq N$$

and hence $N = (R : M)$.

(ii) The invertible ideals of R from a group for multiplication with R as unit element and the inverse of M is $(R : M)$.

Proposition 5.3.12 *Let M be a fractionary ideal of R. The following conditions are equivalent.*

(i) *M is invertible.*

(ii) *M is finitely generated and M_P is invertible in R_P for every prime ideal P of R.*

(iii) *M is finitely generated and $M_\mathfrak{m}$ is invertible in $R_\mathfrak{m}$ for every maximal ideal $\mathfrak{m}$ of R.*

Proof $(i) \Rightarrow (ii)$. Let M be an invertible ideal so that $M(R : M) = R$. This implies that $1 = \Sigma_i a_i b_i$, $a_i \in M$, $b_i \in (R : M)$. If $x \in M$, then $x = \Sigma_i (b_i x) a_i$, where $b_i x \in R$. Hence M is generated over R by the set $\{a_i\}$. Moreover, $M(R : M) = R$ implies

$$M_P(R : M)_P = R_P.$$

Since M, being invertible, is finitely generated, it follows that $M_P(R_P : M_P) = R_P$ for every prime ideal P. Hence M_P is invertible.

$(ii) \Rightarrow (iii)$ is clear.

$(iii) \Rightarrow (i)$. Let $M(R : M) = I$, an integral ideal of R. Then $M_{\mathfrak{m}}(R_{\mathfrak{m}} : M_{\mathfrak{m}}) = I_{\mathfrak{m}}$ for every maximal ideal $\mathfrak{m}$ of R.
By assumption, $M_{\mathfrak{m}}(R_{\mathfrak{m}} : M_{\mathfrak{m}}) = R_{\mathfrak{m}}$, so that $I_{\mathfrak{m}} = R_{\mathfrak{m}}$ for each maximal ideal $\mathfrak{m}$. Since $I \subseteq R$, this implies $I = R$ and M is invertible. □

Proposition 5.3.13 *Let R be a local domain. Every non-zero fractionary ideal of R is invertible if and only if R is a* DVR.

Proof Let $\mathfrak{m}$ be the unique maximal ideal of R. If R is a DVR, $\mathfrak{m} = (t)$ for some $t \in R$. If M is a fractionary ideal, then $aM \subseteq R$ for some $a \in R$, $a \neq 0$. Then $aM = (t^r)$, so that $M = R\alpha$ is principal, where $\alpha = t^r/a$. Hence M is invertible.

Conversely, assume that every fractionary ideal of R is invertible. In particular, every integral ideal is invertible and hence finitely generated. Thus R is Noetherian. By Theorem 5.2.8, it is sufficient to show that every non-zero ideal of R is a power of $\mathfrak{m}$. Suppose this is not true. Let $\Sigma \neq \emptyset$ be the collection of non-zero ideals of R which are not powers of $\mathfrak{m}$. Let I be a maximal element of Σ. Then $I \neq \mathfrak{m}$ and hence $I \subsetneqq \mathfrak{m}$. Since $\mathfrak{m}$ is invertible, this implies $\mathfrak{m}^{-1}I \subsetneqq R$, i.e., $\mathfrak{m}^{-1}I$ is a proper integral ideal of R. Now, $I \subseteq \mathfrak{m}^{-1}I$, as $x \in I$ can be written as $x = \alpha^{-1}(\alpha x)$, $\alpha \in \mathfrak{m}$. If $I = \mathfrak{m}^{-1}I$, then $I = \mathfrak{m}I$, and by Nakayama Lemma, $I = 0$. Hence $I \subsetneqq \mathfrak{m}^{-1}I$ and by maximality, $\mathfrak{m}^{-1}I$ is a power of $\mathfrak{m}$, i.e., I is a power of $\mathfrak{m}$, which is a contradiction. □

Theorem 5.3.14 *Let R be a domain. Then R is a Dedekind domain if and only if every non-zero fractionary ideal of R is invertible.*

Proof Assume that R is a Dedekind domain and let M be a non-zero fractionary ideal of R. Then, for any prime ideal $P \neq 0$ of R, M_P is a fractionary ideal of R_P and since R_P is a DVR, M_P is invertible. Moreover, M is finitely generated, as R is Noetherian. Hence, by Proposition 5.3.12, M is invertible.

Conversely, assume that every non-zero fractionary ideal of R is invertible. In particular, every integral ideal is invertible and hence finitely generated. Thus, R is Noetherian. We now show that for every prime ideal $P \neq 0$ in R, R_P is a DVR. By Proposition 5.3.13, it is sufficient to show that every integral ideal of R_P is invertible. Let J be an integral ideal of R_P and $I = R \cap J$. Then $I_P = J$ and since by assumption I is invertible, $I_P = J$ is also invertible and thus R_P is a DVR. It remains to show that every non-zero prime ideal P of R is maximal. Let P be a non-zero prime ideal of R and $\mathfrak{m}$ a maximal ideal of R containing P. Then $PR_{\mathfrak{m}}$ is a non-zero prime ideal of $R_{\mathfrak{m}}$ and since $R_{\mathfrak{m}}$ is a DVR, $PR_{\mathfrak{m}} = \mathfrak{m}R_{\mathfrak{m}}$, i.e., $P = \mathfrak{m}$. □

Definition 5.3.15 If R is a Dedekind domain, the non-zero fractionary ideals of R form a group under multiplication. The quotient of this group by the subgroup of principal fractionary ideals is called the *ideal class group* of R. This group is important in number theory.

Proposition 5.3.16 *Let R be a Dedekind domain with only finitely many prime ideals. Then R is a principal ideal domain.*

Proof Let $P_1, P_2, \ldots, P_s$ be the non-zero prime ideals of R. Fix i, $1 \leq i \leq s$. If $P_i = P_i^2$, then $\mathfrak{m} = \mathfrak{m}^2$ in R_{P_i}, where $\mathfrak{m} = P_i R_{P_i}$, and by Nakayama Lemma, $\mathfrak{m} = 0$, a contradiction. Choose $t_i \in P_i - P_i^2$ so that $\mathfrak{m} = t_i R_{P_i}$.

Let $I \neq 0$ be an ideal of R and write $I = P_1^{n_1} \cdots P_s^{n_s}$. By Proposition 2.1.21, there exists some $a \in R$ with

$$a \equiv t_i^{n_i} \pmod{P_i^{n_i+1}}, \ 1 \leq i \leq s.$$

Then $a \in P_i^{n_i} - P_i^{n_i+1}$. If the ideal (a) is decomposed as a product $(a) = P_1^{r_1} \cdots P_s^{r_s}$, then clearly $r_i = n_i$ for each i. Hence $I = (a)$ is principal. □

Corollary 5.3.17 *Let R be a Dedekind domain and I a non-zero ideal of R. Then every ideal of R/I is principal.*

Proof Let $I = P_1^{n_1} \cdots P_r^{n_r}$, P_i prime ideal of R and let $S = R - \cup_{i=1}^{r} P_i$. Since $I \neq 0$, we have $S \neq R^*$ and R_S is a Dedekind domain (Exercise 1, Section 5.3). Since R_S has only finitely many maximal ideals $P_i R_S$ $(1 \leq i \leq r)$, it is a PID. Hence every ideal of R_S/IR_S is principal. But

$$\frac{R_S}{IR_S} \simeq (R/I)_{\overline{S}}, \quad \overline{S} = (S+I)/I.$$

Now, $\overline{S}$ consists of units in R/I as for any $s \in S$, $(s) + I$ is not contained in any prime ideal of R. Hence $R/I \simeq (R/I)_{\overline{S}}$ and R/I is a principal ideal ring. □

Corollary 5.3.18 *Let R be a Dedekind domain and I a non-zero ideal of R. Then I is generated by at most 2 elements.*

Proof Choose $a \in I$, $a \neq 0$, and consider $R/(a)$. By the above corollary, $R/(a)$ is a principal ideal ring, so that the ideal $\overline{I} = I/(a)$ is generated by some $\overline{b} = b + (a)$. Then I is generated by a and b. □

Theorem 5.3.19 *Let R be a Dedekind domain with quotient field K and L a finite extension of K. If S is the integral closure of R in L, then S is a Dedekind domain.*

Proof We first assume that L/K is separable. By Proposition 4.4.1, S is a finitely generated R-module and since R is Noetherian, S is also Noetherian. If Q is a non-zero prime ideal of S, then $Q \cap R$ is a non-zero prime ideal of R (Corollary 4.2.3, Proposition 4.2.1). Hence Q is maximal in S. Since S is integrally closed in L, S is a Dedekind domain.

We now consider the general case. There exists a subfield L' of L such that $K \subset L' \subset L$ with L'/K separable and L/L' purely inseparable. If R' is the integral closure of R in L', then R' is a Dedekind domain. Since the integral closure of R' in L is the same as the integral closure of R in L, we may assume without loss of generality that L/K is purely inseparable and $\operatorname{ch} K = p > 0$. Let S be the integral closure of R in L. If $\alpha \in S$, then α satisfies the minimum polynomial of the type $X^{p^e} - a \in K[X]$ $(e \geq 1)$ and by Corollary 4.3.8, $a \in R$. Since L/K is finite, $[L : K] = p^n = q$ for some $n \geq 1$ and clearly, $S = \{\alpha \in L \mid \alpha^q \in R\}$.

Let $K^{q^{-1}} \supseteq L$ be the field obtained by taking the q-th roots of elements of K. Since the mapping $x \mapsto x^q$ is an isomorphism of $K^{q^{-1}}$ onto K, mapping the set $S_1 = \{\alpha \in K^{q^{-1}} \mid \alpha^q \in R\}$ isomorphically onto R, S_1 is a Dedekind domain. Clearly, $S_1 \cap L = S$. It remains to show that S is a Dedekind domain.

Let I be a non-zero integral ideal of S. Since S_1 is a Dedekind domain, IS_1 is invertible. Hence $1 = \Sigma_i a_i b_i$, $a_i \in I, b_i \in (S_1 : IS_1)$. This implies $1 = \Sigma_i a_i^q b_i^q$, with $b_i^q \in K \subset L$. Rewriting the above relation, as $1 = \Sigma_i a_i c_i$, where $c_i = a_i^{q-1} b_i^q$, we have $c_i I \subset b_i^q I^q \subset S_1$ as $b_i I \subset S_1$. Also $c_i I \subset L$ as $b_i^q \in L$. Hence $c_i I \subset S_1 \cap L = S$, i.e., $c_i \in (S : I)$. This implies that $I(S : I) = S$, as $1 = \Sigma a_i c_i$. Hence I is invertible in S and S is a Dedekind domain. □

Let R be a Dedekind domain with quotient field K, L a finite separable extension of K and S the integral closure of R in L. By Proposition 4.4.1, it follows that S is a finitely generated R-module. Let P be a non-zero prime ideal of R and Q a prime ideal of S lying above P. Then S/Q is finite field extension of R/P and its degree is called the degree of Q over P and is denoted by $f(Q/P)$. The ramification index of Q over P denoted by $e(Q/P)$ is defined by $e(Q/P) = v_Q(PS_Q)$, where v_Q is the discrete valuation of L corresponding to S_Q. By Theorem 5.3.6, if we write $PS = Q_1^{e_1} Q_2^{e_r} \cdots Q_r^{e_r}$, where Q_i's are prime. Then $Q_1, \ldots, Q_r$ are precisely the prime ideals of S lying above P and $e(Q_i/P) = e_i, 1 \leq i \leq r$.

Theorem 5.3.20 (Ramification Formula) *Let R be a Dedekind domain with quotient field K, L a finite separable extension of K and S the integral closure of R in L. Let P be a non-zero prime ideal of R, $Q_i(1 \leq i \leq r)$ the prime ideals of S lying above P. Then $\Sigma_i e_i f_i - [L . K]$, where $e_i = e(Q_i/P)$ and $f_i = f(Q_i/P)$.*

We first prove the following proposition.

Proposition 5.3.21 *With the same notation as in Theorem 5.3.20, assume further that $T = R - P$. Then*

(i) S_T is the integral closure of R_T in L.

(ii) $Q_1S_T, \ldots, Q_rS_T$ *are the prime ideals of* S_T *lying above* PR_T.

(iii) $e\left(\frac{Q_iS_T}{PR_T}\right) = e(Q_i/P)$ *and* $f\left(\frac{Q_iS_T}{PR_T}\right) = f(Q_i/P), 1 \le i \le r$.

Proof (i) Since S is the integral closure of R in L, S_T is the integral closure of R_T in L by Proposition 4.3.3.

(ii) If Q' is a prime ideal of S_T, then $Q' = QS_T$, where Q is a prime ideal of S and Q' lies above PR_T if and only if Q lies above P.

(iii) $R_T/PR_T \simeq (R/P)_{\overline{T}}$, where $\overline{T}$ is the image of $R-P$ under the projection $R \to R/P$. Since P is maximal, $\overline{T}$ consists of units of R/P so that $R_T/PR_T \simeq R/P$. Similarly,

$$\frac{S_T}{Q_iS_T} \simeq \frac{S}{Q_i}, \quad 1 \le i \le r.$$

Let $PS = Q_1^{e_1}Q_2^{e_2}\cdots Q_r^{e_r}$ be the unique decomposition of PS as a product of prime ideals. Then $PS_T = (Q_1S_T)^{e_1}\cdots(Q_rS_T)^{e_r}$ and hence

$$e\left(\frac{Q_iS_T}{PR_T}\right) = e_i = e(Q_i/P).$$

Moreover,

$$f\left(\frac{Q_iS_T}{PR_T}\right) = \left[\frac{S_T}{Q_iS_T} : \frac{R_T}{PR_T}\right] = \left[\frac{S}{Q_i} : \frac{R}{P}\right] = f(Q_i/P).$$

□

Proof of Theorem 5.3.20: By Proposition 5.3.21, it is sufficient to prove the result after localising at P. We may therefore assume without loss of generality that R is a DVR with maximal ideal P. Since R is a PID by Corollary 4.4.3, S is a free R-module of rank $[L : K]$. Hence S/PS is also free over $R/P = k$ of rank $[L : K]$. It is therefore sufficient to show that $\Sigma_i e_i f_i = \dim_k(S/PS)$. We have a well defined homomorphism $S \to S/PS \to S/Q_i^{e_i}$ and therefore a homomorphism $S \to \Pi_{i=1}^r(S/Q_i^{e_i})$. This map is surjective by the Chinese Remainder Theorem. Its kernel is $\cap_i Q_i^{e_i} = PS$. Hence $S/PS \simeq \Pi_{i=1}^r S/Q_i^{e_i}$. This is also an isomorphism as k-spaces. It is therefore sufficient to show that $\dim_k S/Q_i^{e_i} = e_if_i$. But

$$S/Q_i^{e_i} \simeq \frac{S_{Q_i}}{Q_i^{e_i}S_{Q_i}} \simeq \frac{S_{Q_i}}{PS_{Q_i}}.$$

But by Proposition 5.2.7,

$$\begin{aligned}\dim_k \frac{S_{Q_i}}{PS_{Q_i}} &= v_{Q_i}(PS_{Q_i})\left[\frac{S_{Q_i}}{Q_iS_{Q_i}} : k\right] \\ &= e(Q_i/P)\left[\frac{S}{Q_i} : k\right] \\ &= e_if_i\end{aligned}$$

and the proof is complete. □

Corollary 5.3.22 *Let k be an algebraically closed field of characteristic 0, t an indeterminate over k and L a finite field extension of $k(t)$. Let Γ be the set of discrete valuations v of L/k such that $v(t) > 0$. Then Γ is a finite set and $\Sigma_{v\in\Gamma}v(t) = [L : k(t)]$.*

Proof Let $K = k(t)$, $R = k[t]_{(t)}$ and $P = (t)R$. Let S be the integral closure of R in L and Q_i $(1 \leq i \leq r)$ the prime ideals of S lying above P, with v_{Q_i} the corresponding discrete valuations on L. Then $v_{Q_i} \in \Gamma$ for $1 \leq i \leq r$.

Conversely, if $v \in \Gamma$, then $k[t] \subset R_v$, the valuation ring corresponding to v and $t \in \mathfrak{m}_v$ the maximal ideal of R_v, so that $R \subset R_v$. Hence $S \subset R_v$. If $\mathfrak{m}' = \mathfrak{m}_v \cap S$, then $\mathfrak{m}'$ is a prime ideal of S containing $(t)S = PS$. Hence $\mathfrak{m}' = Q_i$ for some i and $R_v = S_{Q_i}$. This shows that $\Gamma = \{v_{Q_1}, v_{Q_2}, \ldots, v_{Q_r}\}$. Now, we have $e(Q_i/P) = v_{Q_i}(t)$ and $f(Q_i/P) = 1$, since k is algebraically closed. It follows from Theorem 5.3.20 that $\sum_{v\in\Gamma} v(t) = [L : k(t)]$. □

5.3. EXERCISES

1. Let R be a Dedekind domain and $S(\neq R^*)$ a multiplicatively closed subset of R. Show that R_S is a Dedekind domain.

2. Let R be a Dedekind domain and $f(X) \in R[X]$ be written as $f(X) = \Sigma_i a_i X^i$. Define the content $c(f)$ of f to be the ideal generated by $(a_0, a_1, \ldots, a_n)$. Show that $c(fg) = c(f)c(g)$, $f, g \in R[X]$.

3. Let R be a Dedekind domain with quotient field K and L a subfield of K such that R is integral over $R \cap L$. Show that $R \cap L$ is a Dedekind domain.

4. Show that in a Dedekind domain the notions of primary ideal and irreducible ideal and prime ideal are equivalent.

5. Let R be a Noetherian domain. Show that the following conditions are equivalent.

   (i) R is a Dedekind domain.

   (ii) For every maximal ideal $\mathfrak{m}$ of R, there exists no ideal I such that $\mathfrak{m}^2 \subsetneqq I \subsetneqq \mathfrak{m}$.

(iii) For every maximal ideal $\mathfrak{m}$ of R, the set of $\mathfrak{m}$-primary ideals is totally ordered by inclusion.

(iv) For every maximal ideal $\mathfrak{m}$ of R, every $\mathfrak{m}$-primary ideal is a power of $\mathfrak{m}$.

6. Let R be a Dedekind domain and $T(\neq R^*)$ a multiplicatively closed subset of R. Show that the map $M \to M_T$ is a homomorphism of the group of fractionary ideals of R onto the group of fractionary ideals of R_T and the kernel consists of fractionary ideals of R which intersect T.

CHAPTER

6

COMPLETIONS

Completion, like localisation is an important tool in geometry. It is an abstraction of the process of obtaining power series from polynomials. By successive application of localisation and completion, it is possible to reduce some questions in geometry to the study of the power series ring.

Completion is well behaved with respect to exact sequences for Noetherian rings and modules. An important result concerning completion is the Krull's intersection theorem, which describes the part of the ring annihilated by completion. Krull's theorem and the preservation of exactness are consequences of a basic result known as Artin-Rees lemma.

Completion is best studied by introducing filtered rings and the corresponding graded rings. A graded ring is an abstraction of the polynomial ring in several variables, where the gradation comes from the degree of monomials. Graded rings and modules are useful in projective algebraic geometry.

6.1 Filtered rings and modules

Definition 6.1.1 A filtered ring R is a ring R together with a family $\{R_n\}_{n\geq 0}$ of subgroups of R satisfying the conditions

(i) $R_0 = R$,

(ii) $R_{n+1} \subseteq R_n$, for all $n \geq 0$,

(iii) $R_n R_m \subseteq R_{m+n}$, for all $m, n \geq 0$.

Example 6.1.2 (i) For any ring R, define a filtration by setting $R_0 = R$ and $R_n = 0$, $n \geq 1$. This filtration on R is called the trivial filtration.

(ii) *Let I be an ideal in R and let $R_n = I^n$, $n \geq 0$. Then $\{R_n\}$ is a filtration on R called I-adic filtration.*

(iii) *If $\{R_n\}$ is a filtration on R and S is a subring of R, then $\{S \cap R_n\}$ is a filtration on S called the induced filtration on S.*

Definition 6.1.3 Let R be a filtered ring. A filtered R-module M is an R-module M together with a family $\{M_n\}_{n \geq 0}$ of R-submodules of M satisfying

(i) $M_0 = M$,

(ii) $M_{n+1} \subseteq M_n$, for all $n \geq 0$,

(iii) $R_m M_n \subseteq M_{m+n}$, for all $m, n \geq 0$.

Example 6.1.4 (i) Let M be an R-module. Corresponding to the trivial filtration on R, M has a filtration defined by $M_0 = M$ and $M_n = 0$, $n \geq 1$, called the trivial filtration on M.

(ii) Let I be an ideal of R and consider the I-adic filtration on R. Define the I-adic filtration on M by setting $M_n = I^n M$. Then M is a filtered R-module.

(iii) Let M be a filtered R-module and N an R-submodule of M. The filtration $\{M_n\}$ on M induces a filtration $\{N_n\}$ on N, where $N_n = N \cap M_n$, $n \geq 0$. It also induces a filtration on the quotient module M/N, where $(M/N)_n = (M_n + N)/N$.

Definition 6.1.5 Let M and N be filtered modules over a filtered ring R. A map $f : M \to N$ is called a homomorphism of filtered modules if

(i) *f is an R-module homomorphism, and*

(ii) *$f(M_n) \subseteq N_n$, for all $n \geq 0$.*

Example 6.1.6 (i) Let I be an ideal in R and let M, N be R-modules. Any R-homomorphism $f : M \to N$ is a homomorphism of filtered modules with respect to the I-adic filtrations on M and N.

(ii) *The natural projection $p : M \to M/N$ is a homomorphism of filtered modules.*

Definition 6.1.7 A graded ring R is a ring which can be expressed as a direct sum of subgroups $\{R_n\}$, i.e., $R = \oplus\Sigma_{n \geq 0} R_n$ such that $R_n R_m \subseteq R_{m+n}$ for all $m, n \geq 0$.

Example 6.1.8 (i) For any ring R, $R_0 = R$, $R_n = 0$, $n \geq 1$ defines a gradation on R called the trivial gradation.

(ii) Let $R = K[X_1, \ldots, X_n]$, K a field and R_d the set of homogeneous polynomials over K in $X_1, \ldots, X_n$ of degree d. Then $R = \oplus\Sigma_{d\geq 0} R_d$ and $R_d R_{d'} \subset R_{d+d'}$, so that $\{R_d\}$, $d \geq 0$, is a gradation.

Definition 6.1.9 Let R be a graded ring. An R-module M is called a graded R-module if M can be expressed as a direct sum of subgroups M_n, i.e., $M = \oplus\Sigma_{n\geq 0} M_n$ such that $R_m M_n \subseteq M_{m+n}, m, n \geq 0$.

Example 6.1.10 (i) Let $R = K[X_1, \ldots, X_n]$, the polynomial ring with gradation defined as in Example (ii) above. Then R considered as an R-module is graded.

(ii) Let M be an R-module. The tensor algebra $T(M)$ of M defined by $T(M) = \oplus\Sigma_{i\geq 0} T^i(M)$, where $T^i(M)$ is the tensor product of M with itself i times and $T^0(M) = R$, is a graded ring. (See Exercise 8, Section 1.3).

Remark 6.1.11 The elements of R_n or M_n in a graded ring or a graded module are called elements of homogeneous degree n. Let M be a graded R-module. A submodule N of M is called a graded submodule if $N = \oplus\Sigma_{n\geq 0} N_n$, where $N_n = N \cap M_n$. In this case the quotient R-module M/N is also graded.

Definition 6.1.12 Let M and N be graded modules over a graded ring R. A map $f : M \to N$ is called a homomorphism of graded modules if

(i) f is an R-homomorphism of modules and

(ii) $f(M_n) \subseteq N_n$, for all $n \geq 0$.

Definition 6.1.13 Let R be a filtered ring with filtration $\{R_n\}_{n\geq 0}$. Let $\text{gr}_n(R) = R_n/R_{n+1}$ and $\text{gr}(R) = \oplus\Sigma_{n\geq 0}\ \text{gr}_n(R)$. Then $\text{gr}(R)$ has a natural multiplication induced from R given by

$$(a + R_{n+1})(b + R_{m+1}) = ab + R_{m+n+1},\ a \in R_n,\ b \in R_m.$$

This makes R into a graded ring. This ring is called the associated graded ring of R.

Example 6.1.14 (i) Let R be any ring and $t \in R$ a non-zero-divisor. Consider the (t)-adic filtration on R. Then $\text{gr}(R)$ is naturally isomorphic to the polynomial ring $\frac{R}{(t)}[X]$, where X is the canonical image of $t \in R$ in $\text{gr}_1(R) = (t)/(t^2)$.

(ii) Let $R = k[[X_1, X_2, \ldots, X_n]]$, the power series ring in $X_1, X_2, \ldots, X_n$ over a field k and $\mathfrak{m} = (X_1, X_2, \ldots, X_n)$. Consider the $\mathfrak{m}$-adic filtration on R. Clearly, $f \in \mathfrak{m}^d$ if and only if order $(f) \geq d$ and R_d/R_{d+1} consists of all homogeneous polynomials of degree d in $X_1, \ldots, X_n$ over k. Hence $\text{gr}(R) \simeq k[X_1, X_2, \ldots, X_n]$.

Definition 6.1.15 Let M be a filtered R-module over a filtered ring R, with filtrations $\{M_n\}_{n\geq 0}$ and $\{R_n\}_{n\geq 0}$ respectively. Let

$$\mathrm{gr}_n(M) = M_n/M_{n+1} \quad \text{and} \quad \mathrm{gr}(M) = \oplus \sum_{n\geq 0} \mathrm{gr}_n(M).$$

Then $\mathrm{gr}(M)$ has a natural $\mathrm{gr}(R)$-module structure given by

$$(a + R_{n+1})(x + M_{m+1}) = ax + M_{m+n+1}, \; a \in R_n, \; x \in M_m.$$

This module is called the associated graded module of M.

Proposition 6.1.16 *Let R be a filtered ring, M, N filtered R-modules and $f : M \to N$, a homomorphism of filtered R-modules. Then f induces a natural map $\mathrm{gr}(f) : \mathrm{gr}(M) \to \mathrm{gr}(N)$, which is a homomorphism of graded $\mathrm{gr}(R)$-modules such that*

(i) $\mathrm{gr}(\mathrm{Id}) = \mathrm{Id}$ *and*

(ii) $\mathrm{gr}(hf) = \mathrm{gr}(h)\mathrm{gr}(f)$, *for another homomorphism of filtered modules $h : N \to K$.*

Proof The R-homomorphism $f : M \to N$ satisfies $f(M_n) \subseteq N_n$, $n \geq 0$, and hence induces a map $\overline{f}_n : M_n/M_{n+1} \to N_n/N_{n+1}$ defined by $\overline{f}_n(x + M_{n+1}) = f(x) + N_{n+1}$, $x \in M_n$. We denote this map by $\mathrm{gr}_n(f)$ and extend it to a map $\mathrm{gr}(f) = \oplus\Sigma_{n\geq 0}\, \mathrm{gr}_n(f) : \mathrm{gr}(M) \to \mathrm{gr}(N)$. Clearly, $\mathrm{gr}(f)$ is a homomorphism of graded $\mathrm{gr}(R)$-modules and satisfies conditions (i) and (ii). □

The following example shows that $\mathrm{gr}(f)$ may be zero without f being zero.

Example 6.1.17 Let $R = \mathbb{Z}$, $I = (n)$, $n > 1$ and $M = N = \mathbb{Z}$ with the I-adic filtrations. Then $f : M \to N$ given by $f(x) = nx$, $x \in \mathbb{Z}$, is non-zero, but $\mathrm{gr}(f) : \mathrm{gr}(M) \to \mathrm{gr}(N)$ is clearly zero.

Proposition 6.1.18 *Let R be a filtered ring, M, N filtered R-modules and $f : M \to N$ a homomorphism of filtered R-modules. If the induced map $\mathrm{gr}(f) : \mathrm{gr}(M) \to \mathrm{gr}(N)$ is injective then f is injective, provided $\cap_{n=0}^{\infty} M_n = (0)$.*

Proof By assumption, the map $\mathrm{gr}_n(f) : M_n/M_{n+1} \to N_n/N_{n+1}$ is injective, for all $n \geq 0$. Hence $M_n \cap f^{-1}(N_{n+1}) \subseteq M_{n+1}$, $n \geq 0$. We prove inductively that $f^{-1}(N_n) \subseteq M_n$, for all $n \geq 0$. Clearly, $f^{-1}(N_0) \subseteq M_0$. Assume that $f^{-1}(N_n) \subseteq M_n$ and let $x \in f^{-1}(N_{n+1})$. Since $N_{n+1} \subseteq N_n$, $x \in f^{-1}(N_n) \subseteq M_n$, so that $x \in f^{-1}(N_{n+1}) \cap M_n \subseteq M_{n+1}$, i.e., $f^{-1}(N_{n+1}) \subseteq M_{n+1}$. Since $\cap_0^{\infty} M_n = (0)$, we have $f^{-1}(0) \subseteq f^{-1}(\cap N_n) \subseteq \cap M_n = \{0\}$, i.e., f is injective. □

Remark 6.1.19 With the same assumptions as in the above proposition, $\text{gr}(f)$ is surjective need not imply that f is surjective without additional assumptions on the filtrations.

Proposition 6.1.20 *Let $R = \oplus\Sigma_{d\geq 0} R_d$ be a graded ring. The following conditions are equivalent.*

(i) R is Noetherian.
(ii) R_0 is Noetherian and R is a finitely generated R_0-algebra.

Proof $(i) \Rightarrow (ii)$. Now, $R_0 \simeq R/I$, where $I = \oplus\Sigma_{d>0}R_d$ is an ideal in R. Hence R_0 is Noetherian. Choose a finite generating set for I consisting of homogeneous elements $\{a_1, \ldots, a_r\}$, where a_i is homogeneous of degree n_i $(1 \leq i \leq r)$. Let R' be the R_0-subalgebra of R generated by $\{a_i\}$. We show by induction on n that $R_n \subseteq R'$, for all $n \geq 0$. Clearly, $R_0 \subseteq R'$ assume that $R_d \subseteq R'$ for all $d \leq n-1$. Let $a \in R_n$ $(n > 0)$, so that $a \in I$. Hence $a = \Sigma\lambda_i a_i$, where λ_i is homogeneous of degree $n - n_i < n$. By induction $\lambda_i \in R'$ $(1 \leq i \leq r)$, so that $a \in R'$. Hence $R = R'$ and the proof is complete.

$(ii) \Rightarrow (i)$. Follows from Corollary 3.1.7. □

Definition 6.1.21 Let M be a filtered R-module with filtration $\{M_n\}$ and I an ideal in R. The filtration $\{M_n\}$ is called an I-filtration if $IM_n \subseteq M_{n+1}$ for all $n \geq 0$.

Definition 6.1.22 Let M be a filtered R-module with an I-filtration $\{M_n\}$. The filtration is called I-stable if there exists some m such that, for all $n \geq m$, $IM_n = M_{n+1}$.

Example 6.1.23 The I-adic filtration on M is I-stable.

Proposition 6.1.24 *Let M be a finitely generated filtered R-module over a Noetherian ring R with an I-filtration. The following conditions are equivalent.*

(i) The filtration on M is I-stable.

(ii) If $R^* = \oplus\Sigma_{n\geq 0}I^n$ and $M^* = \oplus\Sigma_{n\geq 0}M_n$, the graded R^*-module M^* is finitely generated.

Proof If $N_n = \oplus\Sigma_0^n M_i$, then N_n is finitely generated over R as each M_i is finitely generated. Define

$$M_n^* = M_0 \oplus M_1 \oplus \cdots \oplus M_n \oplus IM_n \oplus I^2M_n \oplus \cdots.$$

Clearly, M_n^* is a finitely generated R^*-module as N_n is a finitely generated R-module. Hence $M^* = \cup_{n\geq 0}M_n^*$ is finitely generated over R^* if and only if $M^* = M_m^*$, for some m, i.e., $M_{m+k} = I^k M_m$ for all $k \geq 1$. This is the same as the condition for the filtration to be I-stable. □

Proposition 6.1.25 (Artin-Rees Lemma) *Let M be a filtered R-module with an I-stable filtration. Assume that R is Noetherian and M is R-finitely generated. Then the filtration induced by M on a submodule N of M is also I-stable.*

Proof If $\{M_n\}$ is the given filtration of M, clearly $\{N \cap M_n\}$ is a filtration on N. Let

$$R^* = \oplus\Sigma_{n\geq 0}I^n, \quad M^* = \oplus\Sigma_{n\geq 0}M_n \text{ and } N^* = \oplus\Sigma_{n\geq 0}N_n,$$

where $N_n = M \cap N_n$. Since R is Noetherian and I is finitely generated, R^* is a finitely generated R-algebra and hence R^* is Noetherian. Since $\{M_n\}$ is I-stable, M^* is finitely R^* generated by Proposition 6.1.24. Hence N^* is also finitely generated over R^*, i.e., $\{N_n\}$ is I stable. □

Corollary 6.1.26 *Let R be a Noetherian ring, I an ideal in R, M a finitely generated R-module and N a submodule of M. Then there exists some m such that $I^{m+k}M \cap N = I^k(I^m M \cap N)$ for all $k \geq 0$.*

Proof Apply the Artin-Rees Lemma for the I-adic filtration on M. □

6.1. EXERCISES

1. Let $R = \oplus\Sigma_{n\geq 0}R_n$ be a graded ring and $R^+ = \oplus\Sigma_{n\geq 1}R_n$. Show that if M is a graded R-module with $R^+M = M$, then $M = 0$.

2. Let M be a graded module over a graded ring R and N an R-submodule of M. Show that the following conditions are equivalent.
   (i) N is a graded submodule of M.
   (ii) If $x \in N$, all the homogeneous components of x are in N.
   (iii) N is generated by a set of homogeneous elements.

3. Let $M = \oplus\Sigma_{n\geq 0}M_n$ be a graded R-module with trivial gradation on R. Show that if M is finitely generated over R, then there exists some m such that $M_k = 0$, for all $k \geq m$.

4. Give an example of a bijective homomorphism $f : M \to N$ of filtered R-modules for which $\text{gr}(f) : \text{gr}(M) \to \text{gr}(N)$ is neither injective nor surjective.

5. Let M be a filtered R-module, N a submodule with induced filtrations on N and M/N respectively. Show that the exact sequence $0 \to N \xrightarrow{i} M \xrightarrow{p} M/N \to 0$, where i is the inclusion and p is the natural projection, induces an exact sequence $0 \to \text{gr}(N) \xrightarrow{\text{gr}(i)} \text{gr}(M) \xrightarrow{\text{gr}(p)} \text{gr}(M/N) \to 0$.

6. Let R be a filtered ring and M a finitely generated R-module. Show that the filtration on R induces a natural filtration on M and that $\mathrm{gr}(M)$ is a finitely generated $\mathrm{gr}(R)$-module.

7. Let $R = \oplus\Sigma_{n\geq 0}R_n$ be a graded ring and $I = \oplus\Sigma_{n\geq 0}I_n$ a graded ideal in R. Show that I is a prime ideal if and only if $x \notin P$ and $y \notin P$ imply $xy \notin P$, for all homogeneous elements x, y in R.

6.2 Completion

The study of completions require some preliminary concepts on inverse limits. We restrict ourselves to countable inverse systems.

Definition 6.2.1 An inverse system of R-modules is a collection of R-modules $\{M_n\}_{n\geq 0}$ and homomorphisms $\{\theta_n\}$, where θ_n is an R-homomorphism $\theta_n : M_n \longrightarrow M_{n-1} \ \ (n \geq 1)$.

Definition 6.2.2 The inverse limit of the system $\{M_n, \theta_n\}$ is an R-module M together with R homomorphisms $\{f_i\}$, where $f_i : M \longrightarrow M_i$ are such that $\theta_{i+1}f_{i+1} = f_i$, $(i \geq 0)$ and M is universal for this property, i.e., if M' is another R-module with natural maps $g_i : M' \longrightarrow M_i$ satisfying $\theta_{i+1}g_{i+1} = g_i \ \ (i \geq 0)$, then there exists a unique R-linear map $\lambda : M' \longrightarrow M$ with $f_i\lambda = g_i \ \ (i \geq 0)$.

Proposition 6.2.3 *The inverse limit of $\{M_n, \theta_n\}$ exists and is unique up to isomorphism.*

Proof The uniqueness is clearly a consequence of the universal property. To prove the existence, consider the product $N = \Pi_i M_i$ and the submodule $M \subset N$ defined by

$$M = \{(x_i) \in N \mid \theta_{i+1}(x_{i+1}) = x_i, i \geq 0\}.$$

Let $f_i : M \longrightarrow M_i$ be the restriction of the i-th projection $N \longrightarrow M_i$ to M. Clearly, $\theta_{i+1}f_{i+1} = f_i \ \ (i \geq 0)$. Consider an R-module M' and maps $g_i : M' \longrightarrow M_i$ satisfying $\theta_{i+1}g_{i+1} = g_i \ \ (i \geq 0)$. Define $\lambda : M' \longrightarrow M$ by setting $\lambda(x) = (g_i(x)), \ \ x \in M'$. Clearly, λ is R-linear and $f_i\lambda = g_i(i \geq 0)$. □

The inverse limit of the system $\{M_n, \theta_n\}$ is denoted by $\varprojlim_n M_n$.

Example 6.2.4 (i) Let $M_n = k[X]/(X^n)$, k a field and $\theta_{n+1} : M_{n+1} \longrightarrow M_n$ the natural map induced by the identity map on $k[X]$. Every element of M_{n+1} is a polynomial over k of degree at most n, and its image in M_n is the same polynomial with the last term truncated. Hence $\varprojlim_n M_n$ can be identified with $k[[X]]$, the ring of formal power series in X over k.

(*ii*) Let $R' = R[X_1, X_2, \ldots X_n]$ and $I = (X_1, \ldots, X_n)$. If $M_n = R'/I^n$ and $\theta_{n+1} : M_{n+1} \to M_n$ is the natural map induced by the identity map, then $\varprojlim_n M_n \simeq R[[X_1, X_2, \ldots, X_n]]$, the ring of formal power series in $X_1, X_2, \ldots, X_n$ over R.

Let M be a filtered R-module. The filtration $\{M_n\}$ on M defines a topology on M compatible with the abelian group structure of M for which $\{M_n\}$ is the fundamental system of neighbourhoods of (0). It is called the topology induced by the filtration $\{M_n\}$.

Proposition 6.2.5 *Let N be a submodule of a filtered module M. In the topology induced by the filtration, we have*

$$\overline{N} = \cap_{n=0}^{\infty}(N + M_n).$$

Proof $x \notin \overline{N}$ if and only if some neighbourhood of x does not intersect N, i.e., $(x + M_n) \cap N = \emptyset$ for some n, i.e., $x \notin (N + M_n)$ for some n. □

Corollary 6.2.6 *The topology defined by the filtration is Hausdorff if and only if $\cap_{n=0}^{\infty} M_n = (0)$.*

Proof The topology is Hausdorff if and only if

$$\{0\} = \{\overline{0}\} = \cap_{n=0}^{\infty} M_n.$$

□

Let M be a filtered R-module with filtration $\{M_n\}_{n \geq 0}$. Under the topology (uniform structure) defined by the filtration, M admits a Hausdorff completion $\widehat{M}$. It is the set of equivalence classes of Cauchy sequences of elements of M modulo the equivalence relation given by $(x_n) \sim (y_n)$ if for each m there exists some n_0 such that $x_n - y_n \in M_m$ for all $n \geq n_0$. The following proposition shows that the completion can be obtained as an inverse limit.

Theorem 6.2.7 *Let M be a filtered R-module with filtration $\{M_n\}$ and completion $\widehat{M}$. Then $\widehat{M} = \varprojlim_n M/M_n$.*

Proof Let $\widetilde{M} = \varprojlim_n M/M_n$. Let $y \in \widetilde{M}$ with $y = (y_n)$. Choose $x_n \in M$ such that its image $\overline{x_n}$ in M/M_n is y_n. Now, $\{x_n\}$ is a Cauchy sequence in M because $x_n - x_m \in M_m$ whenever $n \geq m$. We map y onto $\{\overline{x}_n\}$; the mapping is well defined because, if we choose $x'_n \in M$ with $\overline{x_n}' = y_n$, then $x_n - x'_n \in M_n, n \geq 0$ and thus $\{x_n\}$ is equivalent to the Cauchy sequence $\{x'_n\}$ in M. Hence the mapping $\alpha : \varprojlim_n M/M_n \to \widehat{M}$, given by $\alpha(y) = \{\overline{x}_n\}$, is well defined. We show that α is an isomorphism. Clearly, α is R-linear. If $\alpha(y) = \{\overline{x}_n\} = 0$, then $x_n \to 0$ and an easy argument shows that $x_n \in M_n$ for all n. This implies that $y_n = 0$ for all n, i.e., $y = 0$.

Consider now any Cauchy sequence $\{z_n\}$ in M. Choose inductively a subsequence $\{x_n\}$ of $\{z_n\}$ such that $x_{n+1} - x_n \in M_n$ for all n. If $y_n = x_n + M_n$ and $y = (y_n)$, then $\alpha(y) = \{\overline{z}_n\}$. Hence α is surjective and the proof is complete. $\square$

Proposition 6.2.8 *Let* $\{M'_n, \theta'_n\}$, $\{M_n, \theta_n\}$ *and* $\{M''_n, \theta''_n\}$ *be inverse systems of* R*-modules such that there exist* R*-linear maps* $\{f_n\}_{n\geq 0}$, $\{g_n\}_{n\geq 0}$ *such that*

$$0 \to M'_n \xrightarrow{f_n} M_n \xrightarrow{g_n} M''_n \to 0$$

is exact for each n *and the diagram*

$$\begin{array}{ccccccccc} 0 & \longrightarrow & M'_{n+1} & \longrightarrow & M_{n+1} & \longrightarrow & M''_{n+1} & \longrightarrow & 0 \\ & & \downarrow{\scriptstyle \theta'_{n+1}} & & \downarrow{\scriptstyle \theta_{n+1}} & & \downarrow{\scriptstyle \theta''_{n+1}} & & \\ 0 & \longrightarrow & M'_n & \longrightarrow & M_n & \longrightarrow & M''_n & \longrightarrow & 0 \end{array}$$

is commutative for all n. *Then the following statements hold.*

(i) The sequence

$$0 \to \left(\varprojlim_n M'_n\right) \to \left(\varprojlim_n M_n\right) \to \left(\varprojlim_n M''_n\right)$$

is exact.

(ii) If θ'_n is surjective for all n, then the sequence

$$0 \to \left(\varprojlim_n M'_n\right) \to \left(\varprojlim_n M_n\right) \to \left(\varprojlim_n M''_n\right) \to 0$$

is exact.

Proof Let $M = \Pi_n M_n$ and $d^M : M \to M$ be defined by $d^M(x_n) = (x_n - \theta_{n+1}(x_{n+1}))$. Then d^M is R-linear and $\operatorname{Ker} d^M = \varprojlim_n M_n$. This implies $\{f_n\}$ and $\{g_n\}$ induces R-linear maps $f = \Pi_n f_n : M' = \Pi_n M'_n \to M$ and $g = \Pi_n g_n : M \to M'' = \Pi_n M''_n$, and it is easy to check that the following diagram is commutative.

$$\begin{array}{ccccccccc} 0 & \longrightarrow & M' & \xrightarrow{f} & M & \xrightarrow{g} & M'' & \longrightarrow & 0 \\ & & \downarrow{\scriptstyle d^{M'}} & & \downarrow{\scriptstyle d^{M}} & & \downarrow{\scriptstyle d^{M''}} & & \\ 0 & \longrightarrow & M' & \xrightarrow{f} & M & \xrightarrow{g} & M'' & \longrightarrow & 0 \end{array}$$

By Proposition 2.2.15, we have an exact sequence

$$0 \to \operatorname{Ker} d^{M'} \to \operatorname{Ker} d^M \to \operatorname{Ker} d^{M''} \to \operatorname{Coker} d^{M'}$$

which proves the relation (i). Assume now that θ'_n is surjective for all n. Then $d^{M'}$ is surjective, i.e., $\operatorname{Coker} d^{M'} = 0$, and (ii) is proved. □

Corollary 6.2.9 *Let*

$$0 \to M' \xrightarrow{f} M \xrightarrow{g} M'' \to 0$$

be an exact sequence of R-modules and $\{M_n\}$ a filtration on M with induced filtration $\{M' \cap f^{-1}(M_n)\}$ on M' and filtration $\{g(M_n)\}$ on M''. If completions are taken with respect to these filtrations, then the sequence

$$0 \to \widehat{M'} \to \widehat{M} \to \widehat{M''} \to 0$$

is exact.

Proof For each n, we have an exact sequence of R-modules

$$0 \to \frac{M'}{M' \cap f^{-1}(M_n)} \to \frac{M}{M_n} \to \frac{M''}{g(M_n)} \to 0$$

such that the following diagram is commutative.

$$\begin{array}{ccccccccc}
0 & \longrightarrow & \frac{M'}{M'\cap f^{-1}(M_{n+1})} & \longrightarrow & \frac{M}{M_{n+1}} & \longrightarrow & \frac{M''}{g(M_{n+1})} & \longrightarrow & 0 \\
 & & \downarrow \theta'_{n+1} & & \downarrow \theta_{n+1} & & \downarrow \theta''_{n+1} & & \\
0 & \longrightarrow & \frac{M'}{M'\cap f^{-1}(M_n)} & \longrightarrow & \frac{M}{M_n} & \longrightarrow & \frac{M''}{g(M_n)} & \longrightarrow & 0.
\end{array}$$

By passing to the inverse limit, the sequence

$$0 \to \widehat{M'} \to \widehat{M} \to \widehat{M''} \to 0$$

is exact by Proposition 6.2.8, as θ'_n is surjective for each n. □

Corollary 6.2.10 *Let M be an R-module with filtration $\{M_n\}$ and completion $\widehat{M}$. Then the completion $\widehat{M_n}$ of M_n with respect to the induced filtration on M_n is a submodule of $\widehat{M}$ and $\widehat{M}/\widehat{M_n} \simeq M/M_n$ for all n.*

Proof Applying Corollary 6.2.9 to $M' = M_n$ and $M'' = M/M_n$, we have the exact sequence

$$0 \to \widehat{M_n} \to \widehat{M} \to \widehat{\left(\frac{M}{M_n}\right)} \to 0$$

i.e.,

$$\frac{\widehat{M}}{\widehat{M_n}} \simeq \widehat{\left(\frac{M}{M_n}\right)} = \widehat{M''}.$$

The induced filtration on M'' is given by $\{M''_m\}$, where $M''_m = (M_n + M_m)/M_n$. Hence $M''_m = 0$ for all $m \geq n$, i.e., M'' has discrete topology. Hence $M'' \simeq \widehat{M''}$ and this implies $M/M_n \simeq \widehat{M}/\widehat{M}_n$. □

Corollary 6.2.11 *Let M be an R-module with filtration $\{M_n\}$. This induces a filtration $\{\widehat{M}_n\}$ on $\widehat{M}$ and $\widehat{\widehat{M}} \simeq \widehat{M}$.*

Proof Clearly, $\{\widehat{M}_n\}$ is a filtration of $\widehat{M}$ as $\{M_n\}$ is a filtration on M. By Corollary 6.2.10, we have $M/M_n \simeq \widehat{M}/\widehat{M}_n$ under a natural isomorphism, making the following diagram commutative

$$\begin{array}{ccc} \dfrac{M}{M_{n+1}} & \rightarrow & \dfrac{\widehat{M}}{\widehat{M}_{n+1}} \\ \downarrow & & \downarrow \\ \dfrac{M}{M_n} & \rightarrow & \dfrac{\widehat{M}}{\widehat{M}_n}. \end{array}$$

Hence $\varprojlim_n M/M_n \simeq \varprojlim_n \widehat{M}/\widehat{M}_n$, i.e., $\widehat{M} \simeq \widehat{\widehat{M}}$. □

6.2. EXERCISES

1. Two I-filtrations $\{M_n\}$ and $\{M'_n\}$ on M are said to be equivalent if there exists an integer k such that $M_{n+k} \subseteq M'_n$ and $M'_{n+k} \subseteq M_n$, for all $n \geq 0$. Show that two such equivalent filtrations define the same topology on M.

2. Let R be a Noetherian ring, M a finitely generated R-module and N a submodule of M. If I is an ideal in R, show that the two filtrations on N, viz. $\{I^nN\}$ and $\{N \cap I^nM\}$ are equivalent. Deduce that the completions of N with respect to these two filtrations are isomorphic.

3. Let p be a fixed prime and let $f_n : \mathbb{Z}/p\mathbb{Z} \to \mathbb{Z}/p^n\mathbb{Z}$ be the natural map defined by $f_n(\overline{1}) = \overline{p^{n-1}}$. Let $f = \oplus\Sigma_n f_n$ be the abelian group map from R to S, where $R = \oplus\Sigma\, \mathbb{Z}/p\mathbb{Z}$ and $S = \oplus\Sigma_n\, \mathbb{Z}/p^n\mathbb{Z}$, R being a countable direct sum of $\mathbb{Z}/p\mathbb{Z}$. Show that the completion of R with respect to the p-adic filtration is isomorphic to R, but the completion of R with respect to the filtration induced on R by the p-adic filtration on S is the direct product $\Pi\ \mathbb{Z}/p\mathbb{Z}$ taken countable number of times.

4. Use Exercise 3 to show that there exist filtered modules M', M, M'' such that $0 \to M'_n \to M_n \to M''_n \to 0$ is exact for all n, but

$$\left(\varprojlim_n M_n\right) \to \left(\varprojlim_n M''_n\right) \to 0$$

is not exact.

5. Let $R = k[[X, Y]]$, the power series ring in X and Y over a field k of ch $\neq 2, \mathfrak{m} = (X, Y)$ and $P = (X^2 - Y^3)$. For the $\mathfrak{m}$-adic filtration on R, show that P is a prime ideal in R closed in the $\mathfrak{m}$-adic topology, but $\mathrm{gr}(P)$ is not a prime ideal in $\mathrm{gr}(R)$.
6. Let M be a filtered module which is Hausdorff and complete with respect to the topology defined by the filtration. Show that a series $\Sigma_n x_n$, $x_n \in M$ converges in M if and only if x_n tends to zero.
7. Let M be a filtered R-module with filtration $\{M_n\}$ such that M is complete with respect to the filtration and N a submodule of M such that N is a closed subspace of M. Show that M/N is complete with respect to tbe induced filtration on M/N.

6.3 I-adic filtration

We consider in this section some special properties of completions with respect to I-adic filtrations. Let M be an R-module and I an ideal in R.

Definition 6.3.1 The topology defined on M by the I-adic filtration is called the I-adic topology and the completion is called the I-adic completion M.

Example 6.3.2 (i) $R = k[X]$, k a field and $I = (X)$. Then the completion $\widehat{R} = \varprojlim_n R/(X^n) \simeq k[[X]]$, the ring of formal power series in X over k.

(ii) Let $R = \mathbb{Z}$ and $I = (p)$, p a prime in $\mathbb{Z}$. The completion $\widehat{R} = \varprojlim_n \mathbb{Z}/p^n\mathbb{Z}$ is called the ring of p-adic integers.

(iii) The completion of $R = k[X_1, \ldots, X_n]$ with respect to the I-adic topology, where $I = (X_1, X_2, \ldots, X_n)$ is the power series ring $k[[X_1, X_2, \ldots, X_n]]$.

Let M be an R-module and I an ideal in R. Let $\widehat{M}$ and $\widehat{R}$ denote respectively the I-adic completions of M and R. Define the scalar operation of $\widehat{R}$ on $\widehat{M}$ by $\overline{\{a_n\}}\,\overline{\{x_n\}} = \overline{\{a_n x_n\}}$, where $\{a_n\}$ is a Cauchy sequence in R and $\{x_n\}$ is a Cauchy sequence in M. This operation is well defined and for this operation $\widehat{M}$ is an $\widehat{R}$-module.

Let M and N be R-modules and $f : M \to N$ an R-linear map. Since $f(I^n M) \subset I^n N$, $n \geq 0$, f induces an R-linear map $f_n : M/I^n M \to N/I^n N$ such that the diagram

$$\begin{array}{ccc} M/I^{n+1}M & \xrightarrow{f_{n+1}} & N/I^{n+1}N \\ \downarrow & & \downarrow \\ M/I^n M & \xrightarrow{f_n} & N/I^n M \end{array}$$

is commutative.

On passage to the inverse limit, we have an $\widehat{R}$-homomorphism $\widehat{f} : \widehat{M} \to \widehat{N}$. Clearly, $\widehat{\mathrm{Id}} = \mathrm{Id}$ and $\widehat{gf} = \widehat{g}\widehat{f}$ for another R-linear map $g : N \to K$.

Proposition 6.3.3 *Let* $0 \to M' \xrightarrow{f} M \xrightarrow{g} M'' \to 0$ *be an exact sequence of finitely generated* R*-modules over a Noetherian ring* R *and* I *an ideal in* R*. Then the* I*-adic completions form an exact sequence of* $\widehat{R}$*-modules, viz.*

$$0 \to \widehat{M'} \xrightarrow{\widehat{f}} \widehat{M} \xrightarrow{\widehat{g}} \widehat{M''} \to 0.$$

Proof By Artin-Rees Lemma, the I-adic completion $\widehat{M'}$ is the same as the completion of M' with respect to the filtration on M' induced by the I-adic filtration on M. (Exercise 2, Section 6.2).

Similarly, the I-adic completion $\widehat{M''}$ is the same as the completion of M'' under the filtration, which is the image of the I-adic filtration on M. The result now follows from Corollary 6.2.9. □

Let M be an R-module and I an ideal in R. Consider the natural map $M \to \widehat{M}$ given by $x \to \{(x + M_n)\}$ and a corresponding map from R to $\widehat{R}$. These maps induce natural R-linear maps $\widehat{R} \otimes_R M \to \widehat{R} \otimes_R \widehat{M} \to \widehat{R} \otimes_{\widehat{R}} \widehat{M} \simeq \widehat{M}$, giving rise to a natural $\widehat{R}$-linear map $\phi_M : \widehat{R} \otimes_R M \to \widehat{M}$. The following proposition gives conditions under which ϕ_M is an isomorphism.

Proposition 6.3.4 *Let* R *be a ring and* M *a finitely generated* R *module. Then the map* $\phi_M : \widehat{R} \otimes_R M \to \widehat{M}$ *is surjective. If, further,* R *is Noetherian, then it is an isomorphism.*

Proof Clearly, $\phi_R : \widehat{R} \otimes_R R \to \widehat{R}$ is an isomorphism, and since tensor product commutes with direct sums, ϕ_M is an isomorphism when M is free of finite rank. If M is a finitely generated R-module, then there exists an exact sequence of R-modules

$$0 \to N \xrightarrow{f} F \xrightarrow{g} M \to 0,$$

where F is a free module of finite rank. Consider the commutative diagram

$$\begin{array}{ccccccccc}
 & & \widehat{R} \otimes_R N & \longrightarrow & \widehat{R} \otimes_R F & \longrightarrow & \widehat{R} \otimes_R M & \longrightarrow & 0 \\
 & & \downarrow{\scriptstyle \phi_N} & & \downarrow{\scriptstyle \phi_F} & & \downarrow{\scriptstyle \phi_M} & & \\
0 & \longrightarrow & \widehat{N} & \xrightarrow{\hat{f}} & \widehat{F} & \xrightarrow{\hat{g}} & \widehat{M} & \longrightarrow & 0
\end{array}$$

with the top row exact. By Corollary 6.2.9 and the second part of the proof of Proposition 6.3.3, $\widehat{g}$ is surjective. Since ϕ_F is an isomorphism, it follows that ϕ_M is surjective.

Assume now that R is Noetherian. Then by Proposition 6.3.3, the bottom row is exact. Since N is also finitely generated, ϕ_N is surjective. This implies that ϕ_M is injective and the proof is complete. □

Corollary 6.3.5 *Let* R *be a Noetherian ring,* I *an ideal in* R*. Then the* I*-adic completion* $\widehat{R}$ *is a flat* R*-algebra.*

Proof Since $\widehat{R} \otimes_R M \simeq \widehat{M}$, the result follows from Proposition 6.3.3. □

Corollary 6.3.6 *Let R be a Noetherian ring, I an ideal in R and $\widehat{R}$ the I-adic completion. If $\alpha \in R$ is a non-zero-divisor of R, then it is also a non-zero-divisor of $\widehat{R}$.*

Proof The homothety given by $\lambda_a : R \to R$, $\lambda_a(b) = ab$, is injective and hence $\widehat{\lambda}_a : \widehat{R} \to \widehat{R}$, which is also a homothety by $a \in \widehat{R}$, is injective. Thus $a \in \widehat{R}$ is a non-zero-divisor. □

Remark 6.3.7 If R is a domain, it is not in general true that $\widehat{R}$ is also a domain (See Exercise 6, Section 6.3).

Proposition 6.3.8 *Let R be a Noetherian ring, I an ideal in R and $\widehat{R}$ the I-adic completion. Then*

(i) $\widehat{I} \simeq \widehat{R} \otimes_R I \simeq \widehat{R}I$,

(ii) $(\widehat{I})^n \simeq (\widehat{I^n})$ for all $n \geq 0$,

(iii) $I^n/I^{n+1} \simeq \widehat{I}^n/\widehat{I}^{n+1}$ for all $n \geq 0$,

(iv) $\widehat{I}$ is contained in the Jacobson radical of $\widehat{R}$.

Proof (i) Since R is Noetherian, I is finitely generated and

$$\phi_I : \widehat{R} \otimes_R I \to \widehat{I}$$

is an isomorphism. Since $\widehat{R}$ is flat over R, the natural map

$$\widehat{R} \otimes_R I \to \widehat{R} \otimes_R R$$

is injective with image isomorphic to $\widehat{R}I$.
(ii) Apply (i) to the ideal I^n. Then

$$(\widehat{I^n}) \simeq \widehat{R}I^n = (\widehat{R}I)^n = (\widehat{I})^n.$$

(iii) By Corollary 6.2.10, we have for each n an isomorphism $\alpha_n : R/I^n \simeq \widehat{R}/\widehat{I}^n$, making the following diagram with exact rows commutative.

$$\begin{array}{ccccccccc}
0 & \longrightarrow & I^n/I^{n+1} & \longrightarrow & R/I^{n+1} & \longrightarrow & R/I^n & \longrightarrow & 0 \\
 & & \downarrow & & \downarrow {\scriptstyle \alpha_{n+1}} & & \downarrow {\scriptstyle \alpha_n} & & \\
0 & \longrightarrow & \widehat{I}^n/\widehat{I}^{n+1} & \longrightarrow & \widehat{R}/\widehat{I}^{n+1} & \longrightarrow & \widehat{R}/\widehat{I}^n & \longrightarrow & 0
\end{array}$$

In particular, $I^n/I^{n+1} \cong \widehat{I}^n/\widehat{I}^{n+1}$.

(iv) By Corollary 6.2.11, $\widehat{R}$ is complete with the $\widehat{I}$-adic topology. Hence for any $a \in \widehat{I}$, the series $1 + a + a^2 + \cdots + a^n + \cdots$ converges in $\widehat{R}$, say to s. Hence $(1 - a)$ is a unit in $\widehat{R}$ with inverse s. This implies $\widehat{I} \subset J(\widehat{R})$ the Jacobson radical of $\widehat{R}$. □

Corollary 6.3.9 *Let R be a local ring with unique maximal ideal $\mathfrak{m}$ and $\widehat{R}$ its $\mathfrak{m}$-adic completion. Then $\widehat{R}$ is a local ring with unique maximal ideal $\widehat{\mathfrak{m}}$.*

Proof $R/\mathfrak{m} \simeq \widehat{R}/\widehat{\mathfrak{m}}$, a field, so that $\widehat{\mathfrak{m}}$ is a maximal ideal in $\widehat{R}$. By Proposition 6.3.8 (iv), $\widehat{\mathfrak{m}}$ is contained in the Jacobson radical of $\widehat{R}$ and hence it is the unique maximal of $\widehat{R}$. □

Theorem 6.3.10 (Krull's Intersection Theorem) *Let M be a finitely generated R-module over a Noetherian ring R and I an ideal in R, $\widehat{M}$ the I-adic completion of M. The natural map $M \to \widehat{M}$ has kernel consisting of those elements of M which are annihilated by some element of $1 + I$.*

Proof Clearly, the kernel of the map $M \to \widehat{M}$ is $\cap_0^\infty I^n M$. If $(1+a)x = 0$ for some $a \in I$, then

$$x = -ax = a^2x = -a^3x = \cdots \in I^n M$$

for all n.

Conversely, let $N = \cap_0^\infty I^n M$. By Artin-Rees Lemma, there exists some n such that $(I^{n+k}M) \cap N = I^k(I^n M \cap N)$ for all $k \geq 0$. Taking $k = 1$, we get $N \subseteq IN$ and hence $N = IN$. Since R is Noetherian, N is finitely generated and by Proposition 2.2.5, there exists some $a \in I$ with $(1 + a)N = 0$. Thus the proof is complete. □

Example 6.3.11 Krull's theorem may be false when R is not Noetherian. Let R be the ring of C^∞-functions on the real line and I the ideal of functions vanishing at 0, the origin. Then, $\mathrm{Ker}(R \to \widehat{R})$ consists of functions whose derivatives vanish at 0 but elements annihilated by $1 + I$ consist of functions vanishing identically in a neighbourhood of 0.

Corollary 6.3.12 *Let R be a Noetherian domain, $I \neq R$ an ideal in R. Then $\cap_{n=0}^\infty I^n = (0)$.*

Proof No non-zero element of R can be annihilated by an element of $1 + I$. □

Corollary 6.3.13 *Let M be a finitely generated module over a Noetherian ring R and $I \subset J(R)$. Then $\cap_{n=0}^\infty I^n M = (0)$, i.e., the I-adic topology on M is Hausdorff.*

Proof Every element of $1 + I$ is a unit. □

Corollary 6.3.14 *Let M be a finitely generated module over a Noetherian local ring R with unique maximal ideal $\mathfrak{m}$. Then*

$$\bigcap_{n=0}^{\infty} \mathfrak{m}^n M = 0$$

i.e., the $\mathfrak{m}$-adic topology on M is Hausdorff. In particular, the $\mathfrak{m}$-adic topology on R is Hausdorff.

Proof Follows from Corollary 6.3.13, as $\mathfrak{m} = J(R)$. □

6.3. EXERCISES

1. Let R be Noetherian ring, I an ideal in R and M a finitely generated R-module. Show that
$$\bigcap_{n=1}^{\infty} I^n M = \bigcap_{\mathfrak{m}\supset I;\mathfrak{m}\in \mathrm{Max R}} \mathrm{Ker}(M \to M_{\mathfrak{m}}).$$

2. Let R be a Noetherian ring, M a finitely generated R-module, I and J ideals in R. Show that the J-adic completion of the I-adic completion of M is isomorphic to the $(I+J)$-adic completion of M.

3. Let $\mathfrak{m}_1, \mathfrak{m}_2, \ldots, \mathfrak{m}_r$ be distinct maximal ideals of a Noetherian ring R and $I = \cap_{i=1}^{r} \mathfrak{m}_i$. Let $\widehat{R}$ be the I-adic completion of R and $\widehat{R}_{\mathfrak{m}_i}$ be the completion of $R_{\mathfrak{m}_i}$ under the $\mathfrak{m}_i R_{\mathfrak{m}_i}$-adic topology. Show that $\widehat{R} \simeq \Pi_{i=1}^{r} \widehat{R}_{\mathfrak{m}_i}$.

4. Let I be an ideal in R. Show that if R is complete for the I-adic topology, then $R[[X]]$ is complete for the (I, X)-adic topology.

5. Let R be a Noetherian ring, I an ideal in R and $S = 1 + I$. Show that R_S can be identified with a subring of $\widehat{R}$.

6. Let $R = \mathbb{C}[X, Y]$, $\mathbb{C}$ the complex number field, and
$$f(X, Y) = X(X^2 + Y^2) + (X^2 - Y^2).$$
Show that the ideal (f) is a prime ideal in R. Let $S = R/(f)$ and $\mathfrak{m} = (X, Y)/(f)$ the maximal ideal of S. Show that $\widehat{S}$, the $\mathfrak{m}$-adic completion of S, is not a domain. (**Hint:** In the ring $C[[X, Y]]$, f decomposes into a product of two power series).

6.4 Associated graded rings

Let R be a ring, M an R-module and I an ideal in R. Consider the I-adic filtrations on M and R and the corresponding associated graded module $\mathrm{gr}_I(M)$ over $\mathrm{gr}_I(R)$. We consider in this section relations between properties of the $\mathrm{gr}_I(R)$-module $\mathrm{gr}_I(M)$ and the R-module M.

Proposition 6.4.1 *Let R be a Noetherian ring, M a finitely generated R-module, and I an ideal in R. Let $\{M_n\}$ be an I-stable filtration on M. Then*

(i) $\mathrm{gr}_I(R) = \oplus\Sigma_{n\geq 0} I^n/I^{n+1}$ *is Noetherian.*

(ii) $\mathrm{gr}_I(R) \simeq \mathrm{gr}_{\widehat{I}}(\widehat{R})$ *as graded rings.*

(iii) $\mathrm{gr}_I(M)$ *is a finitely generated* $\mathrm{gr}_I(R)$*-module.*

Proof (i) Let I be generated by $\{a_1, a_2, \ldots, a_n\}$ and let $\overline{a}_i (1 \leq i \leq n)$ be the image of a_i in I/I^2. Then the set $\{\overline{a}_1, \overline{a}_2, \ldots, \overline{a}_n\}$ generates $\mathrm{gr}_I(R)$ as an algebra over R/I. Since R/I is Noetherian, $\mathrm{gr}_I(R)$ is Noetherian.

(ii) By Proposition 6.3.8(ii), $I^n/I^{n+1} \simeq \widehat{I}^n/\widehat{I}^{n+1}$, $n \geq 0$ under a natural isomorphism, which can be extended to an isomorphism of $\mathrm{gr}_I(R)$ with $\mathrm{gr}_{\widehat{I}}(\widehat{R})$ as graded rings.

(iii) Since $\{M_n\}$ is an I-stable filtration, there exists an m such that $IM_n = M_{n+1}$ for all $n \geq m$, i.e., $I^r M_m = M_{m+r}$ for all $r \geq 1$. Hence, $\mathrm{gr}_I(M)$ is generated by $\oplus\Sigma_{n=0}^{m} M_n/M_{n+1}$ over $\mathrm{gr}_I(R)$. Since M_n/M_{n+1} is Noetherian and is annihilated by I for all n, it is Noetherian as an R/I-module and hence finitely generated over R/I. Hence $\oplus\Sigma_{n=0}^{m} M_n/M_{n+1}$ is finitely generated over R/I. This implies that $\mathrm{gr}_I(M)$ is finitely generated over $\mathrm{gr}_I(R)$. □

Proposition 6.4.2 *Let M and N be filtered R-modules with I-filtrations and $\phi : M \to N$ a homomorphism of filtered modules. Let $\mathrm{gr}(\phi) : \mathrm{gr}(M) \to \mathrm{gr}(N)$ and $\widehat{\phi} : \widehat{M} \to \widehat{N}$ be the homomorphisms induced respectively on the associated graded modules and the completions. Then $\mathrm{gr}(\phi)$ is injective (surjective) implies that $\widehat{\phi}$ is injective (surjective).*

Proof Since $\phi : M \to N$ is a homomorphism of filtered modules, $\phi(M_n) \subseteq N_n \ (n \geq 0)$. Now, ϕ induces natural R-linear maps $\phi_n : M/M_n \to N/N_n$ and $f_n : M_n/M_{n+1} \to N_n/N_{n+1} \ (n \geq 0)$ making the following diagram commutative.

$$\begin{array}{ccccccccc}
0 & \longrightarrow & \dfrac{M_n}{M_{n+1}} & \longrightarrow & \dfrac{M}{M_{n+1}} & \longrightarrow & \dfrac{M}{M_n} & \longrightarrow & 0 \\
 & & \Big\downarrow{\scriptstyle f_n} & & \Big\downarrow{\scriptstyle \phi_{n+1}} & & \Big\downarrow{\scriptstyle \phi_n} & & \\
0 & \longrightarrow & \dfrac{N_n}{N_{n+1}} & \longrightarrow & \dfrac{N}{N_{n+1}} & \longrightarrow & \dfrac{N}{N_n} & \longrightarrow & 0
\end{array}$$

By Proposition 2.2.15 we have an exact sequence

$$\mathrm{Ker}\, f_n \to \mathrm{Ker}\, \phi_{n+1} \to \mathrm{Ker}\, \phi_n \to \mathrm{Coker} f_n \to \mathrm{Coker}\, \phi_{n+1} \to \mathrm{Coker}\, \phi_n.$$

Suppose $\mathrm{gr}(\phi)$ is injective (surjective). We prove by induction on n that each ϕ_n is injective (surjective). Note that $\phi_1 : M_0/M_1 \to N_0/N_1$ is the same as f_0 and thus injective (surjective). Now, by induction hypothesis, $\mathrm{Ker}\, \phi_n = 0$ ($\mathrm{Coker}\, \phi_n = 0$). We also have

Ker $f_n = 0$ (Coker $f_n = 0$). Hence from the exact sequence, Ker $\phi_{n+1} = 0$ (Coker $\phi_{n+1} = 0$). Thus ϕ_n is injective (surjective) for all n. This implies from Proposition 6.2.8 that $\widehat{\phi}$ is injective (surjective). □

Theorem 6.4.3 *Let R be a ring, I an ideal in R, M a filtered R-module with I-filtration $\{M_n\}$. Assume that R is complete in the I-adic topology and $\cap_{n=1}^{\infty} M_n = (0)$. If $\mathrm{gr}_I(M)$ is a finitely generated $\mathrm{gr}_I(R)$ module, then M is a finitely generated R-module.*

Proof Choose a finite generating set of $\mathrm{gr}_I(M)$ over $\mathrm{gr}_I(R)$ consisting of homogeneous elements $\{y_1, \cdots, y_t\}$, where $\deg y_i = n_i \quad (1 \leq i \leq t)$, and choose $x_i \in M_{n_i}$ with $y_i = x_i + M_{n_{i+1}}$. Let

$$F = R \oplus R \oplus \ldots \oplus R$$

t times and

$$F_n = \{(a_i) \mid a_i \in I^{n-n_i}, 1 \leq i \leq t\},$$

where $I^k = R$ if $k \leq 0$. This defines a filtration on F and the map $\phi : F \to M$, given by $\phi[(a_i)] = \Sigma_i a_i x_i$, is a homomorphism of filtered R-modules. The associated graded homomorphism $\mathrm{gr}(\phi) : \mathrm{gr}(F) \to \mathrm{gr}(M)$ is surjective as the $\{y_i\}$ generate $\mathrm{gr}_I(M)$. Hence, by Proposition 6.4.2, $\widehat{\phi} : \widehat{F} \to \widehat{M}$ is surjective. Consider the commutative diagram

$$\begin{array}{ccc} F & \xrightarrow{\phi} & M \\ f \downarrow & & \downarrow g \\ \widehat{F} & \xrightarrow{\widehat{\phi}} & \widehat{M} \end{array}.$$

Since R is complete and F is a free R-module of finite rank, f is an isomorphism. Since $\cap_{n=0}^{\infty} M_n = (0)$, g is injective. This implies that ϕ is surjective as $\widehat{\phi}$ is surjective. Hence M is finitely generated over R. □

Corollary 6.4.4 *With the same assumptions as above, if $\mathrm{gr}_I(M)$ is a Noetherian $\mathrm{gr}_I(R)$-module, then M is a Noetherian R-module.*

Proof Let N be a submodule of M and $N_n = N \cap M_n$. Then $\{N_n\}$ is an I-filtration on N and the inclusion $N_n \subseteq M_n$ induces an injection $N_n/N_{n+1} \to M_n/M_{n+1}$, $n \geq 0$, and hence an injection $\mathrm{gr}_I(N) \to \mathrm{gr}_I(M)$. Since $\mathrm{gr}_I(M)$ is a Noetherian $\mathrm{gr}_I(R)$-module, $\mathrm{gr}_I(N)$ is finitely generated over $\mathrm{gr}_I(R)$ and hence by Theorem 6.4.3, N is finitely generated over R. This implies that M is a Noetherian R-module. □

Corollary 6.4.5 *Let R be a Noetherian ring and I an ideal in R. Then the I-adic completion $\widehat{R}$ is also Noetherian.*

Proof Since R is a Noetherian ring, $\mathrm{gr}_I(R)$ is also Noetherian by Proposition 6.4.1. Since $\mathrm{gr}_I(R) \simeq \mathrm{gr}_{\widehat{I}}(\widehat{R})$, it follows that $\mathrm{gr}_{\widehat{I}}(\widehat{R})$ is Noetherian. Since $\widehat{R}$ is complete, $\cap_{n=0}^{\infty} \widehat{I}^n = 0$.

By applying Corollary 6.4.4 to the ring $\widehat{R}$ and the module $M = \widehat{R}$, we get that $\widehat{R}$ is Noetherian. □

Corollary 6.4.6 *If R is Noetherian, then the power series ring $R[[X_1, X_2, \ldots, X_n]]$ is Noetherian.*

Proof Since $R[X_1, X_2, \ldots, X_n]$ is Noetherian, its completion for the $(X_1, X_2, \ldots, X_n)$-adic topology $R[[X_1 X_2, \ldots, X_n]]$ is Noetherian. □

Proposition 6.4.7 *Let R be a Noetherian ring, I an ideal contained in the Jacobson radical of R. If $\operatorname{gr}_I(R)$ is a domain, then R is a domain.*

Proof Let $a, b \in R$, $a \neq 0$, $b \neq 0$. By Corollary 6.3.13, $\cap_{n=1}^{\infty} I^n = 0$. Hence there exist m and n such that $a \in I^m, a \notin I^{m+1}, b \in I^n, b \notin I^{n+1}$. Let $\overline{a} = a + I^{m+1} \in \operatorname{gr}_I(R)$ and $\overline{b} = b + I^{n+1} \in \operatorname{gr}_I(R)$. Then $\overline{a} \neq 0, \overline{b} \neq 0$, and hence $\overline{ab} = \overline{a} \cdot \overline{b} \neq 0$. This implies $ab \neq 0$, i.e., R is a domain. □

Theorem 6.4.8 (Hensel's Lemma) *Let R be a local ring with maximal ideal $\mathfrak{m}$ such that R is complete with respect to the $\mathfrak{m}$-adic topology. For $f(X) = \Sigma_{i=0}^{d} a_i X^i \in R[X]$ and $\overline{a}_i = a_i + \mathfrak{m} \in R/\mathfrak{m} = k$, let $\overline{f}(X)$ denote the polynomial $\overline{f}(X) = \Sigma_{i=0}^{d} \overline{a}_i X^i \in k[X]$. Assume $f(X) \in R[X]$ is a monic polynomial such that there exist $\alpha(X)$ and $\beta(X)$ relatively prime monic polynomials in $k[X]$ of degrees r and $(d-r)$ respectively with $\overline{f}(X) = \alpha(X)\,\beta(X)$. Then there exist monic polynomials $g(X), h(X) \in R[X]$ of degrees r and $(d-r)$ respectively such that $\overline{g}(X) = \alpha(X)$, $\overline{h}(X) = \beta(X)$ and $f(X) = g(X)h(X)$.*

Proof We construct, inductively, sequences of polynomials

$$\{g_n(X)\} \text{ and } \{h_n(X)\}$$

in $R[X]$ of degrees at most r and $(d-r)$ respectively, such that

$$\overline{g}_n(X) = \alpha(X), \quad \overline{h}_n(X) = \beta(X)$$

and

$$f(X) - g_n(X)h_n(X) \in \mathfrak{m}^n R[X],\ n \geq 1.$$

Let $n = 1$. Since $\overline{f}(X) = \alpha(X)\beta(X)$, there exist

$$g_1(X),\ h_1(X) \in R[X]$$

such that

$$\overline{g}_1(X) = \alpha(X),\ \overline{h}_1(X) = \beta(X),\ f(X) - g_1(X)h_1(X) \in \mathfrak{m}R[X]$$

and

$$\deg g_1(X) = r,\ \deg h_1(X) = d - r.$$

Assume now that $g_n(X)$ and $h_n(X)$ have been constructed satisfying the required conditions. Write

$$f(X) - g_n(X)h_n(X) = \Sigma_{i=0}^{d}\lambda_i X^i, \lambda_i \in \mathfrak{m}^n.$$

Since $\alpha(X), \beta(X)$ are relatively prime, there exist polynomials $\phi_i(X)$ and $\psi_i(X)$ in $R[X]$ such that

$$\overline{\phi}_i(X),\ \overline{\psi}_i(X) \in k[X]$$

are of degrees at most $d-r$ and r respectively and

$$X^i = \overline{\phi}_i(X)\alpha(X) + \overline{\psi}_i(X)\beta(X) \ \ (0 \le i \le d).$$

Since $\overline{g}_n(X) = \alpha(X)$ and $\overline{h}_n(X) = \beta(X)$ and $\phi_i(X),\ \psi_i(X) \in R[X]$ are of degrees at most $d-r$ and r respectively, we have

$$X^i - \phi_i(X)g_n(X) - \psi_i(X)h_n(X) \in \mathfrak{m}R[X].$$

Let

$$g_{n+1}(X) = g_n(X) + \sum_i \lambda_i \psi_i(X)$$

and

$$h_{n+1}(X) = h_n(X) + \sum_i \lambda_i \phi_i(X).$$

Clearly, g_{n+1} and h_{n+1} are of degrees at most r and $(d-r)$ and

$$\overline{g}_{n+1} = \overline{g}_n = \alpha,\ \ \overline{h}_{n+1} = \overline{h}_n = \beta,$$

as $\lambda_i \in \mathfrak{m}^n$. Moreover,

$$\begin{aligned}
f - g_{n+1}h_{n+1} &= f - (g_n + \sum_i \lambda_i\psi_i)(h_n + \sum_i \lambda_i\phi_i)\\
&= f - g_n h_n - \sum_i \lambda_i X^i + \sum_i \lambda_i X^i - g_n \sum_i \lambda_i \phi_i\\
&\quad - h_n \sum_i \lambda_i \psi^i - \sum_{i,j} \lambda_i\lambda_j \psi_i\phi_j\\
&= \sum_i \lambda_i (X^i - g_n\phi_i - h_n\psi_i) - \Sigma_{i,j}\lambda_i\lambda_j\psi_i\phi_j.
\end{aligned}$$

Since $X^i - g_n\phi_i - h_n\psi_i \in \mathfrak{m}R[X]$ and $\lambda_i \in \mathfrak{m}^n$, it follows that

$$f - g_{n+1}h_{n+1} \in \mathfrak{m}^{n+1}R[X].$$

Since $g_{n+1} - g_n \in \mathfrak{m}^n R[X]$, the coefficients of $\{g_n(X)\}$ is a Cauchy sequence and since R is complete, $\{g_n(X)\} \to g(X) \in R[X]$. Similarly, $\{h_n(X)\} \to h(X) \in R[X]$. Since $f - g_n h_n \in \mathfrak{m}^n R[X]$ and R is complete, we have, by passing to the limit, $f(X) = g(X)h(X), \overline{g}(X) = \alpha(X)$ and $\overline{h}(X) = \beta(X)$.

The highest degree coefficients of $g(X)$ and $h(X)$ are respectively of the type $(1+a)X^r$ and $(1+a)^{-1}X^{d-r}$, $a \in \mathfrak{m}$, as $\alpha(X)$ and $f(X)$ are monic. By replacing g and h respectively by $(1+a)^{-1}g$ and $(1+a)h$, we can make both $g(X)$ and $h(X)$ monic and the proof is complete. □

Corollary 6.4.9 *Let R be a local ring with maximal ideal $\mathfrak{m}$, complete with respect to the $\mathfrak{m}$-adic topology. Let $f(X) \in R[X]$ be a monic polynomial such that $\overline{f}(X) \in (R/\mathfrak{m})[X]$ has a simple root $\xi \in R/\mathfrak{m}$. Then $f(X)$ has a simple root $a \in R$ such that $\overline{a} = \xi$.*

Proof Let $\overline{f}[X] = (X - \xi)\beta(X)$, where $X - \xi$ and $\beta(X) \in (R/\mathfrak{m})[X]$ are relatively prime. Then there exists a monic polynomial $X - a \in R[X]$ with $\overline{a} = \xi$, which divides $f(X)$. Hence a is a root of $f(x)$ and it has to be a simple root of $f(X)$ as ξ is a simple root of $\overline{f}(X)$. □

Corollary 6.4.10 (Implicit Function Theorem) *Let $R = k[[X_1, \ldots, X_i]]$, the power series ring in $X_1, X_2, \ldots, X_i$ over a field k. Let*

$$f(z) = z^n + a_1(X)z^{n-1} + \cdots + a_n(X)$$

be a monic polynomial, $a_i(X) \in R$, such that

$$z^n + a_1(0)z^{n-1} + \cdots + a_n(0)$$

admits a simple root $\xi \in k$. Then there exists some $g(X) \in R$ with $g(0) = \xi$ and $f(g(X)) = 0$.

Proof Apply Corollary 6.4.9 to the local ring $R[[X_1, \ldots, X_t]]$, which is $\mathfrak{m}$-adically complete for $\mathfrak{m} = (X_1, X_2, \ldots, X_t)$. □

Corollary 6.4.11 *Let $R = k[[X_1, X_2, \ldots, X_t]]$, the power series ring in $X_1, \ldots, X_t$ over a field k, and d an integer relatively prime to* ch *k. If $a(X) \in R$ has constant term $a(0) \neq 0$ and is a d-th power in k, then $a(X)$ itself is a d-th power in R.*

Proof Apply Corollary 6.4.10 to the polynomial $f(z) = z^d - a(X)$. □

6.4. EXERCISES

1. Let R be a ring. Suppose that there exists a finitely generated ideal I of R such that R/I is Noetherian and R is complete for the I-adic topology. Show that R is Noetherian.

2. Let R be a Noetherian semi-local ring, I its nil radical and J the Jacobson radical. Show that if R with J-adic topology is such that R/I is complete, then R is complete.

3. Let R be a Noetherian ring, I an ideal of R, $J(R)$ the Jacobson radical of R. Show that $I \subseteq J(R)$ if and only if every maximal ideal of R is closed for the I-adic topology. (Such a ring is called a Zariski ring).

4. Let R be a ring with I-adic completion $\widehat{R}$. Show that $\widehat{R}$ is faithfully flat over R if and only if R is a Zariski ring.

5. Let R be a Zariski ring with I-adic topology, $I \subseteq J(R)$. Let M be a finitely generated R-module. Show that if the I-adic completion $\widehat{M}$ is a free $\widehat{R}$-module, then M is a free R-module. Deduce that if $\widehat{R}$ is a domain, J an ideal of R with $J\widehat{R}$ principal, then J is principal.

6. In the proof of Theorem 6.4.8, justify the step "we have, by passing to the limit $f(X) = g(X)h(X), \ldots$." algebraically by showing directly that $f(X) - g(X)h(X) \in \mathfrak{m}^n R[x]$ for all $n \geq 1$, etc.

CHAPTER

7

HOMOLOGY

In the first chapter we observed that an exact sequence of R-modules may fail to be exact on the left when tensored with another R-module. Evaluating the "kernel" of the tensored sequence at the first as well as successive stages leads to the notion of the "Tor" functor. Similarly, an exact sequence of R-modules may fail to be exact on the right after applying the "Hom" functor and the evaluation of the cokernel at the first as well as successive stages leads to the notion of the 'Ext' functor. It is possible to develop a unified theory of derived functors and deduce the 'Tor' and 'Ext' functor as their special cases. We outline this development and derive some elementary properties of the "Tor" and "Ext" functors needed in the subsequent chapters.

7.1 Complexes

Let R be a commutative ring with 1. A (positive) complex X of R-modules is a sequence $\{X_n\}_{n\geq 0}$ of R-modules together with a sequence $\{d_n\}_{n\geq 0}$ of R-linear maps, where $d_n : X_n \to X_{n-1}$ satisfying $d_n\, d_{n+1} = 0$ $(n \geq 0)$. We denote such a complex by the symbol

$$X : \cdots \longrightarrow X_n \xrightarrow{d_n} X_{n-1} \longrightarrow \cdots \longrightarrow X_1 \xrightarrow{d_1} X_0 \longrightarrow 0$$

Let $X = (X_n, d_n)_{n\geq 0}$ and $X' = (X'_n, d'_n)_{n\geq 0}$ be two complexes. A homomorphism $F : X \to X'$ of complexes is a collection $\{F_n\}_{n\geq 0}$ of R-linear maps, $F_n : X_n \to X'_n$, such that the diagram

$$\begin{array}{ccc} X_{n+1} & \xrightarrow{F_{n+1}} & X'_{n+1} \\ \Big\downarrow{\scriptstyle d_{n+1}} & & \Big\downarrow{\scriptstyle d'_{n+1}} \\ X_n & \xrightarrow{F_n} & X'_n \end{array}$$

is commutative for all n.

Definition 7.1.1 Let X be a complex of R-modules. The R-submodules $Z_n(X) = \text{Ker } d_n$ and $B_n(X) = \text{Im } d_{n+1}$ are called respectively the n-cycles and the n-boundaries of X. Since $d_n d_{n+1} = 0,\ B_n(X) \subset Z_n(X),\ n \geq 0$, and the n-th homology module $H_n(X)$ is defined to be $\frac{Z_n(X)}{B_n(X)}$.

Proposition 7.1.2 *Let $F : X \to X'$ be a mapping of complexes. Then F induces an R-linear map $F_n^* : H_n(X) \to H_n(X')$ for all n. Moreover,*

1. $\text{Id}^* = \text{Id}$ *and*

2. *If $G : X' \to X''$ is another mapping of complexes, then $(GF)_n^* = G_n^*\ F_n^*$ for all n.*

Proof Since F is a mapping of complexes, we have the commutative diagram

$$\begin{array}{ccc} X_{n+1} & \xrightarrow{F_{n+1}} & X'_{n+1} \\ \Big\downarrow{\scriptstyle d_{n+1}} & & \Big\downarrow{\scriptstyle d'_{n+1}} \\ X_n & \xrightarrow{F_n} & X'_n \end{array}$$

Since $d'_{n+1}F_{n+1} = F_n d_{n+1}$, we have $F_{n+1}(Z_{n+1}(X)) \subseteq Z_{n+1}(X')$ and $F_n(B_n(X)) \subseteq B_n(X')$ for all n. Hence F_n defines a map

$$F_n^* : \frac{Z_n(X)}{B_n(X)} \longrightarrow \frac{Z_n(X')}{B_n(X')}$$

given by $F_n^*(\alpha + B_n(X)) = F_n(\alpha) + B_n(X')$, $\alpha \in Z_n(X)$. Then F_n^* is well defined and satisfies (1) and (2).

Definition 7.1.3 Let $F, G : X \to X'$ be mappings of complexes. F and G are said to be homotopic if there exists a collection $s = \{s_n\}_{n \geq 0}$ of R-linear maps, where $s_n : X_n \to X'_{n+1}$ such that $d'_{n+1}s_n + s_{n-1}d_n = F_n - G_n$ for all n.

The notation $F \overset{s}{\simeq} G$ denotes that F and G are homotopic under the homotopy s.

Proposition 7.1.4 *Let $F, G : X \longrightarrow X'$ be mappings of complexes such that $F \simeq G$. Then $F_n^* = G_n^*$ for all n.*

Proof Let $s = \{s_n\}_{n \geq 0}$ be a homotopy between F and G. Then $d'_{n+1}\, s_n + s_{n-1}\, d_n = F_n - G_n$. If $\alpha \in \mathbb{Z}_n(X)$, then $d_n \alpha = 0$, so that

$$F_n(\alpha) - G_n(\alpha) = d'_{n+1}\, s_n(\alpha) \in B_n(X').$$

Hence $F_n^*(\alpha + B_n(X)) = G_n^*(\alpha + B_n(X))$ for all n. □

Definition 7.1.5 Two complexes X and X' are said to be homotopically equivalent if there exist mappings $F : X \to X'$ and $G : X' \to X$ such that $G \circ F \simeq \mathrm{Id}_X$ and $F \circ G \simeq \mathrm{Id}_{X'}$.

Corollary 7.1.6 *If $F : X \to X'$ is a homotopy equivalence, then*

$$F_n^* : H_n(X) \to H_n(X')$$

is an isomorphism for all n.

Proof Inverse of F_n^* is $G_n^* : H_n(X') \to H_n(X)$. □

Definition 7.1.7 A complex X is said to be acyclic if $H_n(X) = 0$ for all n.

Definition 7.1.8 A projective resolution of an R-module M is a complex X together with an R-linear map $X_0 \xrightarrow{\varepsilon} M$ such that the sequence

$$\cdots \to X_n \xrightarrow{d_n} X_{n-1} \xrightarrow{d_{n-1}} \cdots \to X_0 \xrightarrow{\varepsilon} M \to 0$$

is exact and all the X_i are R-projective.

If X is a projective resolution of M we denote it by the symbol $X \xrightarrow{\varepsilon} M \to 0$. Clearly, X has the property that $H_n(X) = 0,\ n > 1$, and $H_0(X) \simeq M$.

Example 7.1.9 (i) If M is projective, choose $X_0 = M,\ \varepsilon = \mathrm{Id}$ and $X_n = 0,\ n \geq 1$. Then X is a projective resolution of M.

(ii) Let $R = \mathbb{Z}$ and M any abelian group. Choose a free abelian group X_0 and a surjection $\varepsilon : X_0 \to M$. Let $X_1 = \operatorname{Ker} \varepsilon$. Then X_1 is also a free abelian group and

$$0 \to X_1 \xrightarrow{i} X_0 \xrightarrow{\varepsilon} M \to 0$$

is a projective resolution of M.

Proposition 7.1.10 *Every R-module M has a projective resolution.*

Proof Given an R-module M, choose a free R-module X_0 and a surjection $\varepsilon : X_0 \to M$ with kernel K_0 giving rise to an exact sequence

$$0 \to K_0 \xrightarrow{g_0} X_0 \xrightarrow{f_0} M \to 0.$$

Now choose a free R-module X_1 together with a surjection $f_1 : X_1 \to K_0$ with kernel K_1. Continuing the process, we have a collection of exact sequences

$$0 \to K_n \xrightarrow{g_n} X_n \xrightarrow{f_n} K_{n-1} \to 0$$

$(n \geq 0)$, $f_0 = \varepsilon$ and $K_{-1} = M$. If $d_n : X_n \to X_{n-1}$ is the map $d_n = g_{n-1}f_n$, then the complex

$$\cdots \to X_n \xrightarrow{d_n} X_{n-1} \to \cdots \longrightarrow X_1 \xrightarrow{d_1} X_0 \xrightarrow{\varepsilon} M \to 0$$

is a projective resolution of M. □

The following theorem shows that if X and X' are two projective resolutions of M, then they are homotopically equivalent.

Theorem 7.1.11 *Let $f : M \to M'$ be a homomorphism of R-modules. Let $X \xrightarrow{\varepsilon} M$ and $X' \xrightarrow{\varepsilon'} M'$ be projective resolutions of M and M' respectively. Then there exists a mapping of complexes $F : X \to X'$ such that $f\varepsilon = \varepsilon' F_0$. Moreover, F is unique up to homotopy.*

Proof We first prove the existence of F. Consider the diagram

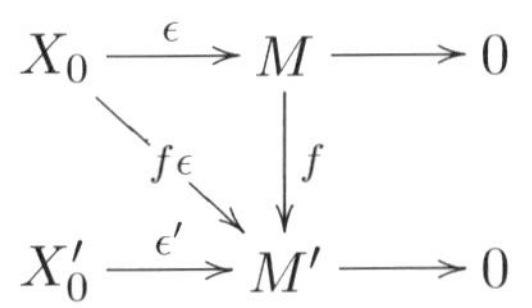

Since X_0 is projective and ε' is surjective, there exists an R-linear map $F_0 : X_0 \to X'_0$ with $\varepsilon' F_0 = f\varepsilon$. Assume inductively that $F_i : X_i \to X'_i$ have been defined for all i, $0 \leq i \leq n$ satisfying $F_{i-1}d_i = d'_i F_i \;\; (i \leq n)$, with $F_{-1} = f,\ d_0 = \varepsilon,\ d'_0 = \varepsilon'$. Consider the diagram

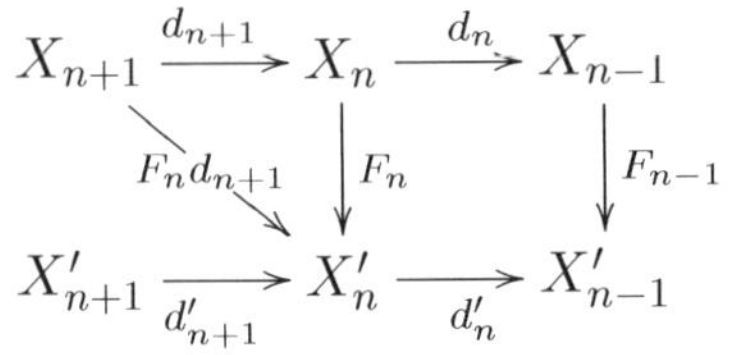

Observe that $\mathrm{Im}\,(F_n d_{n+1}) \subset \mathrm{Im}\, d'_{n+1} = \mathrm{Ker}\, d'_n$, because $d'_n F_n d_{n+1} = F_{n-1}\, d_n\, d_{n+1} = 0$. Consider the diagram

$$\begin{array}{ccccc} & & X_{n+1} & & \\ & & \big\downarrow{\scriptstyle F_n d_{n+1}} & & \\ X'_{n+1} & \xrightarrow{d'_{n+1}} & d'_{n+1}(X'_{n+1}) & \longrightarrow & 0 \end{array}$$

The projectivity of X_{n+1} gives an R-linear map $F_{n+1} : X_{n+1} \to X'_{n+1}$ such that $d'_{n+1} F_{n+1} = F_n d_{n+1}$, and the proof is complete by induction.

We now prove the uniqueness of F up to homotopy. Suppose we have another mapping of complexes $G : X \to X'$ such that $f\varepsilon = \varepsilon' G_0$. We construct a homotopy $s = \{s_n\}_{n \geq 0}$ between F and G inductively as follows. Consider the diagram

$$\begin{array}{ccccccc} X_1 & \xrightarrow{d_1} & X_0 & \xrightarrow{\epsilon} & M & \longrightarrow & 0 \\ \big\downarrow & & \big\downarrow{\scriptstyle F_0 - G_0} & & \big\downarrow{\scriptstyle f} & & \\ X'_1 & \xrightarrow{d'_1} & X'_0 & \xrightarrow{\epsilon'} & M' & \longrightarrow & 0 \end{array}$$

Since $\varepsilon'(F_0 - G_0) = \varepsilon' F_0 - \varepsilon' G_0 = f\varepsilon - f\varepsilon = 0$, we have

$$\mathrm{Im}\, d'_1 = \mathrm{Ker}\, \varepsilon' \supset \mathrm{Im}\,(F_0 - G_0).$$

Since X_0 is projective, there exists an R-linear map $s_0 : X_0 \to X'_1$ such that $d'_1 s_0 = F_0 - G_0$.

Assume inductively that $\{s_i\}$ have been defined, for i, $0 \leq i \leq n$, satisfying $d'_{i+1} s_i + s_{i-1} d_i = F_i - G_i \ (i \leq n)$ with $s_{-1} = 0$. To define s_{n+1}, consider the diagram

$$\begin{array}{ccccc} X_{n+2} & \xrightarrow{d_{n+2}} & X_{n+1} & \xrightarrow{d_{n+1}} & X_n \\ & & {\scriptstyle F_{n+1} - G_{n+1}}\big\downarrow & \swarrow{\scriptstyle s_n} & \big\downarrow{\scriptstyle F_n - G_n} \\ X'_{n+2} & \xrightarrow[d'_{n+2}]{} & X'_{n+1} & \xrightarrow[d'_{n+1}]{} & X'_n \end{array}$$

We observe that $\mathrm{Im}\ d'_{n+2} = \mathrm{Ker}\, d'_{n+1} \supseteq \mathrm{Im}\,(F_{n+1} - G_{n+1} - s_n d_{n+1})$ because

$$\begin{aligned} & d'_{n+1}(F_{n+1} - G_{n+1} - s_n d_{n+1}) \\ &= (F_n - G_n) d_{n+1} - (F_n - G_n - s_{n-1} d_n) d_{n+1} \\ &= 0. \end{aligned}$$

Since X_{n+1} is projective, there exists an R-linear map $s_{n+1} : X_{n+1} \to X'_{n+2}$ such that $d'_{n+2} s_{n+1} = F_{n+1} - G_{n+1} - s_n d_{n+1}$. Thus s_n is defined for all n and $s = \{s_n\}_{n \geq 0}$ is the required homotopy between F and G. $\square$

Remark 7.1.12 The map $F : X \to X'$ is called a lifting of the map $f : M \to M'$ to the corresponding projective resolutions. If $f = \text{Id}$, F can be chosen to be identity and if $g : M' \to M''$ is another R-linear map with a lifting $G : X' \to X''$, then GF is a lifting of gf.

Corollary 7.1.13 *If X and Y are two projective resolutions of M, then X and Y are homotopically equivalent.*

Proof Follows from Theorem 7.1.11 and Remark 7.1.12. □

Theorem 7.1.14 *Let $0 \to X' \xrightarrow{F} X \xrightarrow{G} X'' \to 0$ be an exact sequences of complexes, i.e., for each n, the sequence of R-modules*

$$0 \to X'_n \xrightarrow{F_n} X_n \xrightarrow{G_n} X''_n \to 0$$

is exact. Then there exists a connecting homomorphism

$$\partial_n : H_n(X'') \to H_{n-1}(X')$$

such that the sequence

$$\cdots \to H_n(X') \xrightarrow{F_n} H_n(X) \xrightarrow{G_n} H_n(X'') \xrightarrow{\partial_n} H_{n-1}(X') \to \cdots$$

is exact.

Proof Since the given sequence of complexes is exact, we have a commutative diagram

$$\begin{array}{ccccccccc} 0 & \longrightarrow & X'_n & \xrightarrow{F_n} & X_n & \xrightarrow{G_n} & X''_n & \longrightarrow & 0 \\ & & \downarrow d'_n & & \downarrow d_n & & \downarrow d''_n & & \\ 0 & \longrightarrow & X'_{n-1} & \xrightarrow{F_{n-1}} & X_{n-1} & \xrightarrow{G_{n-1}} & X''_{n-1} & \longrightarrow & 0 \end{array}$$

with exact rows. This induces a commutative diagram

$$\begin{array}{ccccccccc} & & \frac{X'_n}{B_n(X')} & \longrightarrow & \frac{X_n}{B_n(X)} & \longrightarrow & \frac{X''_n}{B_n(X'')} & \longrightarrow & 0 \\ & & \downarrow \overline{d'}_n & & \downarrow \overline{d}_n & & \downarrow \overline{d''}_n & & \\ 0 & \longrightarrow & Z_{n-1}(X') & \longrightarrow & Z_{n-1}(X) & \longrightarrow & Z_{n-1}(X'') & & \end{array}$$

with exact rows, where the vertical maps are induced by the boundary operators, i.e.,

$$\overline{d}_n : \frac{X_n}{B_n(X)} \longrightarrow Z_{n-1}(X)$$

is given by $\overline{d}_n((\alpha+B_n(X)) = d_n\alpha$. Clearly, $\operatorname{Ker}\overline{d}_n = H_n(X)$ and $\operatorname{Coker}\overline{d}_n = H_{n-1}(X)$. By Proposition 2.2.15, we have an exact sequence

$$\operatorname{Ker}\overline{d'}_n \to \operatorname{Ker}\overline{d}_n \to \operatorname{Ker}\overline{d''}_n \to \operatorname{Coker}\overline{d'}_n \to \operatorname{Coker}\overline{d}_n \to \operatorname{Coker}\overline{d''}_n$$

i.e.,

$$H_n(X') \to H_n(X) \to H_n(X'') \xrightarrow{\partial_n} H_{n-1}(X') \to H_{n-1}(X) \to H_{n-1}(X'')$$

is exact for all $n \geq 0$. □

Remark 7.1.15 The exact homology sequence derived above satisfies the naturality condition (see Exercise 4, Section 7.1).

Theorem 7.1.16 *Let $0 \to M' \xrightarrow{f} M \xrightarrow{g} M'' \longrightarrow 0$ be an exact sequence of R-modules. Let $X' \xrightarrow{\varepsilon'} M' \to 0$ and $X'' \xrightarrow{\varepsilon''} M'' \to 0$ be projective resolutions of M' and M'' respectively. Then there exists a projective resolution $X \xrightarrow{\varepsilon} M \to 0$ of M and an exact sequence of complexes $0 \to X' \xrightarrow{F} X \xrightarrow{G} X'' \to 0$ such that the diagram*

$$\begin{array}{ccccccccc} 0 & \longrightarrow & X_0' & \xrightarrow{F_0} & X_0 & \xrightarrow{G_0} & X_0'' & \longrightarrow & 0 \\ & & \downarrow{\scriptstyle \epsilon'} & & \downarrow{\scriptstyle \epsilon} & & \downarrow{\scriptstyle \epsilon''} & & \\ 0 & \longrightarrow & M' & \xrightarrow{f} & M & \xrightarrow{g} & M'' & \longrightarrow & 0 \end{array} \tag{7.1}$$

is commutative.

Proof Let $X_n = X_n' \oplus X_n''$, $F_n : X_n' \to X_n$, the natural inclusion $x' \to (x', 0)$ and $G_n : X_n \to X_n''$ the natural projection $(x', x'') \to x''$. Clearly, the sequence $0 \to X_n' \to X_n \to X_n'' \to 0$ is exact and X_n is projective for all n.

To define $\varepsilon : X_0 \to M$ and $d_n : X_n \to X_{n-1}$ satisfying the required conditions, we first construct a family of R-linear maps $\{\alpha_n\}_{n\geq 0}$, $\alpha_n : X_n'' \to X_{n-1}'$, $\alpha_0 : X_0'' \to M$ satisfying suitable conditions. We then define

$$\varepsilon : X_0 \to M, \quad d_n : X_n \to X_{n-1}$$

respectively by

$$\varepsilon(x_0', x_0'') = f\varepsilon' x_0' + \alpha_0 x_0'', \quad d_n(x_n', x_n'') = (d_n' x_n' + \alpha_n x_n'', d_n'' x_n'').$$

The conditions $d_{n-1}d_n = 0$, $n \geq 1$ and $\varepsilon d_1 = 0$ impose the following conditions on the α_n.

(i) $-d_{n-1}'\alpha_n = \alpha_{n-1}d_n''$, $n \geq 2$,

(ii) $-f\varepsilon'\alpha_1 = \alpha_0 d_1''$.

Since the left half of the diagram 7.1 is commutative, the commutativity of the right half imposes the condition

(iii) $\varepsilon'' = g\alpha_0$.

We show by induction that $\{\alpha_n\}$ can be chosen to satisfy (i), (ii) and (iii).

Since X_0'' is projective, α_0 can be chosen to satisfy (iii). To define α_1, consider the diagram

$$\begin{array}{ccccc} & & X_1'' & & \\ & & \big\downarrow{\scriptstyle \alpha_0 d_1''} & & \\ X_0' & \xrightarrow{-f\epsilon'} & M & \xrightarrow{g} & M'' \end{array}$$

Since $\varepsilon'' d_1'' = g\alpha_0 d_1'' = 0$, we have $\mathrm{Im}\,(\alpha_0 d_1'') \subseteq \mathrm{Ker}\, g = \mathrm{Im}\,(-f\varepsilon')$ and the projectivity of X_1'' gives an R-linear map $\alpha_1 : X_1'' \to X_0'$ with $-f\varepsilon'\alpha_1 = \alpha_0 d_1''$. Using the same type of argument, $\{\alpha_n\}_{n\geq 2}$ can be defined inductively satisfying conditions in (i). Hence ε and d_n, $n \geq 1$, are defined giving rise to the complex

$$X_M : \cdots X_n \xrightarrow{d_n} X_{n-1} \to \cdots \to X_0 \xrightarrow{\varepsilon} M \to 0.$$

The exactness of the sequence of complexes $0 \to X' \to X \to X'' \to 0$ gives the exactness of $0 \to X_M' \to X_M \to X_M'' \to 0$. Since X' and X'' are projective resolutions of M' and M'' respectively the complexes at the extremes are acyclic. This implies that X_M is also acyclic (see Exercise 2, Section 7.1) showing that X is a projective resolution of M. □

Definition 7.1.17 A projective resolution of an exact sequence of R-modules $0 \to M' \to M \to M'' \to 0$ is an exact sequence of complexes $0 \to X' \to X \to X'' \to 0$, where X', X, X'' are projective resolutions of M', M and M'' respectively, such that the diagram

$$\begin{array}{ccccccccc} 0 & \longrightarrow & X_0' & \longrightarrow & X_0 & \longrightarrow & X_0'' & \longrightarrow & 0 \\ & & \big\downarrow{\scriptstyle \epsilon'} & & \big\downarrow{\scriptstyle \epsilon} & & \big\downarrow{\scriptstyle \epsilon''} & & \\ 0 & \longrightarrow & M' & \longrightarrow & M & \longrightarrow & M'' & \longrightarrow & 0 \end{array}$$

is commutative.

Remark 7.1.18 Theorem 7.1.16 shows the existence of such a projective resolution.

7.1. EXERCISES

1. Show that a complex is acyclic if and only if the identity map of X in X is homotopic to the zero map.

2. Let $0 \to X' \to X \to X'' \to 0$ be an exact sequence of complexes. If any two of the three complexes X', X, X'' are acyclic, show that the third is also acyclic.

3. Show that a finitely generated module over a Noetherian ring admits a projective resolution by finitely generated projective modules.

4. Prove the following naturality condition for exact homology sequence. Given a commutative diagram

$$\begin{array}{ccccccccc} 0 & \longrightarrow & X' & \xrightarrow{F} & X & \xrightarrow{G} & X'' & \longrightarrow & 0 \\ & & \downarrow{\scriptstyle \alpha'} & & \downarrow{\scriptstyle \alpha} & & \downarrow{\scriptstyle \alpha''} & & \\ 0 & \longrightarrow & Y' & \xrightarrow{\phi} & Y & \xrightarrow{\psi} & Y'' & \longrightarrow & 0 \end{array}$$

where the rows are exact sequences of complexes and the vertical maps are mappings of complexes, the induced diagram

$$\begin{array}{ccccccccccc} \cdots & \longrightarrow & H_n(X') & \xrightarrow{F_n^*} & H_n(X) & \xrightarrow{G_n^*} & H_n(X'') & \xrightarrow{\delta_n} & H_{n-1}(X') & \longrightarrow & \cdots \\ & & \downarrow{\scriptstyle {\alpha'_n}^*} & & \downarrow{\scriptstyle \alpha_n^*} & & \downarrow{\scriptstyle {\alpha''}_n^*} & & \downarrow{\scriptstyle {\alpha'}_{n-1}^*} & & \\ \cdots & \longrightarrow & H_n(Y') & \xrightarrow[\phi_n^*]{} & H_n(Y) & \xrightarrow[\psi_n^*]{} & H_n(Y'') & \xrightarrow[\delta'_n]{} & H_{n-1}(Y') & \longrightarrow & \cdots \end{array}$$

is commutative.

7.2 Derived functors

An additive covariant (contravariant) functor is a correspondence which associates to each R-module M an R-module $T(M)$ and to each R-linear map $f : M \to M'$, an R-linear map

$$T(f) : T(M) \to T(M') \qquad (T(f) : T(M') \to T(M))$$

satisfying

(i) $T(\mathrm{Id}) = \mathrm{Id}$,

(ii) $T(gf) = T(g)T(f) \quad (T(gf) = T(f)T(g))$,

(iii) $T(f+g) = T(f) + T(g)$.

Example 7.2.1 (i) Let N be a fixed R-module. The functor

$$T(M) = M \otimes_R N \quad \text{and } T(f) = f \otimes \mathrm{Id}_N,$$

for $f : M \to M'$, is a covariant additive functor.

(ii) For a fixed R-module N, the functor $T'(M) = \operatorname{Hom}_R(M, N)$ and $T'(f) = \operatorname{Hom}_R(f, \operatorname{Id}_N)$, $f : M \to M'$ is an additive contravariant functor.

We now define the left derived functors L_nT $(n \geq 0)$ of an additive covariant functor T. Let M be an R-module and $X \xrightarrow{\varepsilon} M$ a projective resolution of M. Applying the functor T on the complex

$$X \ :\cdots \to X_n \xrightarrow{d_n} X_{n-1} \cdots X_1 \xrightarrow{d_1} X_0 \to 0$$

we have the complex

$$T(X) : \cdots \to T(X_n) \xrightarrow{T(d_n)} T(X_{n-1}) \to \cdots \to T(X_1) \xrightarrow{T(d_1)} T(X_0) \to 0.$$

The n-th homology of the complex $T(X)$

$$H_n(T(X)) = \frac{\operatorname{Ker} T(d_n)}{\operatorname{Im} T(d_{n+1})}$$

is defined to be $L_nT(M)$. Let $f : M \to M'$ be an R-linear map and X, X' projective resolutions of M and M' respectively. By Theorem 7.1.11, f can be lifted to a mapping of complexes $F : X \to X'$, which induces a mapping of complexes $T(F) : T(X) \to T(X')$. By Proposition 7.1.2, $T(F)$ induces maps on the homology

$$T(F)_n^* : H_n(T(X)) \to H_n(T(X')) \ (n \geq 0)$$

i.e.,

$$L_nT(f) : L_nT(M) \to L_nT(M') \ (n \geq 0).$$

Proposition 7.2.2 *$L_nT(M)$ and $L_nT(f)$ are well defined and are independent of the projective resolutions. Moreover, $L_nT(n \geq 0)$ are additive covariant functors.*

Proof Let X and Y be two projective resolutions of M. By Corollary 7.1.13, X and Y are homotopically equivalent. Hence $T(X)$ and $T(Y)$ are also homotopically equivalent (Exercise 2, 7.2). This implies by Corollary 7.1.6 that $H_n(T(X)) \simeq H_n(T(Y))$ for all n, i.e., $L_nT(M)$ is unique up to isomorphism. Let $f : M \to M'$ be an R-linear map and X, X' projective resolutions of M, M' respectively. Choose a lifting $F : X \to X'$ of f. Then F is unique up to homotopy, i.e, if $G : X \to X'$ is another lifting of f then $F \overset{s}{\simeq} G$. Thus $T(F) \overset{T(s)}{\simeq} T(G)$ and, by Proposition 7.1.4, $T(F)_n^* = T(G)_n^*$ for all n, i.e., $L_nT(f)$ is well defined. It remains to verify that L_nT $(n \geq 0)$ are covariant functors.

Now $L_nT(\mathrm{Id}) = \mathrm{Id}$, for if $f = \mathrm{Id} : M \to M$, we can choose $X' = X$ and $F = \mathrm{Id}$. Let X', X and X'' be projective resolutions of M', M and M'' respectively. If $F : X \to X'$ is a lifting of $f : M \to M'$ and $G : X' \to X''$ is a lifting of $g : M' \to M''$, then $GF : X \to X''$ is a lifting of $gf : M \to M''$ and $L_nT(gf) = T(GF)_n^* = T(G)_n^* T(F)_n^*$ (T covariant) $= L_nT(g)L_nT(f)$.

The verification that the functors $L_nT \quad (n \geq 0)$ are additive is routine and is left as an exercise (Exercise 3, Section 7.2). □

Theorem 7.2.3 *Let T be a covariant additive functor. Then the following statements hold:*

(i) *If M is projective, then $L_nT(M) = 0, \quad n \geq 1$.*

(ii) *If T is right exact, then $L_0T \simeq T$.*

(iii) *For any exact sequence of R-modules,*

$$0 \to M' \xrightarrow{f} M \xrightarrow{g} M'' \to 0,$$

there is an exact sequence

$$\to L_nT(M') \xrightarrow{L_nT(f)} L_nT(M) \xrightarrow{L_nT(g)}$$

$$L_nT(M'') \xrightarrow{\partial_n} L_{n-1}T(M') \to \cdots$$

which has the naturality property.

Proof (i) If M is projective, $0 \to X_0 \xrightarrow{\text{Id}} M \to 0$ is a projective resolution of M, where $X_0 = M$ and $X_i = 0$, $i \geq 1$. If $f : N \to K$ is the zero map, condition (iii) of the functor shows that $T(f) = 0$. In particular, if $N = 0$, $\text{Id}_N = 0$, and hence $T(\text{Id}_N) = \text{Id}_{T(N)} = 0$, i.e., $T(N) = 0$. Hence $X_i = 0$, for $i \geq 1$, implies $T(X_i) = 0$, $i \geq 1$, i.e., $L_nT(M) = 0, \quad n \geq 1$.

(ii) Assume now that T is right exact and let

$$\to X_n \to \cdots \to X_1 \xrightarrow{d_1} X_0 \xrightarrow{\varepsilon} M \to 0$$

be a projective resolution of M. Since T is right exact, the sequence

$$T(X_1) \xrightarrow{T(d_1)} T(X_0) \xrightarrow{T(\varepsilon)} T(M) \to 0$$

is exact. Hence

$$T(M) \simeq \frac{T(X_0)}{\text{Ker } T(\varepsilon)} = \frac{T(X_0)}{\text{Im } T(d_1)} = L_0T(M).$$

(iii) Let $0 \to M' \xrightarrow{f} M \xrightarrow{g} M'' \to 0$ be an exact sequence of R-modules. By Theorem 7.1.16, it has a projective resolution

$$0 \to X' \xrightarrow{F} X \xrightarrow{G} X'' \to 0.$$

For each n,

$$0 \to X'_n \xrightarrow{F_n} X_n \xrightarrow{G_n} X''_n \to 0$$

is split exact. By Exercise 8, 7.2, the sequence

$$0 \to T(X'_n) \xrightarrow{T(F_n)} T(X_n) \xrightarrow{T(G_n)} T(X''_n) \to 0$$

is exact and splits, i.e., we have an exact sequence of complexes

$$0 \to T(X') \xrightarrow{T(F)} T(X) \xrightarrow{T(G)} T(X'') \to 0.$$

This induces an exact homology sequence

$$\cdots \to H_nT(X') \to H_nT(X) \to H_nT(X'') \xrightarrow{\partial_n} H_{n-1}T(X') \to \cdots,$$

i.e.,

$$\cdots \to L_nT(M') \to L_nT(M) \to L_nT(M'') \xrightarrow{\partial_n} L_{n-1}T(M') \to \cdots.$$

The naturality condition is a consequence of the naturality of the homology exact sequence. □

We now define the right derived functors $R^nT (n \geq 0)$ of an additive contravariant functor. Let M be an R-module and $X \xrightarrow{\varepsilon} M$ a projective resolution of M. Applying T on the complex X

$$X : \cdots \to X_n \xrightarrow{d_n} X_{n-1} \to \cdots X_1 \xrightarrow{d_1} X_0 \longrightarrow 0$$

we have the complex $T(X)$

$$T(X) : 0 \to T(X_0) \xrightarrow{T(d_1)} T(X_1) \to \cdots \to T(X_{n-1}) \xrightarrow{T(d_n)} T(X_n) \to \cdots$$

The n-th homology of the complex $T(X)$, i.e.,

$$H_n(T(X)) = \frac{\text{Ker}\, T(d_{n+1})}{\text{Im}\, T(d_n)}$$

is defined to be $R^nT(M)$. Let $f : M \to M'$ be an R-linear map, and X, X' projective resolutions of M and M' respectively. Then f can be lifted to a mapping $F : X \to X'$ of complexes which induces a map $T(F) : T(X') \to T(X)$. Since $T(F)$ is also a mapping of complexes, it induces mappings on the homology

$$T(F)^*_n : H_n(T(X')) \to H_n(T(X)) \; (n \geq 0),$$

i.e.,

$$R^nT(f) : R^nT(M') \to R^nT(M) \; (n \geq 0).$$

The following are analogues of Proposition 7.2.2, Theorem 7.2.3, and can be proved on the same lines.

Proposition 7.2.4 *$R^nT(M)$ and $R^nT(f)$ are well defined and are independent of the projective resolutions. Moreover, R^nT $(n \geq 0)$ are additive contravariant functors.*

Theorem 7.2.5 *Let T be a contravariant additive functor. Then the following statements hold:*
(i) *If M is projective, then $R^nT(M) = 0, n \geq 1$.*
(ii) *$R^0T \simeq T$ if T is left exact.*
(iii) *For any exact sequence of R-modules*

$$0 \to M' \xrightarrow{f} M \xrightarrow{g} M'' \to 0,$$

there is an exact sequence

$$\to R^nT(M'') \xrightarrow{R^nT(g)} R^nT(M) \xrightarrow{R^nT(f)} R^nT(M') \to R^{n+1}T(M') \to \cdots,$$

which satisfies the naturality condition.

We now consider the left derived functors of the tensor product functor.

Let N be a fixed R-module. Consider the functor T given by $T(M) = M \otimes_R N$ and $T(f) = f \otimes \mathrm{Id}_N$ for $f : M \to M'$.

Since T is a covariant additive functor, L_nT $(n \geq 0)$ are defined and we denote L_nT by $\mathrm{Tor}_n^R(\ , N)$. From the definition of L_nT, it is clear that if M is an R-module, $X \xrightarrow{\varepsilon} M \to 0$, a projective resolution of M, and $X \otimes_R N$ is the complex

$$X \otimes_R N : \cdots \to X_{n+1} \otimes_R N \to X_n \otimes_R N \to \cdots \to X_0 \otimes_R N \to 0,$$

then $\mathrm{Tor}_n^R(M, N) = H_n(X \otimes_R N)$. If $f : M \to M'$ is an R-linear map, X, X' projective resolution of M and M' respectively, f can be lifted to a mapping of complexes $F : X \to X'$, which induces a map

$$\overline{F} = F \otimes_R N : X \otimes_R N \to X' \otimes_R N.$$

The induced map on the homology

$$\overline{F}_n^* \cdot \mathrm{Tor}_n^R(M, N) \to \mathrm{Tor}_n^R(M', N)$$

is the map $\mathrm{Tor}_n^R(f, N)$.

From Theorem 7.2.3, it is clear that when M is projective, $\mathrm{Tor}_n^R(M, N) = 0$ for all R-modules N and all $n \geq 1$, and that $\mathrm{Tor}_0^R(M, N) \simeq M \otimes_R N$ for any two R-modules M and N. We will now show that for any R-linear map $\phi : N \to N'$, there exists an R-linear map

$$\mathrm{Tor}_n^R(M, \phi) : \mathrm{Tor}_n^R(M, N) \to \mathrm{Tor}_n^R(M, N').$$

If X is a projective resolution of M, ϕ induces a mapping of complexes $G : X \otimes_R N \to X \otimes N'$, which in turn induces a mapping on the homology

$$G_n^* : H_n(X \otimes_R N) \to H_n(X \otimes_R N'),$$

i.e.,

$$\mathrm{Tor}_n^R(M, \phi) : \mathrm{Tor}_n^R(M, N) \to \mathrm{Tor}_n^R(M, N').$$

Clearly, the correspondence $\phi \to \mathrm{Tor}_n^R(M, \phi)$ satisfies the conditions (i), (ii) and (iii) of a covariant functor.

Proposition 7.2.6 *Let*

$$0 \to M' \xrightarrow{f} M \xrightarrow{g} M'' \to 0$$

and

$$0 \to N' \xrightarrow{\phi} N \xrightarrow{\psi} N'' \to 0$$

be exact sequences of R-modules. Then there exist exact sequences

$$\cdots \to \mathrm{Tor}_n^R(M', N) \overset{\mathrm{Tor}_n(f,N)}{\longrightarrow} \mathrm{Tor}_n^R(M, N)$$

$$\overset{\mathrm{Tor}_n(g,N)}{\longrightarrow} \mathrm{Tor}_n^R(M'', N) \xrightarrow{\partial_n} \mathrm{Tor}_{n-1}^R(M', N) \to \cdots$$

and

$$\cdots \to \mathrm{Tor}_n^R(M, N') \overset{\mathrm{Tor}_n(M,\phi)}{\longrightarrow} \mathrm{Tor}_n^R(M, N)$$

$$\overset{\mathrm{Tor}_n(M,\psi)}{\longrightarrow} \mathrm{Tor}_n^R(M, N'') \xrightarrow{\partial_n} \mathrm{Tor}_{n-1}^R(M, N) \to \cdots$$

which have the naturality property.

Proof The first exact sequence is a consequence of Theorem 7.2.3, as $\mathrm{Tor}_n^R(M, N) = L_nT(M)$ for the functor $T = \ \otimes_R N$.

To establish the second exact sequence, consider a projective resolution X of M and the exact sequence of complexes

$$0 \to X \otimes_R N' \overset{\mathrm{Id}_X \otimes \phi}{\longrightarrow} X \otimes_R N \overset{\mathrm{Id}_X \otimes \psi}{\longrightarrow} X \otimes_R N'' \to 0$$

induced by the sequence $0 \to N' \xrightarrow{\phi} N \xrightarrow{\psi} N'' \to 0$.

By Theorem 7.1.14, there is a natural exact homology sequence

$$\cdots \to H_n(X \otimes_R N') \to H_n(X \otimes_R N) \to$$

$$H_n(X \otimes_R N'') \xrightarrow{\partial_n} H_{n-1}(X \otimes_R N') \to \cdots$$

i.e.,

$$\to \mathrm{Tor}_n^R(M, N') \to \mathrm{Tor}_n^R(M, N) \to$$

$$\mathrm{Tor}_n^R(M, N'') \to \mathrm{Tor}_{n-1}^R(M, N') \to \cdots$$

is exact. □

We now consider the right derived functors of the Hom functor. Let N be a fixed R-module. Consider the functor T defined by $M \to T(M) = \mathrm{Hom}_R(M, N)$ and for an R-linear map $f : M \to M', T(f)$ is given by

$$T(f) = \mathrm{Hom}_R(f, I_N) : \mathrm{Hom}_R(M', N) \to \mathrm{Hom}_R(M, N).$$

Since T is a contravariant additive functor, $R^nT \quad (n \geq 0)$ are defined and we denote R^nT by $\mathrm{Ext}_R^n(\ , N)$. From the definition of R^nT it is clear that if M is an R-module and $X \xrightarrow{\varepsilon} M \to 0$, a projective resolution of M, and $\mathrm{Hom}_R(X, N)$ is the complex

$$0 \to \mathrm{Hom}_R(X_0, N) \to \mathrm{Hom}_R(X_1, N) \to \cdots \mathrm{Hom}_R(X_n, N) \to \cdots$$

then $\mathrm{Ext}_R^n(M, N) = H_n(\mathrm{Hom}_R(X, N))$. If $f : M \to M'$ is an R-linear map, X, X' projective resolutions of M and M' respectively, then f can be lifted to a mapping of complexes $F : X \to X'$, which induces a map $\overline{F} : \mathrm{Hom}_R(X', N) \to \mathrm{Hom}_R(X, N)$, which in turn defines a map on homology

$$\overline{F}_n^* : \mathrm{Ext}_R^n(M', N) \to \mathrm{Ext}_R^n(M, N).$$

This map is denoted by $\mathrm{Ext}_R^n(f, N)$. The correspondence $f \mapsto \mathrm{Ext}_R^n(f, N)$ satisfies conditions (i), (ii) and (iii) of a contravariant additive functor. Clearly, we have $\mathrm{Ext}_R^n(M, N) = 0$ for $n \geq 1$ if M is projective and $\mathrm{Ext}_R^0(M, N) \simeq \mathrm{Hom}_R(M, N)$ for all M and N (Theorem 7.2.5).

We will now show that the Ext functor has the property that for any R-linear map $\phi : N \to N'$, there is an induced R-linear map

$$\mathrm{Ext}_R^n(M, \phi) : \mathrm{Ext}_R^n(M, N) \to \mathrm{Ext}_R^n(M, N').$$

We have only to take a projective resolution X of M and consider the mapping F induced by ϕ on the complexes

$$\mathrm{Hom}_R(X, N) \xrightarrow{F} \mathrm{Hom}_R(X, N'),$$

which in turn induces a mapping on the homology

$$\mathrm{Ext}_R^n(M, \phi) : \mathrm{Ext}_R^n(M, N) \to \mathrm{Ext}_R^n(M, N').$$

The correspondence $\phi \to \mathrm{Ext}_R^n(M,\phi)$ satisfies the conditions (i), (ii) and (iii) of a covariant functor. The following proposition on Ext can be proved on lines similar to Proposition 7.2.6.

Proposition 7.2.7 *Let* $0 \to M' \xrightarrow{f} M \xrightarrow{g} M'' \to 0$ *and* $0 \to N' \xrightarrow{\phi} N \xrightarrow{\psi} N'' \to 0$ *be exact sequences of R-modules. Then there exist exact sequences*

$$\cdots \to \mathrm{Ext}_R^n(M'',N) \xrightarrow{\mathrm{Ext}^n(g,N)} \mathrm{Ext}_R^n(M,N) \xrightarrow{\mathrm{Ext}^n(f,N)} \mathrm{Ext}_R^n(M',N) \xrightarrow{\partial_n} \mathrm{Ext}_R^{n+1}(M'',N) \to \cdots$$

and

$$\cdots \to \mathrm{Ext}_R^n(M,N') \xrightarrow{\mathrm{Ext}^n(M,\phi)} \mathrm{Ext}_R^n(M,N) \xrightarrow{\mathrm{Ext}^n(M,\psi)} \mathrm{Ext}_R^n(M,N'') \xrightarrow{\partial_n} \mathrm{Ext}_R^{n+1}(M,N') \to \cdots$$

which have the naturality property.

7.2. EXERCISES

1. Show that functor given by $T(M) = M \otimes_R M$ and $T(f) = f \otimes f$ is covariant but not additive.
2. Show that an additive functor carries homotopically equivalent complexes to homotopically equivalent complexes.
3. Show that L_nT and R^nT $(n \geq 0)$ are additive functors.
4. Show that for any covariant functor T, the functor L_0T is right exact.
5. Prove the naturality condition for long exact sequences, viz., given a commutative diagram

$$\begin{array}{ccccccccc} 0 & \longrightarrow & M' & \xrightarrow{f} & M & \xrightarrow{g} & M'' & \longrightarrow & 0 \\ & & \downarrow{\scriptstyle \alpha'} & & \downarrow{\scriptstyle \alpha} & & \downarrow{\scriptstyle \alpha''} & & \\ 0 & \longrightarrow & N' & \xrightarrow{\phi} & N & \xrightarrow{\psi} & N'' & \longrightarrow & 0 \end{array}$$

of R-modules with exact rows, the induced diagram

$$\begin{array}{ccccccccccc} \cdots & \longrightarrow & L_nT(M') & \xrightarrow{L_nT(f)} & L_nT(M) & \xrightarrow{L_nT(g)} & L_nT(M'') & \xrightarrow{\delta_n} & L_{n-1}T(M') & \longrightarrow & \cdots \\ & & \downarrow{\scriptstyle L_nT(\alpha')} & & \downarrow{\scriptstyle L_nT(\alpha)} & & \downarrow{\scriptstyle L_nT(\alpha'')} & & \downarrow{\scriptstyle L_{n-1}T(\alpha')} & & \\ \cdots & \longrightarrow & L_nT(N') & \xrightarrow[L_nT(\phi)]{} & L_nT(N) & \xrightarrow[L_nT(\psi)]{} & L_nT(N'') & \xrightarrow[\delta_n]{} & L_{n-1}T(N') & \longrightarrow & \cdots \end{array}$$

is commutative. State and prove a corresponding result for R^nT $(n \geq 0)$.

6. Show that $\mathrm{Ext}_R^n(M, N)$ and $\mathrm{Tor}_n^R(M, N)$ are finitely generated R-modules if M and N are finitely generated modules over a Noetherian ring.

7. Let M and N be R-modules and $a \in R$. Show that the scalar multiplication $\lambda_a : M \to M$ defined by $\lambda_a(x) = ax$, $x \in M$, induces

$$\lambda_a^* : \mathrm{Ext}_R^n(M, N) \to \mathrm{Ext}_R^n(M, N),$$

which is also scalar multiplication by a. State and prove a corresponding result for the scalar multiplication on the second variable and also for the Tor functor.

8. Show that an additive functor carries a split exact sequence to a split exact sequence.

7.3 Homological dimension

We give some applications of the Ext and Tor functors in the study of homological dimension of rings and modules. Homological dimension is closely related to the dimension of rings and modules to be studied in the next chapter.

Let M be an R-module. The projective dimension of M over R, denoted as $\mathrm{pd}_R\, M$, is defined as follows.

Definition 7.3.1 $\mathrm{pd}_R\, M = n$ if X has a projective resolution of length n, i.e.,

$$X_i = 0 \quad \text{for} \quad i > n, \quad X_n \neq 0,$$

and n is the least with this property. Otherwise $pd_R M$, is defined to be ∞.

Example 7.3.2 (i) If M is R-projective, then the sequence

$$0 \to X_0 \xrightarrow{I} M \to 0, \quad X_0 = M,$$

is a projective resolution of M of length 0. Hence $\mathrm{pd}_R\, M = 0$. The converse is also true, i.e., if $0 \to X_0 \to M \to 0$ is a projective resolution of M of length 0, then $M \simeq X_0$ is projective.

(ii) Let $R = \mathbb{Z}$. Since R is a PID, every projective R-module, being a submodule of a free module, is free. Let M be an abelian group which has non-zero torsion. Express M as the quotient of a free abelian group X_0 with kernel X_1. The sequence

$$0 \to X_1 \xrightarrow{i} X_0 \xrightarrow{\varepsilon} M \to 0$$

is exact, with X_0, X_1 free so that $\mathrm{pd}_R\, M \leq 1$. If $\mathrm{pd}_R\, M = 0$, then M is projective and hence free over $\mathbb{Z}$. Hence $\mathrm{pd}_R\, M = 1$ if M has torsion elements.

(iii) Let $R = \mathbb{Z}/4\mathbb{Z}$ and $M = \overline{2}(\mathbb{Z}/4\mathbb{Z})$. Consider the map

$$p : \mathbb{Z}/4\mathbb{Z} \to \mathbb{Z}/4\mathbb{Z}$$

given by $p(\alpha) = \overline{2}\alpha$, $\alpha \in \mathbb{Z}/4\mathbb{Z}$. Clearly, $\operatorname{Im} p = \overline{2}(\mathbb{Z}/4\mathbb{Z}) = \operatorname{Ker} p$. This gives an exact sequence

$$0 \to \overline{2}(\mathbb{Z}/4\mathbb{Z}) \xrightarrow{i} \mathbb{Z}/4\mathbb{Z} \xrightarrow{p} \overline{2}(\mathbb{Z}/4\mathbb{Z}) \to 0.$$

Consider the complex

$$\cdots \to \mathbb{Z}/4\mathbb{Z} \xrightarrow{\lambda} \mathbb{Z}/4\mathbb{Z} \xrightarrow{\lambda} \cdots \to \mathbb{Z}/4\mathbb{Z}$$

$$\xrightarrow{\lambda} \mathbb{Z}/4\mathbb{Z} \xrightarrow{\lambda} \mathbb{Z}/4\mathbb{Z} \xrightarrow{p} \overline{2}(\mathbb{Z}/4\mathbb{Z}) \to 0,$$

where $\lambda = ip$. Clearly, it is a projective resolution of $M = \overline{2}(\mathbb{Z}/4\mathbb{Z})$ over $R = \mathbb{Z}/4\mathbb{Z}$ of infinite length. It can be shown that $\operatorname{pd}_R M = \infty$.

The global dimension of a ring R, denoted by $\operatorname{gl.dim} R$, is defined as follows.

Definition 7.3.3 $\operatorname{gl.dim} R = \sup_M \operatorname{pd}_R M$, where the supremum is taken over all R-modules M.

Example 7.3.4 (i) If R is a field, then $\operatorname{pd}_R M = 0$ for all R-modules M, as every R-module is free. Hence $\operatorname{gl.dim} R = 0$.

(ii) If $R = \mathbb{Z}$, we have seen in Example 7.3.2 (ii) that $\operatorname{pd}_R M \leq 1$ for any R-module M, and $\operatorname{pd}_R M = 1$ if M has torsion elements. Hence $\operatorname{gl.dim} R = 1$. Similarly, if R is a PID which is not a field, then $\operatorname{gl.dim} R = 1$. In particular, $\operatorname{gl.dim} k[X] = 1$, k a field.

(iii) If $R = \mathbb{Z}/4\mathbb{Z}$ and $M = \overline{2}\mathbb{Z}/4\mathbb{Z}$, $\operatorname{pd}_R M = \infty$ and $\operatorname{gl.dim} \mathbb{Z}/4\mathbb{Z} = \infty$.

Proposition 7.3.5 *Let M be an R-module. Then M is R-projective if and only if $\operatorname{Ext}^1_R(M, N) = 0$ for all R-modules N.*

Proof If M is R-projective, clearly $\operatorname{Ext}^n_R(M, N) = 0$, $n \geq 1$ for all N (Theorem 7.2.5). Conversely, assume that $\operatorname{Ext}^1_R(M, N) = 0$ for all N. To show that M is R-projective, consider an exact sequence

$$0 \to N' \xrightarrow{\phi} N \xrightarrow{\psi} N'' \to 0$$

and an R-linear map $f : M \to N''$. It is sufficient to find some $g \in \operatorname{Hom}_R(M, N)$ with $\psi g = f$. Consider the exact sequence (Proposition 7.2.7)

$$0 \to \operatorname{Hom}_R(M, N') \xrightarrow{\overline{\phi}} \operatorname{Hom}_R(M, N)$$

$$\xrightarrow{\overline{\psi}} \operatorname{Hom}_R(M, N'') \xrightarrow{\delta} \operatorname{Ext}^1_R(M, N) = 0.$$

It follows that $\overline{\psi}$ is surjective, i.e., there exists some $g \in \operatorname{Hom}_R(M, N)$ with $\overline{\psi}(g) = \psi g = f$. □

Theorem 7.3.6 *Let M be an R-module. The following conditions are equivalent.*

(i) $\operatorname{pd}_R M \leq n$.

(ii) $\operatorname{Ext}^{n+i}_R(M, N) = 0$, $i \geq 1$ for all R-modules N.

(iii) $\operatorname{Ext}^{n+1}_R(M, N) = 0$ for all R-modules N.

(iv) If $0 \to K_{n-1} \to X_{n-1} \to X_{n-2} \to \cdots \to X_0 \to M \to 0$ is exact with X_i projective $(0 \leq i \leq n-1)$, then K_{n-1} is projective.

Proof $(i) \Rightarrow (ii)$. If $\operatorname{pd}_R(M) \leq n$, then M has a projective resolution X of length n and using this resolution to compute Ext, we have, for $i \geq 1$, $\operatorname{Ext}^{n+i}_R(M, N) = 0$ for all R-modules N, as $X_{n+i} = 0$.

$(ii) \Rightarrow (iii)$ is clear.

$(iii) \Rightarrow (iv)$. Assume that $\operatorname{Ext}^{n+1}_R(M, N) = 0$, for all R-modules N. Consider an exact sequence

$$0 \to K_{n-1} \xrightarrow{i_{n-1}} X_{n-1} \xrightarrow{d_{n-1}} X_{n-2} \to \cdots \to X_1 \xrightarrow{d_1} X_0 \xrightarrow{\varepsilon} M \to 0,$$

with X_i $(0 \leq i \leq n-1)$ projective. It is sufficient to show that K_{n-1} is projective, i.e., $\operatorname{Ext}^1_R(K_{n-1}, N) = 0$ for all N. We show that $\operatorname{Ext}^1_R(K_{n-1}, N) \simeq \operatorname{Ext}^{n+1}_R(M, N)$.

The given exact sequence splits into short exact sequences

$$0 \to K_0 \xrightarrow{i_0} X_0 \xrightarrow{\varepsilon} M \to 0$$

$$0 \to K_1 \xrightarrow{i_1} X_1 \xrightarrow{d_1} K_0 \to 0$$

$$\cdots\cdots\cdots\cdots\cdots\cdots\cdots\cdots\cdots\cdots$$

$$0 \to K_{n-1} \xrightarrow{i_{n-1}} X_{n-1} \xrightarrow{d_{n-1}} K_{n-2} \to 0,$$

where

$$K_0 = \operatorname{Ker} \varepsilon = \operatorname{Im} d_1, \quad \text{and} \quad K_i = \operatorname{Ker} d_i \ (1 \leq i \leq n-1).$$

The exactness of

$$0 \to K_0 \xrightarrow{i_0} X_0 \xrightarrow{\varepsilon} M \to 0$$

implies the exactness of

$$\to \operatorname{Ext}^n_R(X_0, N) \to \operatorname{Ext}^n_R(K_0, N) \to \operatorname{Ext}^{n+1}_R(M, N)$$

$$\to \operatorname{Ext}_R^{n+1}(X_0, N) \to \cdots$$

and the two terms on the extremes vanish as X_0 is projective $(n \geq 1)$. Hence $\operatorname{Ext}_R^n(K_0, N) \simeq \operatorname{Ext}_R^{n+1}(M, N)$ $(n \geq 1)$. Using the same argument with the second exact sequence

$$0 \to K_1 \to X_1 \to K_0 \to 0$$

we have $\operatorname{Ext}_R^n(K_0, N) \simeq \operatorname{Ext}_R^{n-1}(K_1, N)$. Continuing this process we have

$$\operatorname{Ext}_R^{n+1}(M, N) \simeq \operatorname{Ext}_R^n(K_0, N) \simeq \operatorname{Ext}_R^{n-1}(K_1, N) \simeq$$

$$\operatorname{Ext}_R^{n-2}(K_2, N) \simeq \cdots \simeq \operatorname{Ext}_R^1(K_{n-1}, N).$$

Hence $\operatorname{Ext}_R^{n+1}(M, N) = 0$ implies $\operatorname{Ext}_R^1(K_{n-1}, N) = 0$ for all R-modules N. Thus K_{n-1} is projective by Proposition 7.3.5.

$(iv) \Rightarrow (i)$. Construct a projective resolution of M as in Proposition 7.1.10 and stop at the $(n-1)$-th stage giving rise to an exact sequence

$$0 \to K_{n-1} \to X_{n-1} \to \cdots \to X_0 \to M \to 0$$

with X_i $(0 \leq i \leq n-1)$ projective. By assumption (iv), K_{n-1} is projective so that we have a projective resolution of M of length n. Hence $\operatorname{pd}_R M \leq n$. □

Corollary 7.3.7 $\operatorname{pd}_R M \leq n$ *if and only if* $\operatorname{Ext}_R^{n+1}(M, N) = 0$ *for all R-modules N.*

Corollary 7.3.8 $\operatorname{gl.dim} R \leq n$ *if and only if* $\operatorname{Ext}_R^{n+1}(M, N) = 0$ *for all R-modules M and N.*

Proof $\operatorname{gl.dim} R \leq n$ if and only if $\operatorname{pd}_R M \leq n$, for all M, i.e., $\operatorname{Ext}_R^{n+1}(M, N) = 0$ for all R-modules M and N. □

The Ext functor can also be developed by using injective resolutions instead of projective resolutions. The notion of an injective module is dual to the notion of a projective module.

Definition 7.3.9 An R-module N is injective if and only if for any exact sequence $0 \to M' \xrightarrow{f} M$ of R-modules the induced sequence

$$\operatorname{Hom}_R(M, N) \xrightarrow{\operatorname{Hom}_R(f, N)} \operatorname{Hom}_R(M', N) \to 0$$

is exact i.e., any R-linear map $M' \to N$ can be extended to an R-linear map $M \to N$.

Example 7.3.10 (i) If R is a field, any R-module N is injective as any R-linear map of a subspace into N can be extended to the whole space.

(ii) $\mathbb{Z}$ is not $\mathbb{Z}$-injective. Consider the homomorphism $0 \to \mathbb{Z} \xrightarrow{f} \mathbb{Q}$. It induces f^* : $\operatorname{Hom}_{\mathbb{Z}}(\mathbb{Q}, \mathbb{Z}) \to \operatorname{Hom}_{\mathbb{Z}}(\mathbb{Z}, \mathbb{Z})$. If f^* is surjective, $\operatorname{Id}_{\mathbb{Z}}$ has a preimage under f^*, i.e., $\mathbb{Z}$ is a direct summand of $\mathbb{Q}$, a contradiction.

We will show that an abelian group G is an injective $\mathbb{Z}$-module if and only if it is divisible, i.e., for every $x \in G, n \in \mathbb{Z}$, $n \neq 0$, there exists some $y \in G$ with $ny = x$. More generally, we have the following:

Theorem 7.3.11 *An R-module N is injective if and only if for any ideal I of R, then any R-homomorphism $f : I \to N$ can be extended to an R-homomorphism $f^* : R \to N$.*

Proof Clearly, if N is injective, then any homomorphism $f : I \to N$ can be extended to $f^* : R \to N$.

Conversely, assume this condition for all ideals I of R and consider a submodule M' of M and an R-homomorphism $f : M' \to N$. We construct a maximal extension for f and show that its domain is the whole of M. Consider $\sum = \{(M_\alpha, f_\alpha)\}$, where M_α is a submodule of M containing M' and f_α is an extension of f. Define $(M_\alpha, f_\alpha) \leq (M_\beta, f_\beta)$ if and only if M_α is a submodule of M_β and $f_\alpha = f_\beta \mid_{M_\alpha}$. By Zorn's lemma, $\sum$ has a maximal element (M^*, f^*). We claim that $M = M^*$. If not, choose $z \in M$, $z \notin M^*$ and consider

$$I = \{a \in R \mid az \in M^*\}.$$

I is an ideal of R and the map $h : I \to N$, given by $h(a) = f^*(az)$, is R-linear. By assumption, h has an extension $h^* : R \to N$. Extend now f^* to $f_1 : M^* + Rz = M_1 \to N$, by defining

$$f_1(x + bz) = f^*(x) + h^*(b), \quad x \in M, \quad b \in R.$$

Clearly, f_1 is a well defined R-linear map and

$$(M^*, f^*) \leq (M_1, f_1), \quad M^* \subsetneq M_1.$$

This contradicts the maximality of (M^*, f^*). □

Corollary 7.3.12 *An abelian group G is an injective $\mathbb{Z}$-module if and only if it is divisible.*

Proof Let G be $\mathbb{Z}$-injective, $x \in G$, and $n \in \mathbb{Z}$, $n \neq 0$. Consider $I = n\mathbb{Z}$ and $f : I \to G$, defined by $f(\lambda n) = \lambda x$, $\lambda \in \mathbb{Z}$. Then f has an extension $f^* : \mathbb{Z} \to G$. If $y = f(1)$, then $ny = f^*(n) = f(n) = x$. Thus G is divisible.

Conversely, assume that G is divisible. Let $I = n\mathbb{Z}$ be an ideal of $\mathbb{Z}$ and $f : I \to G$ any $\mathbb{Z}$-homomorphism. If $f(n) = x$, then there exists some $y \in G$ with $ny = x$. Define $f^* : \mathbb{Z} \to G$ by $f(\lambda) = \lambda y$, $\lambda \in \mathbb{Z}$. Then $f^*(n) = nf^*(1) = ny = x = f(n)$, so that f^* is an extension of f. Hence G is $\mathbb{Z}$-injective. □

Proposition 7.3.13 *Any R-module M can be embedded in an injective R-module.*

Proof Consider first the case where $R = \mathbb{Z}$ and M a free abelian group. Now $\mathbb{Z} \subset \mathbb{Q}$, the additive group of rationals. $\mathbb{Q}$ is $\mathbb{Z}$-injective as it is divisible. Hence M, being a direct sum of copies of $\mathbb{Z}$, can be embedded in a direct sum of copies of $\mathbb{Q}$, which is divisible, i.e., injective. Now, any abelian group M is isomorphic to F/K, with F free. Choose an injective $\mathbb{Z}$-module N with $F \subset N$ so that $M \simeq F/K \subset N/K$ which is injective as the quotient of a divisible group is divisible.

We now consider the general case. Let M be an R-module and consider $M^* = \operatorname{Hom}_{\mathbb{Z}}(M, \mathbb{Q}/\mathbb{Z})$. Then M^* is an R-module for the operation $(af)(x) = f(ax)$, $a \in R, x \in M, f \in M^*$. The natural map $i_M : M \to M^{**}$, given by $i_M(x)(f) = f(x)$, $x \in M, f \in M^*$ is R-linear. We claim that i_M is a one-one map. If $x \neq 0$, there exists a $\mathbb{Z}$-homomorphism $h : Rx \to \mathbb{Q}/\mathbb{Z}$, with $h(x) \neq 0$, by assigning any non-zero value to x if the order of x is infinite or the value $1/n + \mathbb{Z}$, if the order of x is n. Since $\mathbb{Q}/\mathbb{Z}$ is $\mathbb{Z}$-injective, h has an extension $h^* : M \to \mathbb{Q}/\mathbb{Z}$, an abelian group homomorphism, with $h^*(x) = h(x) \neq 0$. Hence

$$i_M(x)(h^*) \neq 0,$$

i.e., i_M is one-one. Express M^* as a quotient of a free R-module F, i.e., $\theta : F \to M^*$ is a surjective R-linear map. Then $\theta^* : M^{**} \to F^*$ is a one-one R-homomorphism and hence M is isomorphic to an R-submodule of F^*. We will show that F^* is R-injective. More generally we will show that P^* is injective for any projective R-module P. Consider the diagram

$$\begin{array}{ccccc} 0 & \longrightarrow & M' & \xrightarrow{\ i\ } & M \\ & & \downarrow{\scriptstyle g} & & \\ & & P^* & & \end{array}$$

where g is R-linear. Since $\mathbb{Q}/\mathbb{Z}$ is $\mathbb{Z}$-injective, $i^* : M^* \to M'^*$ is surjective. Consider the diagram

$$\begin{array}{ccccc} M^* & \xrightarrow{\ i^*\ } & M'^* & \longrightarrow & 0 \\ & & \uparrow{\scriptstyle g^*} & & \\ P & \xrightarrow{\ i_P\ } & P^{**} & & \end{array}$$

Since P is R-projective, there exists an R-linear map $h : P \to M^*$ with $i^*h = g^*i_P$. If $f = h^*i_M : M \to P^*$, then f is an extension of g. Hence P^* is injective. □

Corollary 7.3.14 *Any R-module M has an injective resolution, i.e., there exists an exact sequence*

$$0 \to M \to Q_0 \xrightarrow{d_0} Q_1 \to \cdots \to Q_n \xrightarrow{d_n} \cdots$$

of R-modules with Q_i injective for all $i \geq 0$.

Proof Embed M in an injective module Q_0, so that we have an exact sequence of R-modules

$$0 \to M \to Q_0 \to Q_0/M \to 0.$$

Embed now Q_0/M in an injective module Q_1 and continue the process as in Proposition 7.1.10 to yield an injective resolution for M. □

Proposition 7.3.15 *The following conditions are equivalent for an R-module N:*

(*i*) *N is injective.*

(*ii*) $\mathrm{Ext}^1_R(M, N) = 0$ *for all R-modules M.*

(*iii*) $\mathrm{Ext}^1_R(R/I, N) = 0$ *for every ideal I of R.*

Proof $(i) \Rightarrow (ii)$. Let M be an R-module. Express M as a quotient of a free R-module F with kernel K, so that we have an exact sequence of R-modules

$$0 \to K \xrightarrow{j} F \xrightarrow{p} M \to 0.$$

By Proposition 7.2.7 we have an exact sequence

$$\to \mathrm{Hom}_R(F, N) \xrightarrow{j^*} \mathrm{Hom}_R(K, N) \xrightarrow{\partial} \mathrm{Ext}^1_R(M, N) \xrightarrow{\overline{p}} \mathrm{Ext}^1_R(F, N) = 0$$

as F is free. Since N is injective, j^* is surjective and hence $\partial = 0$. Now the condition $\mathrm{Ext}^1_R(M, N) = 0$ follows, as $\mathrm{Ker}\ \overline{p} = \mathrm{Im}\ \partial = 0$.

$(ii) \Rightarrow (iii)$ is trivial.

$(iii) \Rightarrow (i)$. Let I be an ideal of R. The exact sequence

$$0 \to I \to R \to R/I \to 0$$

gives an exact sequence

$$\to \mathrm{Hom}_R(R, N) \xrightarrow{i^*} \mathrm{Hom}_R(I, N) \to \mathrm{Ext}^1_R(R/I, N) = 0.$$

By Theorem 7.3.11, i^* is surjective. Hence N is injective. □

Lemma 7.3.16 *If N is an injective R-module, then $\mathrm{Ext}^n_R(M, N) = 0$ for all $n \geq 1$ and for all R-modules M.*

Proof We prove the result by induction on n. This is true for $n = 1$ by Proposition 7.3.15. Assume the result for $n - 1$ $(n \geq 2)$ and consider an exact sequence

$$0 \to K \to F \to M \to 0$$

with F free. This induces an exact sequence

$$\operatorname{Ext}_R^{n-1}(K, N) \to \operatorname{Ext}_R^n(M, N) \to \operatorname{Ext}_R^n(F, N) = 0.$$

By induction, $\operatorname{Ext}_R^{n-1}(K, N) = 0$ and hence $\operatorname{Ext}_R^n(M, N) = 0$. □

Proposition 7.3.17 *Let* $0 \to M' \to Q \to M'' \to 0$ *be an exact sequence of* R*-modules with* Q *injective. Then*

$$\operatorname{Ext}_R^{n+1}(M, M') \simeq \operatorname{Ext}_R^n(M, M'') \ (n \geq 1)$$

for all R*-modules* M.

Proof The given exact sequence

$$0 \to M' \to Q \to M'' \to 0$$

induces an exact sequence

$$\to \operatorname{Ext}_R^n(M, Q) \to \operatorname{Ext}_R^n(M, M'') \to \operatorname{Ext}_R^{n+1}(M, M') \to \operatorname{Ext}_R^{n+1}(M, Q) \to .$$

Since, for $n \geq 1$, the terms at the extremes are zero, by Lemma 7.3.16, the middle terms are isomorphic for $n \geq 1$. □

We now define the injective dimension of an R-module M.

Definition 7.3.18 Let N be an R-module. The injective dimension of N over R, denoted by $id_R N$, is n if N has an injective resolution of length n and n is the least integer with this property. Otherwise, $id_R(M) = \infty$.

Theorem 7.3.19 *Let* N *be an* R*-module. The following conditions are equivalent.*

(i) $id_R(N) \leq n$.

(ii) $\operatorname{Ext}_R^{n+i}(M, N) = 0, i \geq 1$ *for all* R*-modules* M.

(iii) $\operatorname{Ext}_R^{n+1}(M, N) = 0$ *for all* R*-modules* M.

(iv) *If* $0 \to N \to Q_0 \to Q_1 \to \cdots \to Q_{n-1} \to T_{n-1} \to 0$ *is exact with* Q_i $(0 \leq i \leq n-1)$ *injective, then* T_{n-1} *is injective.*

Proof $(i) \Rightarrow (ii)$. Let

$$0 \to N \to Q_0 \xrightarrow{d_0} \cdots \longrightarrow Q_{n-1} \xrightarrow{d_{n-1}} Q_n \to 0$$

be an injective resolution of N of length n. It splits into short exact sequences

$$0 \to N \to Q_0 \xrightarrow{d_0} K_0 \to 0$$

$$0 \to K_0 \to Q_1 \xrightarrow{d_1} K_1 \to 0$$

$$\cdots\cdots\cdots\cdots\cdots\cdots\cdots\cdots$$

$$0 \to K_{n-2} \to Q_{n-1} \xrightarrow{d_{n-1}} Q_n \to 0$$

where $K_i = \operatorname{Im} d_i \ \ (0 \leq i \leq n-2)$. Repeated use of Proposition 7.3.17 gives

$$\operatorname{Ext}_R^{n+i}(M, N) \simeq \operatorname{Ext}_R^{n+i-1}(M, K_0) \simeq \operatorname{Ext}_R^{n+i-2}(M, K_1)$$

$$\cdots \simeq \operatorname{Ext}_R^{i+1}(M, K_{n-2}) \simeq \operatorname{Ext}_R^{i}(M, Q_n) = 0 \quad (i \geq 1)$$

by Lemma 7.3.16 as Q_n is injective.

$(ii) \Rightarrow (iii)$ is clear.

$(iii) \Rightarrow (iv)$. The given exact sequence induces an isomorphism, as in (ii) above, viz.,

$$\operatorname{Ext}_R^{n+1}(M, N) \simeq \operatorname{Ext}_R^{1}(M, T_{n-1})$$

for all R-modules M. By assumption (iii) we have $\operatorname{Ext}_R^{n+1}(M, N) = 0$ for all R-modules M. This implies $\operatorname{Ext}_R^{1}(M, T_{n-1}) = 0$ for all R-modules M. Hence by Proposition 7.3.15, T_{n-1} is injective.

$(iv) \Rightarrow (i)$. Construct an injective resolution of N and stop at the n-th stage yielding an exact sequence

$$0 \to N \to Q_0 \to Q_1 \to \cdots \to Q_{n-1} \to T_{n-1} \to 0$$

with $Q_i \ \ (0 \leq i \leq n-1)$ injective. By (iv), T_{n-1} is injective and hence $id_R N \leq n$. □

Corollary 7.3.20 $id_R N \leq n \Leftrightarrow \operatorname{Ext}_R^{n+1}(M, N) = 0$ *for all R-modules M.*

Corollary 7.3.21 $\operatorname{gl.dim} R = \sup_N id_R(N)$ *over all R-modules N.*

Proof $\operatorname{gl.dim} R \leq n \Leftrightarrow \operatorname{pd}_R(M) \leq n$ for all R-modules $M \Leftrightarrow \operatorname{Ext}_R^{n+1}(M, N) = 0$ for all R-modules M, N, $\Leftrightarrow id_R(N) \leq n$ for all R-modules N. □

Corollary 7.3.22 $\operatorname{gl.dim} R = \sup_I \operatorname{pd}_R R/I$, *where the supremum is taken over all ideals I of R.*

Proof If $\sup_I \operatorname{pd}_R R/I = \infty$, then $\operatorname{gl.dim} R = \infty$. Assume therefore that $\sup_I \operatorname{pd}_R R/I \leq n$. It is sufficient to show, in view of Corollary 7.3.21, that $id_R(N) \leq n$ for any R-module N. Consider an exact sequence

$$0 \to N \to Q_0 \to \cdots \to Q_{n-1} \to T_{n-1} \to 0$$

with Q_i injective $(0 \leq i \leq n-1)$. Now, as in Theorem 7.3.19, repeated use of Proposition 7.3.17 gives

$$\mathrm{Ext}^1_R(R/I, T_{n-1}) \simeq \mathrm{Ext}^{n+1}_R(R/I, N) = 0$$

as $\mathrm{pd}_R\, R/I \leq n$. This implies that T_{n-1} is injective by Proposition 7.3.15. Hence $id_R N \leq n$. □

We now study $\mathrm{gl.dim}\, R$ when R is a local ring and show that the Tor functor can be used to compute the global dimension of local rings.

Proposition 7.3.23 *For R-modules M and N, there exists an isomorphism* $\mathrm{Tor}^R_n(M, N) \simeq \mathrm{Tor}^R_n(N, M)$, *which is functorial in M and N.*

Proof We first observe that if P is projective, then $\mathrm{Tor}^R_n(P, N) = 0$ and $\mathrm{Tor}^R_n(N, P) = 0$ for all R-modules N and $n \geq 1$. The first relation follows by computing Tor using the resolution

$$0 \to X_0 \xrightarrow{I} P \to 0, \quad X_0 = P.$$

To prove the second relation, consider a projective resolution X of N

$$X : \cdots \to X_n \xrightarrow{d_n} \cdots \to X_1 \xrightarrow{d_1} X_0 \xrightarrow{\varepsilon} N \to 0.$$

Since P is projective, the complex

$$X \otimes_R P : \cdots \to X_n \otimes_R P \to \cdots \to X_0 \otimes_R P \to N \otimes_R P \to 0$$

is exact. Hence $\mathrm{Tor}^R_n(N, P) = H_n(X \otimes_R P) = 0$, $n \geq 1$.

We now prove the proposition by induction on n. For $n = 0$, the result is true as

$$M \otimes_R N \simeq N \otimes_R M.$$

Let $n \geq 1$ and assume the result for $(n-1)$. Express M as the quotient of a free module F with kernel K so that we have an exact sequence

$$0 \to K \to F \to M \to 0.$$

This gives rise to the following exact sequences

$$0 = \mathrm{Tor}^R_n(F, N) \to \mathrm{Tor}^R_n(M, N) \xrightarrow{\partial_n} \mathrm{Tor}^R_{n-1}(K, N) \xrightarrow{i_{n-1}} \mathrm{Tor}^R_{n-1}(F, N)$$

$$0 = \mathrm{Tor}^R_n(N, F) \to \mathrm{Tor}^R_n(N, M) \xrightarrow{\overline{\partial_n}} \mathrm{Tor}^R_{n-1}(N, K) \xrightarrow{\overline{i_{n-1}}} \mathrm{Tor}^R_{n-1}(N, F).$$

By induction, there exist isomorphisms α and β making the following diagram commutative:

$$\begin{array}{ccc} \mathrm{Tor}_{n-1}^R(K,N) & \xrightarrow{i_{n-1}} & \mathrm{Tor}_{n-1}^R(F,N) \\ \downarrow{\scriptstyle\alpha} & & \downarrow{\scriptstyle\beta} \\ \mathrm{Tor}_{n-1}^R(N,K) & \xrightarrow{\overline{i_{n-1}}} & \mathrm{Tor}_{n-1}^R(N,F) \end{array}$$

β induces therefore an isomorphism $\gamma : \mathrm{Im}\ \partial_n \to \mathrm{Im}\ \overline{\partial}_n$, i.e.,

$$\gamma : \mathrm{Tor}_n^R(M,N) \to \mathrm{Tor}_n^R(N,M)$$

is an isomorphism. □

Proposition 7.3.24 *An R-module M is flat if and only if*

$$\mathrm{Tor}_1^R(M,N) = 0$$

for all R-modules N.

Proof Assume $\mathrm{Tor}_1^R(M,N) = 0$ for all R-modules N and consider an exact sequence

$$0 \to N' \xrightarrow{\phi} N \xrightarrow{\psi} N'' \to 0.$$

By Proposition 7.2.6, this gives an exact sequence

$$0 = \mathrm{Tor}_1^R(M,N'') \to M \otimes N' \xrightarrow{\overline{\phi}} M \otimes N \xrightarrow{\overline{\psi}} M \otimes N'' \to 0.$$

Hence $\mathrm{Ker}\,\overline{\phi} = 0$, and M is R-flat.

Conversely, assume that M is R-flat, and let N be any R-module. Consider an exact sequence

$$0 \to K \xrightarrow{\phi} F \to N \to 0$$

with F free. This gives an exact sequence

$$0 = \mathrm{Tor}_1^R(M,F) \to \mathrm{Tor}_1^R(M,N) \to$$

$$K \otimes_R N \xrightarrow{\overline{\phi}} F \otimes_R N \xrightarrow{\overline{\psi}} M \otimes_R N \to 0.$$

Since N is flat, $\mathrm{Ker}\,\overline{\phi} = 0$. Hence

$$\mathrm{Tor}_1^R(M,N) \simeq \mathrm{Im}\,\mathrm{Tor}_1^R(M,N) = \mathrm{Ker}\,\overline{\phi} = 0. \qquad \square$$

Corollary 7.3.25 *Let R be a Noetherian local ring and M a finitely generated R-module. Then M is free if and only if $\mathrm{Tor}_1^R(M,N) = 0$ for all R-modules N.*

Proof Follows from Corollary 2.2.17. □

Theorem 7.3.26 *Let R be a Noetherian local ring with maximal ideal $\mathfrak{m}$ and $k = R/\mathfrak{m}$. If M is a finitely generated R-module, then M is free if and only if $\operatorname{Tor}_1^R(M,k) = 0$.*

Proof If M is free, $\operatorname{Tor}_1^R(M,k) = 0$ by Proposition 7.3.24. Conversely, assume that $\operatorname{Tor}_1^R(M,k) = 0$. Let $\{x_1, \dots, x_n\}$ be a minimal generating set of M and let $u : F \to M$ be the surjective R-linear map given by $u(e_i) = x_i$, where F is free with basis $\{e_i\}(1 \leq i \leq n)$.

Let $K = \operatorname{Ker} u$. We have an exact sequence

$$0 \to K \to F \xrightarrow{u} M \to 0.$$

This gives an exact sequence

$$0 = \operatorname{Tor}_1^R(M,k) \to K \otimes_R k \to F \otimes_R k \xrightarrow{\overline{u}} M \otimes_R k \to 0.$$

Since $\{x_1, \dots, x_n\}$ is a minimal generating set for M, $\overline{u}$ is an isomorphism. Hence $\operatorname{Ker} \overline{u} \simeq K \otimes_R k = 0$, i.e., $K/\mathfrak{m}K = 0$. Since R is Noetherian and K is finitely generated, by Nakayama Lemma, $K = 0$. Thus M is free. □

Proposition 7.3.27 *Let R be a Noetherian local ring with maximal ideal $\mathfrak{m}$ and $k = R/\mathfrak{m}$. Let M be a finitely generated R-module. Then $\operatorname{pd}_R M \leq n$ if and only if $\operatorname{Tor}_{n+1}^R(M,N) = 0$ for all R-modules N.*

Proof If $\operatorname{pd}_R M \leq n$, then M has a projective resolution of length at most n which can be used to compute $\operatorname{Tor}_{n+1}^R(M,N)$. This implies $\operatorname{Tor}_{n+1}^R(M,N) = 0$.

Conversely, assume that $\operatorname{Tor}_{n+1}^R(M,N) = 0$ for all R-modules N and consider an exact sequence

$$0 \to K_{n-1} \to X_{n-1} \to \cdots \to X_0 \to M \to 0 \tag{7.2}$$

with $X_i \ (0 \leq i \leq n-1)$ projective. By Corollary 7.3.25, it is sufficient to show that K_{n-1} is projective, i.e., $\operatorname{Tor}_1^R(K_{n-1},N) = 0$. The exact sequence 7.2 splits into short exact sequences. Using the argument as in Theorem 7.3.6 on the Tor functor instead of the Ext functor, we have

$$\operatorname{Tor}_1^R(K_{n-1},N) \simeq \operatorname{Tor}_{n+1}^R(M,N) = 0.$$

□

Corollary 7.3.28 *With the same assumptions as above,*

$$\operatorname{pd}_R M \leq n \text{ if and only if } \operatorname{Tor}_{n+1}^R(M,k) = 0.$$

Proof If $\operatorname{pd}_R M \leq n$, then $\operatorname{Tor}_{n+1}^R(M,k) = 0$. Conversely, if $\operatorname{Tor}_{n+1}^R(M,k) = 0$, then

$$\operatorname{Tor}_1^R(K_{n-1};k) \simeq \operatorname{Tor}_{n+1}^R(M,k) = 0$$

and this implies by Theorem 7.3.26 that K_{n-1} is projective. Hence $\mathrm{pd}_R M \leq n$. □

Theorem 7.3.29 *Let R be a Noetherian local ring with maximal ideal $\mathfrak{m}$ and $k = R/\mathfrak{m}$. The following conditions are equivalent.*
(i) $\mathrm{gl.dim}\, R \leq n$.
(ii) $\mathrm{Tor}_{n+1}^R(M, N) = 0$ *for all R-modules M, N.*
(iii) $\mathrm{Tor}_{n+1}^R(k, k) = 0$.

Proof $(i) \Rightarrow (ii)$. If $\mathrm{gl.dim}\, R \leq n$, then $\mathrm{pd}_R M \leq n$ for any R-module M and this implies (ii) as in the proof Proposition 7.3.27.

$(ii) \Rightarrow (iii)$ is clear.

$(iii) \Rightarrow (i)$ Assume that $\mathrm{Tor}_{n+1}^R(k, k) = 0$. This implies by Corollary 7.3.28 that $\mathrm{pd}_R k \leq n$. Hence $\mathrm{Tor}_{n+1}^R(k, M) = 0$ for all R-modules M. By Proposition 7.3.23, $\mathrm{Tor}_{n+1}^R(M, k) \simeq \mathrm{Tor}_{n+1}^R(k, M) = 0$.

Hence by Corollary 7.3.28, $\mathrm{pd}_R M \leq n$, i.e., $\mathrm{gl.dim}\, R \leq n$. □

Corollary 7.3.30 *Let R be a Noetherian local ring with maximal ideal $\mathfrak{m}$ and $k = R/\mathfrak{m}$. Then* $\mathrm{gl.dim}\, R = \mathrm{pd}_R k$.

Proof If $\mathrm{pd}_R k = \infty$, then $\mathrm{Tor}_{n+1}^R(k, k) \neq 0$ for each n and $\mathrm{gl.dim}\, R = \infty$. If $\mathrm{pd}_R k < \infty$, then $\mathrm{pd}_R k \leq n \Leftrightarrow \mathrm{Tor}_{n+1}^R(k, k) = 0 \Leftrightarrow \mathrm{gl.dim}\, R \leq n$. □

Remark 7.3.31 If $\mathrm{gl.dim}\, R = n$, then k has a projective resolution of length n. An explicit resolution of k over R is given by the Koszul complex which will be described in the next chapter.

As an application of the Tor functor, we shall now derive local criterion for flatness. We first prove a proposition.

Proposition 7.3.32 *Let $f : R \to S$ be a homomorphism of rings and M an R-module. The following conditions are equivalent.*
(i) $\mathrm{Tor}_1^R(M, N) = 0$ *for any S-module N.*
(ii) $\mathrm{Tor}_1^R(M, S) = 0$ *and $M \otimes S$ is a flat S-module.*

Proof $(i) \Rightarrow (ii)$. If $N = S$ in (i), we have $\mathrm{Tor}_1^R(M, S) = 0$. If

$$0 \to N' \to N \to N'' \to 0$$

is an exact sequence of S-modules, we have an exact sequence

$$0 = \mathrm{Tor}_1^R(M, N'') \to M \otimes_R N' \to M \otimes_R N \to M \otimes_R N'' \to 0$$

i.e.,

$$0 \to M \otimes_R N' \to M \otimes_R N$$

is exact. Since $M \otimes_R S \otimes_S N \simeq M \otimes_R N$, this implies $M \otimes_R S$ is S-flat.

$(ii) \Rightarrow (i)$. Since $\mathrm{Tor}_1^R(M, S) = 0$, $\mathrm{Tor}_1^R(M, F) = 0$ for any free S-module F. Express N as a quotient of a free S-module so that we have an exact sequence

$$0 \to K \xrightarrow{i} F \to N \to 0.$$

This induces an exact sequence

$$0 = \mathrm{Tor}_1^R(M, F) \to \mathrm{Tor}_1^R(M, N) \to$$

$$M \otimes_R K \xrightarrow{i^*} M \otimes_R F \to M \otimes_R N \to 0.$$

Since $M \otimes_R S$ is S-flat, $M \otimes_R S \otimes_S K \to M \otimes_R S \otimes_S F$ is injective, i.e., i^* is injective. This implies $\mathrm{Tor}_1^R(M, N) = 0$. □

Theorem 7.3.33 (Local criterion for flatness) *Let $f : R \to S$ be a homomorphism of Noetherian rings, I an ideal of R such that its extension I^e under f is contained in the Jacobson radical of S. Let M be a finitely generated S-module. The following conditions are equivalent.*

(i) *M is R-flat.*

(ii) *$M \otimes_R R/I$ is R/I-flat and $\mathrm{Tor}_1^R(M, R/I) = 0$.*

(iii) *$\mathrm{Tor}_1^R(M, N) = 0$ for all R-modules N annihilated by I.*

(iv) *$\mathrm{Tor}_1^R(M, N) = 0$ for all R-modules N annihilated by I^s for some $s \geq 1$.*

(v) *$M \otimes_R R/I^s$ is R/I^s flat for all $s \geq 1$.*

Proof $(i) \Rightarrow (ii)$ follows from Proposition 1.4.7 and Proposition 7.3.24.

$(ii) \Rightarrow (iii)$. Take $S = R/I$ in Proposition 7.3.32.

$(iii) \Rightarrow (iv)$. We prove this by induction on s. For $s = 1$, it is true by (iii). Assume (iv) for all R-modules L annihilated by I^k $(1 \leq k \leq s-1)$ and let N be annihilated by I^s. The exact sequence

$$0 \to IN \to N \to N/IN \to 0$$

induces an exact sequence

$$\mathrm{Tor}_1^R(M, IN) \to \mathrm{Tor}_1^R(M, N) \to \mathrm{Tor}_1^R(M, N/IN).$$

The terms at the extremes are zero by induction and hence $\mathrm{Tor}_1^R(M, N) = 0$.

$(iv) \Rightarrow (v)$. Take $S = R/I^s$ in Proposition 7.3.32.

$(v) \Rightarrow (i)$. Let $0 \to N' \to N$ be an exact sequence of R-modules with N' and N finitely generated. It is sufficient to show that

$$0 \to N' \otimes_R M \to N \otimes_R M \text{ is exact.}$$

The given sequence induces an exact sequence

$$0 \to \frac{N'}{N' \cap I^k N} \to \frac{N}{I^k N} \text{ for all } k.$$

Since $M \otimes_R R/I^k$ is R/I^k-flat for all k, the sequence

$$0 \to \frac{N'}{N' \cap I^k N} \otimes_{\frac{R}{I^k}} M \otimes_R \frac{R}{I^k} \longrightarrow \frac{N}{I^k N} \otimes_{\frac{R}{I^k}} M \otimes_R \frac{R}{I^k}$$

is exact, i.e.,

$$0 \to \frac{N'}{N' \cap I^k N} \otimes_R M \to \frac{N}{I^k N} \otimes_R M \tag{7.3}$$

is exact for all k. By Artin Rees Lemma, there exists an integer m such that

$$I^{k-m}(N' \cap I^m N) = N' \cap I^k N \text{ for all } k \geq m.$$

Let P_m denote the image of $(N' \cap I^m N) \otimes_R M \to N' \otimes_R M$. Then $\frac{N'}{N' \cap I^k N} \otimes_R M$ can be identified with the quotient of $N' \otimes_R M$ by P_k. Hence, the sequence (7.3) becomes the exact sequence

$$0 \longrightarrow \frac{N' \otimes M}{I^{k-m} P_m} \to \frac{N \otimes M}{I^k (N \otimes M)} \quad (k \geq m). \tag{7.4}$$

Since $I^{k-m}(N' \cap I^m N) = N' \cap I^k N$ for all $k \geq m$, $\{I^{k-m} P_m\}$ and $I^k(N' \otimes_R M)$ induce the same topology on $N' \otimes_R M$.

Taking I^e-adic completions, we have the exact sequence

$$0 \to \widehat{N' \otimes M} \to \widehat{N \otimes M}.$$

Since $I^e \subset J(S)$, the natural maps $N' \otimes M \to \widehat{N' \otimes M}$ and $N \otimes M \to \widehat{N \otimes M}$ are injective. Hence $0 \to N' \otimes_R M \to N \otimes_R M$ is exact. □

Corollary 7.3.34 *Let $f : R \to S$ be a local homomorphism of Noetherian local rings R and S with maximal ideals $\mathfrak{m}$ and $\mathfrak{n}$ respectively, i.e., f satisfies $f(\mathfrak{m}) \subseteq \mathfrak{n}$. Let M be a finitely generated S-module and $t \in R$ a non-zero-divisor. Then M is R-flat if and only if M/tM is $R/(t)$-flat and t is a non-zero-divisor of M.*

Proof Consider R/I where $I = (t)$. Consider the exact sequence

$$0 \to R \xrightarrow{\lambda} R \to R/I \to 0,$$

where λ is multiplication by t. This gives an exact sequence

$$0 = \operatorname{Tor}_1^R(M, R) \to \operatorname{Tor}_1^R(M, R/I) \to M \otimes R \xrightarrow{\lambda^*} M \otimes R.$$

Hence $\operatorname{Tor}_1^R(M, R/I) = 0$ if and only if $\operatorname{Ker} \lambda^* = 0$, i.e., t is a non-zero-divisor of M. The result now follows by the equivalence of (i) and (ii). □

7.3. EXERCISES

1. Show that if M is a finitely generated module over a Noetherian ring R and if $\operatorname{pd}_R M = n$, then $\operatorname{Ext}_R^n(M, R) \neq 0$.

2. Let $0 \to M' \to M \to M'' \to 0$ be an exact sequence of R-modules. Show that $\operatorname{pd}_R M \leq \max\{\operatorname{pd}_R M', \operatorname{pd}_R M''\}$.

3. Show that a direct product $\prod_\alpha N_\alpha = N$ of R-modules is injective if and only if each N_α is injective.

4. Two R-modules M and N are said to be projectively equivalent if $M \oplus P \simeq N \oplus Q$ for some projective R-modules P and Q. Show that for two such modules, $\operatorname{pd}_R M = \operatorname{pd}_R N$.

   Illustrate by an example that the converse is not true.

5. Let M be a finitely generated module over a Noetherian ring R. Show that $\operatorname{pd}_R M = \sup_P \operatorname{pd}_{R_P} M_P$, where the supremum is taken over all prime ideals P of R and can be restricted to all maximal ideals $\mathfrak{m}$ of R. Deduce that $\operatorname{gl.dim} R = \sup \operatorname{gl.dim} R_P = \sup_{\mathfrak{m}} \operatorname{gl.dim} R_{\mathfrak{m}}$.

6. Let M be a finitely generated module over a Noetherian ring R with Jacobson radical J. If M/JM is a projective R/J module and $\operatorname{Tor}_1^R(M, R/J) = 0$, show that M is R-projective.

7. Let $f : R \to S$ be a ring homomorphism such that S is R-flat. If M and N are R-modules, show that

$$\operatorname{Tor}_n^R(M, N) \otimes_R S \simeq \operatorname{Tor}_n^S(M \otimes_R S, N \otimes_R S) \ (n \geq 0).$$

8. Let R be a local ring with maximal ideal $\mathfrak{m}$. Show that

$$\operatorname{Tor}_1^R(k, k) \simeq \mathfrak{m}/\mathfrak{m}^2.$$

9. Show that $R = \dfrac{k[[X]]}{(X^2)}$ is a local ring with $\mathrm{gl.dim}\, R = \infty$.

10. Let $0 \to M' \to M \to M'' \to 0$ be an exact sequence of R-modules with M projective and M'' not projective. Show that $\mathrm{pd}_R\, M'' = 1 + \mathrm{pd}_R\, M'$.

11. Let R be a ring, $a \in R$ a non-unit and a non-zero-divisor and $R^* = R/(a)$. Show that if M is a non-zero R^*-module with $\mathrm{pd}_{R^*}\, M < \infty$, then $\mathrm{pd}_R\, M = \mathrm{pd}_{R^*}\, M + 1$.

12. Let $0 \to M' \to M \to M'' \to 0$ be an exact sequence of R-modules. Show that

    $(i)\ id_R M > id_R M'' \Rightarrow id_R M' = id_R M,$

    $(ii)\ id_R M = id_R M'' \Rightarrow id_R M' \leq id_R M + 1,$

    $(iii)\ id_R M < id_R M'' \Rightarrow id_R M' = id_R M'' + 1.$

13. For R-modules M, N and K, consider the adjoint isomorphism (Exercise 6, Section 1.3)
$$\phi : \mathrm{Hom}_R(M, \mathrm{Hom}_R(N, K)) \simeq \mathrm{Hom}_R(M \otimes_R N, K).$$
By replacing M by a projective resolution of M and passing to homology, show that
$$\mathrm{Ext}^i_R(M, \mathrm{Hom}_R(N, K)) \simeq \mathrm{Hom}_R(\mathrm{Tor}_i^R(M, N), K),$$
$i \geq 0$, when K is R-injective.

CHAPTER
8

DIMENSION

Dimension of a variety is a basic numerical invariant. Its algebraic analogue leads to the notion of dimension for rings and modules. There are three different formulations of dimension. We outline all of them and prove their equivalence for Noetherian modules over a local ring. One of these is developed by using Hilbert Samuel polynomial which is less intuitive but more useful in proving properties about dimension. Geometric invariants like degree of a variety or arithmetic genus can be made precise by using Hilbert Samuel polynomial. Dimension of affine algebras have special properties and these are studied in detail. Dimension is a coarse invariant as varieties of a given dimension are highly non-isomorphic. We introduce another numerical invariant called depth, which is closely related to dimension and investigate the relations between them. Rings and modules for which the two coincide form a special class called Cohen–Macaulay rings and modules. We study properties of rings and modules which are Cohen–Macaulay.

8.1 Hilbert Samuel polynomial

Definition 8.1.1 A *polynomial function* f is a map $f : \mathbb{N} \to \mathbb{Q}$ such that there exists a polynomial $g(X) \in \mathbb{Q}[X]$ with $f(n) = g(n)$ for all n greater than or equal to some fixed n_0.

Clearly, such a polynomial $g(X)$ is unique. Its degree and leading coefficient are called the *degree* and *leading* coefficient of the polynomial function f.

Proposition 8.1.2 *Let $f : \mathbb{N} \to \mathbb{Q}$ be a mapping. Then f is a polynomial function of degree r if and only if $\Delta f(n) : \mathbb{N} \to \mathbb{Q}$ defined by $\Delta f = f(n+1) - f(n)$ is a polynomial function of degree $(r-1)$.*

Proof If f is a polynomial function of degree r, then there exists some $g(X) \in \mathbb{Q}[X]$ of degree r such that $f(n) = g(n)$, $n \gg 0$ (i.e., for all n greater than or equal to some fixed n_0). If $h(X) = g(X+1) - g(X)$, $\Delta f(n) = h(n)$, $n \gg 0$, showing that Δf is a polynomial function of degree $(r-1)$.

The converse is proved by induction on r. If $r = 1$, then

$$\Delta f(n) = f(n+1) - f(n) = q_0 \in \mathbb{Q} \quad (n \gg 0),$$

i.e., $f(n+1) - q_0(n+1) = f(n) - q_0 n = \alpha$ $(n \gg 0)$. If $g(X) = q_0 X + \alpha$, then $f(n) = g(n)$ $(n \gg 0)$.
Let $r > 1$ and

$$g(X) = a_0 X^{r-1} + a_1 X^{r-2} + \cdots + a_{r-1} \in \mathbb{Q}[X],$$

$a_0 \neq 0$ be such that $\Delta f(n) = g(n)$ $(n \gg 0)$, i.e.,

$$f(n+1) - f(n) = \frac{a_0}{r}[(n+1)^r - n^r] + h(n),$$

where $h(X) \in Q[X]$ is of degree $\leq (r-2)$. If $\phi(n) = f(n) - \dfrac{a_0}{r}$, then $\Delta\phi$ is a polynomial function of degree $\leq r-2$. By induction, ϕ is a polynomial function of degree $r-1$. Hence

$$f(n) = \phi(n) + \frac{a_0 n^r}{r}$$

$(a_0 \neq 0)$ is polynomial function of degree r. □

Remark 8.1.3 The above result is also true for $r = 0$ if we assign the degree -1 to the zero polynomial.

We now consider a special type of polynomial function.

Let $R = \oplus\Sigma_{n\geq 0} R_n$ be a graded ring such that R_0 is Artinian and R is finitely generated as an R_0-algebra generated by r elements $a_1, a_2, \ldots, a_r$ in R_1. Let $M = \oplus\Sigma_{n\geq 0} M_n$ be a finitely generated graded R-module.

Proposition 8.1.4 *With the assumptions as above, each M_n is a finitely generated R_0-module.*

Proof Since R_0 is Noetherian and R is a finitely generated R_0 algebra, R is Noetherian. Let $N = \oplus\Sigma_{m\geq n} M_m \subseteq M$. Since M is finitely generated over R, so is N. Choose a generating set $x_1, x_2, \ldots, x_s$ of N over R and write $x_i = y_i + z_i$, $y_i \in M_n$ and $z_i \in \oplus\Sigma_{m>n} M_m$. We claim that $y_1, y_2, \ldots, y_s$ is a generating set of M_n over R_0. Let $y \in M_n$. Write $y = \Sigma_{i=1}^s a_i x_i$, where $a_i = b_i + c_i$, $b_i \in R_0$, $c_i \in \oplus\Sigma_{i>0} R_i$. Then $y = \Sigma_{i=1}^s b_i y_i$. □

Corollary 8.1.5 $\ell_{R_0}(M_n) < \infty$ *for each n.*

Theorem 8.1.6 *Let R be a graded ring with R_0 Artinian and R a finitely generated algebra over R_0 generated by r elements in R_1. Let M be a finitely generated graded R-module. The map*

$$\chi(M,\ ) : \mathbb{N} \to \mathbb{Z}, \quad \text{given by } \chi(M, n) = \ell_{R_0}(M_n),$$

is a polynomial function of degree $\leq r - 1$.

Proof The proof is by induction on r. If $r = 0$, then $R = R_0$. Choose a homogeneous generating set of M over $R = R_0$. If d is the maximum of their degrees, then $M_n = 0$ for $n > d$. Hence

$$\chi(M, n) = 0 \ (n \gg 0),$$

i.e., $\chi(M,\ )$ is a polynomial function of degree -1.

Let $r > 0$ and assume the result for all $r' < r$. Choose a generating set $a_1, a_2, \ldots, a_r$ in R_1 of R over R_0. If $\lambda_{a_r} : M \to M$ is the scalar multiplication defined by a_r, then $\lambda_{a_r}(M_n) \subseteq M_{n+1}$. Let C_{n+1} and K_n be the corresponding Coker and the Ker respectively, so that there is an exact sequence

$$0 \to K_n \to M_n \overset{\lambda_{a_r}}{\to} M_{n+1} \to C_{n+1} \to 0. \tag{8.1}$$

If $K = \oplus\Sigma_{n\geq 0}K_n$ and $C = \oplus\Sigma_{n\geq 0}C_n$, then K and C are Noetherian graded R-modules being, respectively, a graded submodule of the Noetherian graded module M and the corresponding quotient module. Hence $\chi(K, n)$ and $\chi(C, n)$ are defined. The exactness of the sequence (8.1) shows (Exercise 1, Section 3.4)

$$\chi(K, n) - \chi(M, n) + \chi(M, n+1) - \chi(C, n+1) = 0,$$

i.e.,

$$\Delta\chi(M, n) = \chi(C, n+1) - \chi(K, n).$$

Since a_r annihilates both K and C, K and C are finitely generated graded R'-modules, where R' is the graded subring of R generated by $a_1, a_2, \ldots, a_{r-1}$ over R_0. By induction, $\chi(K, n)$ and $\chi(C, n)$ are polynomial functions of degree $\leq r - 2$, i.e., $\Delta\chi(M, n)$ is a polynomial function of degree $\leq r - 2$. Hence by Proposition 8.1.2, $\chi(M, n)$ is a polynomial function of degree $\leq r - 1$. □

Corollary 8.1.7 *Let R be a Noetherian local ring with maximal ideal $\mathfrak{m}$ generated by r elements and $k = R/\mathfrak{m}$. Then*

$$\chi(\mathrm{gr}_{\mathfrak{m}}(R), n) = \ell_k(\mathfrak{m}^n/\mathfrak{m}^{n+1})$$

is a polynomial function of degree $\leq r - 1$.

Proof Apply Theorem 8.1.6 to the graded ring

$$\mathrm{gr}_{\mathfrak{m}}(R) = \oplus \sum_{i \geq 0} \frac{\mathfrak{m}^i}{\mathfrak{m}^{i+1}}$$

as a graded module over itself. □

Definition 8.1.8 The polynomial defining the polynomial function $\chi(M, n)$ is called the *Hilbert polynomial* of M.

We now consider a special case of the Hilbert polynomial.

Definition 8.1.9 Let R be a Noetherian local ring with maximal ideal $\mathfrak{m}$. An ideal I of R is called an *ideal of definition* of R if $\mathfrak{m}^n \subseteq I \subseteq \mathfrak{m}$ for some $n \geq 1$.

Clearly, I is an ideal of definition of R if and only if R/I is an Artinian ring.

If M is a finitely generated R-module, then M/IM is a finitely generated R/I-module and has finite length over R/I. Consider the I-adic filtrations defined on R and M respectively. If we denote the corresponding associated graded ring and module by

$$\mathrm{gr}_I(R) = \oplus \sum_{n \geq 0} \frac{I^n}{I^{n+1}}, \quad \mathrm{gr}_I(M) = \oplus \sum_{n \geq 0} \frac{I^n M}{I^{n+1} M},$$

then $\mathrm{gr}_I(M)$ is a finitely generated $\mathrm{gr}_I(R)$-module. If I is generated by $a_1, \ldots, a_r$ over R, their images $\overline{a}_1, \overline{a}_2, \ldots, \overline{a}_r$ in I/I^2 generate $\mathrm{gr}_I(R)$ over R/I. Hence $\chi(\mathrm{gr}_I(M), n)$ is well defined, where

$$\chi(\mathrm{gr}_I(M), n) = \ell_{R/I}\left(\frac{I^n M}{I^{n+1} M}\right).$$

Since Supp $(M/I^n M) = \{\mathfrak{m}\}$, $\ell_R(M/I^n M) < \infty$. If we denote $P_I(M, n) = \ell_R(M/I^n M)$, the exactness of the sequence

$$0 \to \frac{I^n M}{I^{n+1} M} \to \frac{M}{I^{n+1} M} \to \frac{M}{I^n M} \to 0$$

shows that

$$\Delta P_I(M, n) = P_I(M, n+1) - P_I(M, n) = \chi(\mathrm{gr}_I(M), n).$$

□

Proposition 8.1.10 *Let R be a Noetherian local ring, M a finitely generated R-module, I an ideal of definition of R generated by r elements. Then $P_I(M, n)$, defined by $P_I(M, n) = \ell_R(M/I^n M)$, is a polynomial function of degree $\leq r$.*

Proof By Theorem 8.1.6, $\chi(\mathrm{gr}_I(M), n)$ is a polynomial function of degree $\leq r - 1$. The result follows from Proposition 8.1.2. □

Proposition 8.1.11 *The degree of the polynomial function $P_I(M, n)$ does not depend on the ideal of definition I.*

Proof It is sufficient to show that $\deg P_I(M, n)$ is equal to $\deg P_{\mathfrak{m}}(M, n)$. Since $\mathfrak{m}^t \subseteq I$ for some $t \geq 1$, $\mathfrak{m}^{tn} \subseteq I^n \subseteq \mathfrak{m}^n$ for all n. Hence $P_{\mathfrak{m}}(M, tn) \geq P_I(M, n) \geq P_{\mathfrak{m}}(M, n)$. This shows that $P_I(M, \ )$ and $P_{\mathfrak{m}}(M, \ )$ have the same degree. □

Definition 8.1.12 The degree of $P_I(M, n)$ for any ideal of definition I is denoted by $d(M)$.

Theorem 8.1.13 *Let R be a Noetherian local ring, I an ideal of definition and $0 \to M' \to M \to M'' \to 0$ an exact sequence of finitely generated R-modules. Then*

$$P_I(M'', n) + P_I(M', n) = P_I(M, n) + R(n),$$

where $R(n)$ is a polynomial function of degree less than $d(M)$ and the leading coefficient of $R(n)$ is non-negative.

Proof The exact sequence

$$0 \to M' \to M \to M'' \to 0$$

induces, for each n, an exact sequence of R-modules

$$0 \to \frac{M'}{M' \cap I^n M} \to \frac{M}{I^n M} \to \frac{M''}{I^n M''} \to 0$$

Let $M' \cap I^n M = M'_n$, so that we have

$$P_I(M, n) - P_I(M'', n) = \ell_R\left(\frac{M'}{M'_n}\right). \tag{8.2}$$

Thus $\ell_R(M'/M'_n)$ is a polynomial function. By Artin Rees Lemma, $IM'_n = M'_{n+1}$, for all $n \geq n_0$, so that

$$I^{n_0+n} M' \subseteq M'_{n+n_0} = I^n M'_{n_0} \subseteq I^n M'$$

which implies

$$\ell_R\left(\frac{M'}{I^{n+n_0} M'}\right) \geq \ell_R\left(\frac{M'}{M'_{n+n_0}}\right) \geq \ell_R\left(\frac{M'}{I^n M'}\right),$$

$$i.e., \qquad P_I(M', n+n_0) \geq \ell_R\left(\frac{M'}{M'_{n+n_0}}\right) \geq P_I(M', n). \tag{8.3}$$

This shows that the polynomial functions $P_I(M', n)$ and $\ell_R(M'/M'_n)$ have the same degree and the same leading coefficient. Hence

$$P_I(M', n) - \ell_R\left(\frac{M'}{M'_n}\right) = R(n)$$

is a polynomial function of degree less than $\deg \ell_R(M'/M'_n)$, which is less than or equal to $d(M) = \deg P_I(M, n)$ by (8.2). Since by (8.3), $R(n) \geq 0$ for $n \gg 0$, the leading coefficient of $R(n)$ is non-negative. □

Corollary 8.1.14 *If M' is a submodule of M, then $d(M') \leq d(M)$.*

Proof If $M/M' = M''$, we have

$$\deg P_I(M'', n) \leq \deg P_I(M, n)$$

always. Hence, by Theorem 8.1.13, we have

$$\deg P_I(M', n) \leq \deg P_I(M, n).$$

□

Let R be a Noetherian local ring with maximal ideal $\mathfrak{m}$ generated by r elements. We saw in Corollary 8.1.7 that

$$\chi(\mathrm{gr}_{\mathfrak{m}}(R), n) = \ell_{R/\mathfrak{m}}\left(\frac{\mathfrak{m}^n}{\mathfrak{m}^{n+1}}\right)$$

is a polynomial function of degree $\leq (r-1)$. The following theorem gives conditions under which the degree is equal to $(r-1)$.

Theorem 8.1.15 *Let R be a Noetherian local ring with maximal ideal $\mathfrak{m}$ generated by $a_1, a_2, \ldots, a_r$ and $R/\mathfrak{m} = k$. The graded k-algebra homomorphism ϕ : $k[X_1, X_2, \ldots, X_r] \to \mathrm{gr}_{\mathfrak{m}}(R)$ given by*

$$\phi(X_i) = \overline{a}_i, = a_i + \mathfrak{m}^2 \ \ (1 \leq i \leq r),$$

is an isomorphism if and only if $\deg \chi(\mathrm{gr}_{\mathfrak{m}}(R), n) = r - 1$.

Proof Assume that ϕ is an isomorphism and let A_n be the n-th homogeneous component of $A = k[X_1, X_2, \ldots, X_r]$. Then $A_n \simeq \mathfrak{m}^n/\mathfrak{m}^{n+1}$ as k-spaces, so that

$$\chi(\mathrm{gr}_{\mathfrak{m}}(R), n) = \ell_k(\mathfrak{m}^n/\mathfrak{m}^{n+1}) = \ell_k(A_n) = \binom{n+r-1}{r-1}$$

(Exercise 1, Section 8.1). Since $n \to \binom{n+r-1}{r-1}$ is a polynomial function of degree $r-1$ (Exercise 2, Section 8.1), the result follows.

Conversely, assume that ϕ is not an isomorphism, so that $\operatorname{Ker}\phi = B \neq 0$. B has an induced gradation $B = \oplus\Sigma B_n$ and we have, for each n, an exact sequence of k-spaces

$$0 \to B_n \to A_n \to \mathfrak{m}^n/\mathfrak{m}^{n+1} \to 0.$$

This implies

$$\chi(\operatorname{gr}_{\mathfrak{m}}(R), n) = \binom{n+r-1}{r-1} - \ell_k(B_n). \tag{8.4}$$

Choose a non-zero homogeneous element $b \in B_d$ of degree d. Then $bA_n \subseteq B_{n+d}$, $n \geq 0$, so that

$$\ell_k(B_{n+d}) \geq \ell_k(bA_n) = \ell_k(A_n) \geq \ell_k(B_n).$$

Thus $\ell_k(B_n)$ and $\ell_k(A_n) = \binom{n+r-1}{r-1}$ are polynomial functions of the same degree $(r-1)$ and the same leading coefficient. Equation (8.4) now gives $\deg \chi(\operatorname{gr}_{\mathfrak{m}}(R), n) < r-1$, a contradiction to the assumption. Hence ϕ is an isomorphism. □

Corollary 8.1.16 *ϕ is an isomorphism if and only if*

$$\deg P_{\mathfrak{m}}(R, n) = r.$$

Proof Since $\chi(\operatorname{gr}_{\mathfrak{m}}(R), n) = \Delta P_{\mathfrak{m}}(R, n)$, the result follows from Proposition 8.1.2 and Theorem 8.1.15. □

8.1. EXERCISES

1. Let $S = R[X_1, \ldots, X_r]$ be the polynomial ring in $X_1, X_2, \ldots, X_r$ over R. Show that if S_n is the n-th homogeneous component of S consisting of homogeneous polynomials of degree n, then S_n is a free R-module of rank $\binom{n+r-1}{r-1}$.

2. Show that if $r \geq 1$, then $n \to \binom{n}{r}$ is a polynomial function of degree r. Deduce that if R is Artinian, then $\ell_R(S_n)$ is a polynomial function of degree $r-1$.

3. Let f be a polynomial function of degree r. Show that there exist $a_0, a_1, \ldots, a_r \in \mathbb{Q}$ such that

$$\begin{aligned} g(n) &= a_0 + a_1\binom{n}{1} + a_2\binom{n}{2} + \cdots + a_r\binom{n}{r} \\ &= f(n) \quad (n \gg 0). \end{aligned}$$

4. Let M be a finitely generated module over a Noetherian local ring R and $a \in \mathfrak{m}$, a non-zero-divisor on M. Show that $d(M/aM) = d(M) - 1$.

5. Let M be a finitely generated module over a Noetherian local ring R with maximal ideal $\mathfrak{m}$ and $\widehat{M}$ its $\mathfrak{m}$-adic completion. Show that $d(M) = d(\widehat{M})$.

8.2 Krull dimension

Definition 8.2.1 Let R be a commutative ring with 1. By a *chain* of prime ideals of R we mean a finite strictly increasing sequence of prime ideals of R of the type $P_0 \subsetneqq P_1 \subsetneqq P_2 \subsetneqq \cdots \subsetneqq P_n$.

The integer n is called the *length* of the chain.

Definition 8.2.2 The *Krull dimension* of R is the supremum of all lengths of chains of prime ideals of R. The Krull dimension of R is denoted by $\dim R$.

Example 8.2.3

(i) If R is Artinian, then $\dim R = 0$. (Proposition 3.3.9)

(ii) If R is a Dedekind domain, then $\dim R = 1$.

(iii) If $R = k[X_1, X_2, \ldots, X_n, \ldots]$, k a field, then $\dim R = \infty$ as $(X_1) \subsetneqq (X_1, X_2) \subsetneqq \cdots \subsetneqq (X_1, X_2, \ldots, X_n)$ $(n \geq 1)$ are chains of prime ideals of arbitrary length.

The following example shows that $\dim R$ can be infinite even if R is Noetherian.

Example 8.2.4
Let $R = k[X_1, X_2, \ldots, X_n, \ldots]$ be the polynomial ring in $X_1, X_2, \ldots, X_n, \ldots$ over a field k and $\{n_i\}$ an increasing sequence of positive integers satisfying

$$n_{i+1} - n_i > n_i - n_{i-1} \quad \text{for all } i.$$

Let $\quad P_i = (X_{n_i}, X_{n_i+1}, \ldots X_{n_{i+1}})R$ and $S = R - \underset{i}{\cup} P_i$.

Then R_S is Noetherian as each localisation of R_S at its maximal ideals is Noetherian and every non-zero element is contained in at most finitely many maximal ideals (Exercise 7, Section 3.1). Clearly, $\dim R_S = \infty$ as there exist in R_S chains of length $n_{i+1} - n_i$, and these are unbounded.

However we shall show later that for a Noetherian local ring R, $\dim R < \infty$.

Definition 8.2.5 Let P be a prime ideal of R. Then the *height* of P, denoted by $\operatorname{ht}(P)$, is $\dim R_P$. The *coheight* of P, denoted by $\operatorname{Coht}(P)$, is $\dim R/P$.

For any ideal I, we define

$$\operatorname{ht} I = \operatorname{Inf} \operatorname{ht}(P) \quad \text{and} \quad \operatorname{Coht} I = \operatorname{Sup} \operatorname{Coht} P,$$

where the Inf and Sup are taken over all prime ideals $P \supseteq I$.

We now generalise the concept of Krull dimension for a finitely generated R-module M.

Definition 8.2.6 $\dim M = \dim\left(\dfrac{R}{\mathrm{Ann}(M)}\right)$ if $M \neq 0$ and $\dim M = -1$ if $M = 0$.

A prime ideal P of R contains $\mathrm{Ann}(M)$ if and only if $P \in \mathrm{Supp}(M)$. Hence

$$\dim M = \operatorname*{Sup}_{P \in \mathrm{Supp}(M)} \mathrm{Coht}(P) = \operatorname*{Sup}_{P \in \mathrm{Ass}(M)} \mathrm{Coht}(P)$$

as the minimal elements of $\mathrm{Supp}(M)$ and $\mathrm{Ass}(M)$ are the same.

Example 8.2.7

(*i*) Let M be a finitely generated R-module. Then $\dim M = 0$ if and only if every $P \in \mathrm{Supp}(M)$ is maximal. This is equivalent to the condition that $\ell_R(M) < \infty$ (Theorem 3.4.13).

(*ii*) Let R be a Noetherian local ring with maximal ideal $\mathfrak{m}$ and M a finitely generated R-module. Since $\mathrm{Supp}(M/\mathfrak{m}M) = \{\mathfrak{m}\}$, $\ell_R(M/\mathfrak{m}M) < \infty$ and there exists a least integer r such that

$$\ell_R(M/(a_1, \ldots, a_r)M) < \infty, \quad a_1, a_2, \ldots, a_r \in \mathfrak{m}.$$

Definition 8.2.8 The *Chevalley dimension* $s(M)$ of M is the smallest integer r for which there exist

$$a_1, a_2, \ldots, a_r \in \mathfrak{m} \text{ with } \ell_R(M/(a_1, \ldots, a_r)M) < \infty.$$

If $M = 0$, $s(M)$ is defined to be -1.

Theorem 8.2.9 (Dimension Theorem) *Let R be a Noetherian local ring and M a finitely generated R-module. Then*

$$\dim(M) = d(M) = s(M).$$

Proof: We first show that $\dim(M) \leq d(M)$. If $d(M) = -1$, then

$$P_{\mathfrak{m}}(M, n) = 0 \text{ for } n \geq n_0.$$

Hence $M = \mathfrak{m}^n M$ for all $n \geq n_0$. By Nakayama Lemma, $M = 0$, i.e., $\dim M = -1$.

Assume now $d(M) \geq 0$. Choose a prime ideal $P \in \mathrm{Ass}(M)$ with $\dim M = \mathrm{Coht}\, P = \dim R/P$. Since R/P is isomorphic to a submodule of M, we have, by Corollary 8.1.14, $d(R/P) \leq d(M)$. It is therefore sufficient to show that $\dim R/P \leq d(R/P)$ for every $P \in \mathrm{Ass}(M)$. Consider any chain of prime ideals

$$P = P_0 \subsetneqq P_1 \subsetneqq \cdots \subsetneqq P_r$$

of R. We must show that $r \leq d(R/P)$. The proof is by induction on r. Since $R/P \neq 0$, $d(R/P) \geq 0$ and the result is true for $r = 0$. Assume the result for $(r-1)$.

Choose $a \in P_1$, $a \notin P$ and a prime ideal $P' \in \mathrm{Ass}(R/Ra + P)$ with $P' \subseteq P_1$. Then there exists a chain of prime ideals

$$P' \subsetneqq P_2 \subsetneqq \cdots \subsetneqq P_r.$$

By induction, $r - 1 \leq d(R/P')$, i.e.,

$$r \leq 1 + d(R/P') \leq 1 + d(R/Ra + P).$$

Consider the exact sequence

$$0 \longrightarrow R/P \xrightarrow{\lambda_a} R/P \longrightarrow R/(Ra + P) \longrightarrow 0$$

where λ_a is scalar multiplication by a. By Theorem 8.1.13,

$$P_{\mathfrak{m}}(R/P, n) + P_{\mathfrak{m}}(R/(Ra + P), n) = P_{\mathfrak{m}}(R/P, n) + R(n),$$

where $R(n)$ is a polynomial function of degree less than $d(R/P)$. Hence $d(R/Ra + P) < d(R/P)$, i.e., $r \leq d(R/P)$ and the proof is complete.

We now show that $d(M) \leq s(M)$. If $s(M) = -1$, then $M = 0$ and $d(M) = -1$.

Assume $s(M) = s \geq 0$ and let $a_1, a_2, \ldots, a_s \in \mathfrak{m}$ be such that $\dfrac{M}{(a_1, a_2, \ldots, a_s)M}$ has finite length. Since

$$\frac{M}{(a_1, a_2, \ldots, a_s)M} \simeq M \otimes_R \frac{R}{(a_1, a_2, \ldots, a_s)}$$

by Proposition 3.2.30 and Theorem 3.4.13,

$$\mathrm{Supp}\left(\frac{R}{(a_1, a_2, \ldots, a_s)}\right) \cap \mathrm{Supp}(M) = \{\mathfrak{m}\}.$$

Let $I = \mathrm{Ann}\,(M)$ and $J = I + (a_1, \ldots, a_s)$. Then $\mathrm{Supp}(R/J) = \{\mathfrak{m}\}$ and, in particular, $\mathrm{Ass}(R/J) = \{\mathfrak{m}\}$, i.e., J is $\mathfrak{m}$-primary. Hence J is an ideal of definition of R. Let $\overline{R} = R/I$ and $\overline{J} = J/I$. Then $\overline{R}$ is a local ring with $\overline{J}$ as an ideal of definition, generated by $\overline{a}_1, \ldots, \overline{a}_s$. Considering M as an $\overline{R}$-module, it follows from Proposition 8.1.10 that $P_{\overline{J}}(M, n)$ has degree at most s. Since

$$\ell_{\overline{R}}\left(\frac{M}{\overline{J}^n M}\right) = \ell_R\left(\frac{M}{J^n M}\right),$$

$P_{\overline{J}}(M, n)$ and $P_J(M, n)$ have the same degree and $d(M) \leq s(M)$.

We now show that $s(M) \leq \dim M$ by induction on $\dim M$. If $\dim M = 0$, then $\ell_R(M) < \infty$, so that $s(M) = 0$.

Assume $\dim M > 0$ and let $P_1, P_2, \ldots, P_k \in \operatorname{Ass}(M)$ for which $\dim M = \operatorname{Coht} P_i$ $(1 \leq i \leq k)$. Since $\dim M > 0$, $P_i \neq \mathfrak{m}$ $(1 \leq i \leq k)$, so that $\mathfrak{m} \not\subseteq \underset{i}{\cup} P_i$. Choose $a \in \mathfrak{m}$, $a \notin \underset{i}{\cup} P_i$ and let $M' = M/aM$. Then

$$\operatorname{Supp}(M') \subseteq \operatorname{Supp}(M) - \{P_1, P_2, \ldots, P_k\}$$

and hence $\dim M' < \dim M$. Choose $a_1, \ldots, a_t \in \mathfrak{m}$, t least, such that $M'/(a_1, \ldots, a_t)M'$ has finite length. This implies

$$M/(a, a_1, \ldots, a_t)M \simeq \frac{M'}{(a_1, \ldots, a_t)M'}$$

also has finite length and hence $s(M) \leq t + 1$. By induction, $t \leq \dim M'$ and hence $s(M) \leq \dim M$. □

Corollary 8.2.10 *Let R be a Noetherian local ring with maximal ideal $\mathfrak{m}$ and M a finitely generated R-module. Then $\dim M < \infty$. In particular, $\dim R < \infty$ and is equal to the minimum number of generators of an ideal of definition.*

Proof $\dim(M) = d(M) < \infty$ and $\dim R = s(R) =$ minimum number of generators of an ideal of definition. □

Corollary 8.2.11 *Let R be a Noetherian local ring with maximal ideal $\mathfrak{m}$ and $k = R/\mathfrak{m}$. Then $\dim R \leq \dim_k(\mathfrak{m}/\mathfrak{m}^2)$.*

Proof Choose $a_1, \ldots, a_r \in \mathfrak{m}$ such that $\overline{a}_1, \ldots, \overline{a}_r$ is a k-basis of $\mathfrak{m}/\mathfrak{m}^2$. Then $a_1, \ldots, a_r$ generate $\mathfrak{m}$ so that $\dim R \leq r = \dim_k(\mathfrak{m}/\mathfrak{m}^2)$. □

Corollary 8.2.12 *Let R be a Noetherian local ring with maximal ideal $\mathfrak{m}$ and $\widehat{R}$ its $\mathfrak{m}$-adic completion. Then $\dim R = \dim \widehat{R}$.*

Proof By Proposition 6.3.8, $R/\mathfrak{m}^n \simeq \widehat{R}/\widehat{\mathfrak{m}}^n$ $(n \geq 1)$ so that $P_{\mathfrak{m}}(R, n) = P_{\widehat{\mathfrak{m}}}(\widehat{R}, n)$. Hence they have the same degree. □

Corollary 8.2.13 *The prime ideals of a Noetherian ring satisfy descending chain condition.*

Proof By localising, we reduce to the case of a local ring. The result in this case follows as $\dim R < \infty$. □

Corollary 8.2.14 *Let R be a Noetherian ring and P a prime ideal of R. The following conditions are equivalent.*

(i) $\text{ht}(P) \leq n$.

(ii) *There exists an ideal I of R generated by n elements such that P is a minimal prime ideal of I.*

Proof $(ii) \Rightarrow (i)$. Under the assumption (ii), IR_P is an ideal of definition of R_P generated by n elements so that $\text{ht}\, P = \dim R_P \leq n$.
$(i) \Rightarrow (ii)$. By (i), $\dim R_P \leq n$, so that there exists an ideal of definition of R_P generated by n elements, say $(a_1/s), \ldots, (a_n/s)$, $s \in R - P$. The ideal I generated by $\{a_1, \ldots, a_n\}$ satisfies (ii). □

Corollary 8.2.15 (Krull's Principal Ideal Theorem). *Let R be a Noetherian ring, $a \in R$ which is neither a unit nor a zero-divisor of R. Then every minimal prime ideal of (a) has height* 1.

Proof Follows from $(ii) \Rightarrow (i)$ of Corollary 8.2.14 for $n = 1$ and observing that if $\text{ht}(P) = 0$, then P consists of zero-divisors of R. □

Corollary 8.2.16 *Let R be a Noetherian local ring with maximal ideal $\mathfrak{m}$ and $a \in \mathfrak{m}$ a non-zero-divisor. If $R' = R/(a)$, then* $\dim R' = \dim R - 1$.

Proof Since $a \in \mathfrak{m}$ is a non-zero-divisor, $\dim R > 0$. Taking $M = R$ in the proof of last implication of Dimension Theorem, we get that $\dim R' < \dim R$, i.e., $1 + \dim R' \leq \dim R$.

Conversely, let $a_1, a_2, \ldots, a_s \in \mathfrak{m}$ be such that their images in R' generate an $\mathfrak{m}' = \mathfrak{m}/(a)$-primary ideal ($s = \dim R'$). Then $a, a_1, a_2, \ldots, a_s$ generate an $\mathfrak{m}$-primary ideal in R. Hence $\dim R \leq 1 + s = 1 + \dim R'$. □

Corollary 8.2.17 *Let P be a prime ideal of a Noetherian ring R and $a \in P$ a non-zero-divisor. Then* $\text{ht}\, P/(a) = \text{ht}(P) - 1$.

Proof Apply Corollary 8.2.16 to the ring R_P. □

Corollary 8.2.18 *If $R = k[[X_1, X_2, \ldots, X_n]]$ is the power series ring in $X_1, X_2, \ldots, X_n$ over a field k, then* $\dim R = n$.

Proof By Corollary 8.2.10, $\dim R \leq n$, as $X_1, \ldots, X_n$ generate the maximal ideal. But $\dim R \geq n$ follows by considering the chain of prime ideals

$$(X_1) \subsetneq (X_1, X_2) \subsetneq \cdots \subsetneq (X_1, X_2, \ldots, X_n) \quad \text{in } R.$$

□

We now study the behaviour of height and coheight under ring extensions.

Proposition 8.2.19 *Let $R \subset S$ be an integral extension, J an ideal of S and $I = J \cap R$. Then*

(i) $\dim R = \dim S$ *and* $\operatorname{Coht} I = \operatorname{Coht} J$

(ii) $\operatorname{ht}(J) \leq \operatorname{ht}(I)$ *and equality holds if further R and S are domains and R is integrally closed.*

Proof (i) Since $R \subset S$ is integral, Theorem 4.2.8 shows that any chain of prime ideals of R can be lifted to a chain of prime ideals of S. Hence $\dim S \geq \dim R$. Now, any chain of prime ideals of S contracted to R gives a chain of prime ideals of R. Hence $\dim R \geq \dim S$, i.e., $\dim R = \dim S$. Since $R/I \subset S/J$ is an integral extension, $\operatorname{Coht} I = \dim R/I = \dim S/J = \operatorname{Coht} J$.

(ii) Let J be a prime ideal of S. Then $I = J \cap R$ is a prime ideal of R and for any chain of prime ideals of S contained in J, its contraction to R is a chain of prime ideals of R contained in I. Hence $\operatorname{ht}(J) \leq \operatorname{ht}\,(I)$. If J is any ideal, and $I = J \cap R$, choose a prime ideal P of R containing I with $\operatorname{ht}(P) = \operatorname{ht}(I)$. Since $R/I \subset S/J$ is an integral extension, there exists a prime ideal Q of S containing J which lies above P. Then $\operatorname{ht}(J) \leq \operatorname{ht}(Q) \leq \operatorname{ht}(P) = \operatorname{ht}(I)$.

To prove equality, assume that R and S are domains and R is integrally closed. Consider first the case when J is a prime ideal of S. If $I = J \cap R$, then I is a prime ideal of R and for any chain $I = P_0 \supsetneqq P_1 \supsetneqq \cdots \supsetneqq P_n$ of prime ideals of R, there exists a chain of prime ideals of S lying above it (Theorem 4.3.16). Hence $\operatorname{ht}(J) \geq \operatorname{ht}(I)$. Now, let J be any ideal of S and Q any prime ideal of S containing J. Then $P = Q \cap R \supset I$ and $\operatorname{ht}(Q) \geq \operatorname{ht}(P) \geq \operatorname{ht}(I)$. Since Q is an arbitrary prime ideal containing J, $\operatorname{ht}(J) \geq \operatorname{ht}(I)$. □

Proposition 8.2.20 *Let $R \subset S$ be an extension of rings such that S is flat over R. Then the going down theorem holds for the extension $R \subset S$.*

Proof Let $P' \subset P$ be prime ideals of R and Q a prime ideal of S lying above P. Then S_Q is flat over R_P and $R_P \to S_Q$ being a local homomorphism (i.e., mapping the unique maximal ideal into the unique maximal ideal), S_Q is faithfully flat over R_P. Hence by Theorem 2.3.26, there exists a prime ideal Q^* of S_Q lying above $P'R_P$. If $Q' = Q^* \cap S$, then Q' is the required prime ideal of S lying above P' and $Q' \subset Q$. □

Corollary 8.2.21 *Let $R \subset S$ be an extension such that S is faithfully flat over R. Then* $\dim S \geq \dim R$.

Proof Consider a chain of prime ideals of R, $P_0 \subsetneq P_1 \subsetneq \cdots \subsetneq P_r$. Since S is faithfully flat over R, there exists a prime ideal Q_r of S lying above P_r, and the above chain can be lifted to a chain of prime ideals of S. Hence $\dim S \geq \dim R$. $\square$

If P is a prime ideal of R, we have clearly the inequality $\operatorname{ht} P + \operatorname{Coht} P \leq \dim R$. We now give an example of a prime ideal P in a Noetherian ring R for which $\operatorname{ht} P + \operatorname{Coht} P < \dim R$.

Example 8.2.22
Let $S = k[[X,Y,Z]]$, the power series ring over a field k and $R = S/I$, where $I = (XY, XZ)$. Let $\overline{X}, \overline{Y}, \overline{Z}$ denote the images modulo I of X, Y, Z in R. Since $(\overline{X}) \subsetneq (\overline{X}, \overline{Y}) \subsetneq (\overline{X}, \overline{Y}, \overline{Z})$ is a chain of prime ideals of R, $\dim R \geq 2$. By Corollary 8.2.16, $\dim R \leq 2$. Hence $\dim R = 2$.

Since $I = (X) \cap (Y, Z)$, we have $(0) = (\overline{X}) \cap (\overline{Y}, \overline{Z})$. Now, $P = (\overline{Y}, \overline{Z})$ is a prime ideal of ht 0 in R and $R/P \simeq k[[X]]$ is of dimension 1. Thus $\operatorname{ht} P + \operatorname{Coht} P = 1 < \dim R$.

Theorem 8.2.23 *Let R be a Noetherian domain. Then R is integrally closed if and only if it satisfies the following two conditions.*

(i) *For every height one prime ideal P, R_P is a* DVR.

(ii) *The associated prime ideals of a non-zero principal ideal are all of height* 1.

Proof Assume R is integrally closed. If P is a prime ideal of ht 1, R_P is one dimensional. If R is integrally closed, so is R_P. Hence R_P is a DVR (Theorem 5.2.8).

To prove (ii), assume that P is an associated prime ideal of $I = Ra$. By localisation, $PR_P = \mathfrak{m}$ is an associated prime ideal of a principal ideal (b) in $R_P = R'$. We show that R' is a DVR.

Let $\lambda \notin (b)$ with $\lambda\mathfrak{m} \subseteq (b)$. Then $\lambda\mathfrak{m}b^{-1} \subset R'$ and $\lambda b^{-1} \notin R'$.

Now, $\lambda\mathfrak{m}b^{-1} \not\subseteq \mathfrak{m}$, for if $\lambda\mathfrak{m}b^{-1} \subseteq \mathfrak{m}$, then λb^{-1} is integral over R' (Lemma 4.1.4) and this is a contradiction as R' is integrally closed. Hence $\lambda\mathfrak{m}b^{-1} = R'$. There exists $t \in \mathfrak{m}$ with $\lambda t b^{-1} = 1$. For any $x \in \mathfrak{m}$, we have $x = 1 \cdot x = (\lambda t b^{-1})x$, i.e., $\mathfrak{m} = (t)$ showing that R' is a DVR.

Since $\dim R' = 1$, $\operatorname{ht} P = 1$.

Conversely, assume that R satisfies conditions (i) and (ii). We show that $R = \cap R_P$. (The intersection being taken over all P with $\operatorname{ht} P = 1$). If $\alpha = a/b \in R_P$ for all P with $\operatorname{ht} P = 1$, then $a \in bR_P$ for all $P \in \operatorname{Ass}(b)$ by (ii). This implies $a \in bR$, i.e., $\alpha \in R$. By (i) each R_P is integrally closed and hence R is integrally closed. $\square$

Corollary 8.2.24 *Let R be a Noetherian integrally closed domain. Then*

$$R = \bigcap_{\operatorname{ht} P = 1} R_P.$$

Proof As R is a Noetherian integrally closed domain, condition (ii) of Theorem 8.2.23 holds. The proof of Theorem 8.2.23 shows that $R = \bigcap_{\text{ht } P=1} R_P$. $\square$

8.2. EXERCISES

1. Let $R \subset S$ be domains, S integral over R and R integrally closed. Let Q be a prime ideal of S and $P = R \cap Q$. Show that $\dim R_P = \dim S_Q$.

2. Let $I \subset J$ be ideals such that J is not contained in any minimal prime ideal of I. Show that $\text{Coht}(I) \geq 1 + \text{Coht}(J)$.

3. Let R be a Noetherian ring such that R has only finitely many height 1 prime ideals. Show that R has only finitely many prime ideals and they are all of height ≤ 1.

4. Let $P \subset Q$ be prime ideals in a Noetherian ring. If there exists one prime ideal P' with $P \subsetneqq P' \subsetneqq Q$, show that there exist infinitely many such prime ideals.

5. Let R be a Noetherian local ring, M a finitely generated R-module, $a \in \mathfrak{m}$. Show that $\dim(M/aM) \geq \dim(M) - 1$ and equality holds if a is not a zero-divisor of M.

6. Let R be a Noetherian local ring. Show that either R is a domain or every principal prime ideal of R has $\text{ht}\, 0$.

7. Let R be a Noetherian ring and I an ideal of $\text{ht}\, 0$. Show that I consists entirely of zero-divisors. Show that the converse is also true if (0) has no embedded primes.

8. Let M be a finitely generated module over a local ring R with $\mathfrak{m}$-adic completion $\widehat{M}$. Show that $\dim_R M = \dim_{\widehat{R}} \widehat{M}$.

9. Let $f : R \to S$ be a local homomorphism of local rings R and S, and $k = R/\mathfrak{m}$. Show that
$$\dim S \leq \dim R + \dim(k \otimes_R S).$$

8.3 Dimension of algebras

We study in this section dimension of affine k-algebras i.e., finitely generated k-algebras, which are domains. We begin with the study of dimension of polynomial algebras.

Proposition 8.3.1 *Let R be a ring, $S = R[X]$ the polynomial ring in X over R. If $Q \subsetneqq Q'$ are prime ideals of S lying above the same prime ideal P of R, then $Q = PS$.*

Proof: By passing to the quotient by PS, we may assume that $P = 0$, i.e., R is a domain. By localising with respect to P, we may assume that R is a field. Then $S = R[X]$ is a PID, and clearly $Q = 0$. □

Proposition 8.3.2 *Let R be a Noetherian ring, $S = R[X]$, I an ideal of R and $J = IS$. If P is a minimal prime ideal of I, then $Q = PS$ is a minimal prime ideal of J.*

Proof By passing to the quotient modulo I, we may assume that $I = 0$. If Q is not a minimal prime ideal of S, then there exists a prime ideal $Q' \subsetneqq Q$. As $Q' \cap R \subseteq Q \cap R = P$, we have $Q' \cap R = P$, and by Proposition 8.3.1, $Q' = PS$, a contradiction as $Q' \neq Q$. □

Proposition 8.3.3 *Let R be a Noetherian ring, $S = R[X]$, P a prime ideal of R and $Q = PS$. Then* $\operatorname{ht} P = \operatorname{ht} Q$.

Proof If $\operatorname{ht} P = n$, then by Corollary 8.2.14, there is an ideal I of R generated by n elements such that P is a minimal prime ideal of I. By Proposition 8.3.2, Q is a minimal prime ideal of $J = IS$ and since J is also generated over S by n elements, $\operatorname{ht} Q \leq n = \operatorname{ht} P$.

Conversely, for any chain of prime ideals

$$P_0 \subsetneqq P_1 \subsetneqq \cdots \subsetneqq P_n = P$$

of R, there is a chain

$$Q_0 \subsetneqq Q_1 \subsetneqq \cdots \subsetneqq Q_n = Q$$

of prime ideals of S where $Q_i = P_i[X]$. Hence $\operatorname{ht} Q \geq \operatorname{ht} P$. □

Proposition 8.3.4 *Let R be a Noetherian ring and $S = R[X]$. Then $\dim S = 1 + \dim R$.*

Proof Let

$$P_0 \subsetneqq P_1 \subsetneqq \cdots \subsetneqq P_n$$

be a chain of prime ideals of R. Define $Q_i = P_i S \ (0 \leq i \leq n)$ and $Q_{n+1} = Q_n + (X)$. Then

$$Q_0 \subsetneqq Q_1 \subsetneqq \cdots \subsetneqq Q_n \subsetneqq Q_{n+1}$$

is a chain of prime ideals of S and hence $\dim S \geq 1 + \dim R$.

To prove the reverse inequality, consider a chain

$$Q_0 \subsetneqq Q_1 \subsetneqq \cdots \subsetneqq Q_n$$

of prime ideals of S and let $P_i = Q_i \cap R \ (0 \leq i \leq n)$. We show that $n \leq \dim R + 1$. If all the P_i are distint, then clearly $n \leq \dim R + 1$. If not, choose largest j such that $P_j = P_{j+1}$.

By Proposition 8.3.1, $Q_j = P_jS$ and by Proposition 8.3.3, $\operatorname{ht} Q_j = \operatorname{ht} P_j$. Since $\operatorname{ht} Q_j \geq j$, we have $\operatorname{ht} P_j \geq j$. But

$$P_j = P_{j+1} \subsetneqq P_{j+2} \subsetneqq \cdots \subsetneqq P_n$$

is a chain of prime ideals of R so that $\operatorname{ht} P_j + n - j - 1 \leq \dim R$, i.e., $n - 1 \leq \dim R$. Thus $n \leq 1 + \dim R$. This implies $\dim S \leq 1 + \dim R$ and the proof is complete. □

Corollary 8.3.5 *If R is a Noetherian ring, then*

$$\dim R[X_1, X_2, \ldots, X_n] = n + \dim R.$$

Corollary 8.3.6 *If K is a field, then* $\dim K[X_1, X_2, \ldots, X_n] = n$.

Corollary 8.3.7 *Let $R = K[X_1, X_2, \ldots, X_n]$, K a field. Then*

$$\operatorname{ht}(X_1, X_2, \ldots, X_i) = i \quad (1 \leq i \leq n).$$

Proof Since $\dim R = n$, $\operatorname{ht}(X_1, X_2, \ldots, X_n) = n$. By Corollary 8.2.17, $\operatorname{ht}(X_1, \ldots, X_{n-1}) = n - 1$ and by iteration we have $\operatorname{ht}(X_1, \ldots, X_i) = i$ $(1 \leq i \leq n)$. □

We saw in Section 8.2 an example of a prime ideal P of a Noetherian ring R such that $\operatorname{ht} P + \operatorname{Coht} P$ is less than $\dim R$. However, we will show that this cannot happen in integral domains which are affine k-algebras. We first prove a general version of Noether's Normalisation Theorem which says that any ideal of height i in the polynomial ring $K[X_1, \ldots, X_n]$, K a field, is of the type as in Corollary 8.3.7 by a change of variables up to an integral extension.

Theorem 8.3.8 *Let $R = K[X_1, \ldots, X_n]$, K a field and I an ideal of R with $\operatorname{ht} I = h$. Then there exist $y_1, y_2, \ldots, y_n \in R$ algebraically independent over K such that*

(i) R is integral over $K[y_1, y_2, \ldots, y_n]$ and

(ii) $I \cap K[y_1, \ldots, y_n] = (y_1, y_2, \ldots, y_h)$.

Proof The proof is by induction on h. If $h = 0$, $I = 0$ and define $y_i = X_i$ $(1 \leq i \leq n)$.

If $h \geq 1$, we claim that there exists an ideal $I' \subset I$ with $\operatorname{ht} I' = h-1$. This is so because there exists a prime ideal $P \supset I$ with $\operatorname{ht} P = h$. If

$$P_0 \subsetneqq P_1 \subsetneqq \cdots \subsetneqq P_{h-1} \subsetneqq P_h = P$$

is a chain of prime ideals in P of length h, then $I' = I \cap P_{h-1}$ has height $(h - 1)$. By induction, there exist $y'_1, y'_2, \ldots, y'_n$ satisfying (i) and (ii) for I'.

Let $R' = K[y'_1, y'_2, \ldots, y'_n]$. Then $I' \cap R' = \sum_{i<h} y'_i R'$. By Proposition 8.2.19, $\operatorname{ht}(I' \cap R') = \operatorname{ht}(I') = h-1$ and $\operatorname{ht}(I \cap R') = \operatorname{ht} I = h$. Hence there exists some $f \in I \cap R'$ with $f \notin I' \cap R'$. By replacing f by $f(0, 0, \ldots, 0, y'_h, y'_{h+1}, \ldots, y'_n)$, we may assume without loss of generality that $f \in K[y'_h, y'_{h+1}, \ldots, y'_n]$. By Exercise 7, Section 4.4, there exist $y_h = f, y_{h+1}, \ldots, y_n$ such that $K[y'_h, y'_{h+1}, \ldots, y'_n]$ is integral over $K[y_h, \ldots, y_n]$.

Define $y_i = y'_i$, $i < h$. We show that $y_1, y_2, \ldots, y_n$ satisfy (i) and (ii) for I. Now, R is clearly integral over $K[y_1, y_2, \ldots, y_n]$. Since $I \cap K[y_1, y_2, \ldots, y_n]$ contains $(y_1, y_2, \ldots, y_h)$ which is of height at least h,

$$I \cap K[y_1, y_2, \ldots, y_n] = (y_1, y_2, \ldots, y_h).$$ □

Proposition 8.3.9 *Let R be an integral domain which is an affine k-algebra over a field k and P a prime ideal of R. Then*

$$\operatorname{ht} P + \operatorname{Coht} P = \dim R.$$

Proof By the Normalisation Theorem (Theorem 4.4.5) R is integral over a polynomial algebra. In view of Proposition 8.2.19, it is sufficient to prove the result for a polynomial algebra over k. Let $R = k[X_1, X_2, \ldots, X_n]$ and P a prime ideal of R of ht h. By Theorem 8.3.8, there exist $y_1, y_2, \ldots, y_n$ algebraically independent over k such that R is integral over $k[y_1, y_2, \ldots, y_n]$ and $Q = P \cap k[y_1, y_2, \ldots, y_n] = (y_1, y_2, \ldots, y_h)$. By Corollary 8.3.7, $\operatorname{ht} Q = h$. Since

$$\frac{k[y_1, y_2, \ldots, y_n]}{Q} \simeq k[y_{h+1}, \ldots, y_n],$$

$$\operatorname{Coht} Q = \dim k[y_{h+1}, \ldots, y_n] = n - h.$$

Hence $\operatorname{Coht} P = \operatorname{Coht} Q = n - h$. Thus $\operatorname{ht} P + \operatorname{Coht} P = n = \dim R$. □

Definition 8.3.10 Let R be an affine k-algebra with quotient field F. Then $\operatorname{Tr.deg}_k R$ is defined to be equal to $\operatorname{Tr.deg}_k F$.

Proposition 8.3.11 *Let R be an affine k-algebra over a field k. Then* $\operatorname{Tr.deg}_k R = \dim R$.

Proof By Normalisation Theorem, R is integral over a polynomial algebra $k[y_1, y_2, \ldots, y_n]$, where $n = \dim R$. Let F and L be the quotient fields of R and $k[y_1, y_2, \ldots, y_n]$ respectively so that F/L is algebraic. Hence $\operatorname{Tr.deg}_k F = \operatorname{Tr.deg}_k L = n$. □

Theorem 8.3.12 *Let R_1 and R_2 be integral domains which are affine k-algebras over a field k and P a minimal prime ideal of $R_1 \otimes_k R_2$. Then*

$$\operatorname{Coht}(P) = \dim R_1 + \dim R_2.$$

Proof By Normalisation Theorem, there exist polynomial algebras S_1 and S_2 over k such that $S_1 \subset R_1$ and $S_2 \subset R_2$ are integral extensions. Let K_1, L_1, K_2, L_2 be the quotient fields of S_1, R_1, S_2 and R_2 respectively. Consider the commutative diagram

$$\begin{array}{ccccc} & & 0 & & 0 \\ & & \downarrow & & \downarrow \\ 0 & \longrightarrow & S_1 \otimes_k S_2 & \longrightarrow & R_1 \otimes_k R_2 \\ & & \downarrow & & \downarrow \\ 0 & \longrightarrow & K_1 \otimes_k K_2 & \longrightarrow & L_1 \otimes_k L_2 \end{array}$$

(Verify that the two rows and the two columns are exact.)

Since L_i is free over K_i $(i = 1, 2)$, $L_1 \otimes_k L_2$ is free over $K_1 \otimes_k K_2$. Hence no non-zero element of the polynomial algebra $S_1 \otimes S_2$ can be a zero-divisor of $R_1 \otimes R_2$. Since P is a minimal prime ideal of $R_1 \otimes R_2$, its contraction Q to $S_1 \otimes S_2$ is zero; for if $Q \neq 0$ there would exist $x \in Q$, $x \neq 0$ such that x is a zero-divisor of $R_1 \otimes R_2$. Hence

$$\begin{aligned} \operatorname{Coht} P = \operatorname{Coht} Q &= \dim S_1 \otimes S_2 = \dim S_1 + \dim S_2 \\ &= \dim R_1 + \dim R_2. \end{aligned}$$

□

We now prove the algebraic analogue of the geometric result concerning the dimension of the irreducible components of intersection of two irreducible affine varieties.

Theorem 8.3.13 *Let P_1 and P_2 be prime ideals of the polynomial ring $R = k[X_1, X_2, \ldots, X_n]$, k a field, and P a minimal prime ideal containing $P_1 + P_2$. Then $\operatorname{Coht}(P) \geq \operatorname{Coht}(P_1) + \operatorname{Coht}(P_2) - n$.*

Proof Consider the map $\phi : R \otimes_k R \to R$ given by $\phi(a \otimes b) = ab$; $a, b \in R$. Clearly, ϕ is a k-algebra homomorphism with $\operatorname{Ker} \phi$ generated by elements of the type $a \otimes 1 - 1 \otimes a$, $a \in R$, for if

$$\sum_i a_i \otimes b_i \in \operatorname{Ker} \phi, \quad \text{then} \quad \sum_i a_i . b_i = 0 \quad \text{and}$$

$$\sum_i a_i \otimes b_i = \sum_i (a_i \otimes 1 - 1 \otimes a_i)(1 \otimes b_i).$$

Let $S = R \otimes_k R$, $\overline{S} = R/P_1 \otimes_k R/P_2$ and $\psi : S \to \overline{S}$ the tensor product of the natural projections $R \to R/P_1$ and $R \to R/P_2$. We have an exact sequence $0 \to I \to S \to \overline{S} \to 0$, where

$$I = \operatorname{Ker} \psi = P_1 \otimes R \oplus R \otimes P_2.$$

Since P is a minimal prime ideal of R containing $P_1 + P_2$, $Q = \phi^{-1}(P)$ is a minimal prime ideal of S containing $I + \operatorname{Ker} \phi$ and its image $\overline{Q}$ under ψ in $\overline{S}$ is a minimal prime ideal of $\overline{S}$ containing $\psi(\operatorname{Ker} \phi)$. Since $\operatorname{Ker} \phi$ is generated by n elements $X_i \otimes 1 - 1 \otimes X_i$, $(1 \leq i \leq n)$, its image under ψ is also generated by n elements, so that $\operatorname{ht}(\overline{Q}) \leq n$. Let $\overline{Q}_0$

be a minimal prime ideal of $\overline{S}$ contained in $\overline{Q}$, so that $\overline{Q}/\overline{Q}_0$ has height at most n in $\overline{S}/\overline{Q}_0$. By Theorem 8.3.12

$$\dim \overline{S}/\overline{Q}_0 = \dim R/P_1 + \dim R/P_2.$$

Since $\mathrm{ht}(\overline{Q}/\overline{Q}_0) = \dim(\overline{S}/\overline{Q}_0) - \dim(\overline{S}/\overline{Q})$, we have

$$n \geq \mathrm{ht}(\overline{Q}/\overline{Q}_0) \geq \dim R/P_1 + \dim R/P_2 - \dim R/P,$$

i.e.,

$$\mathrm{Coht}\, P \geq \mathrm{Coht}\, P_1 + \mathrm{Coht}\, P_2 - n.$$

□

8.3. EXERCISES

1. Let P be a prime ideal of ht h in $R = k[X_1, X_2, \ldots, X_n]$, k a field. Show that PR_P is generated by h elements in R_P. Show also that if P is a maximal ideal of R, then P is generated by n elements.

2. Let R be an affine k-algebra, k a field. Show that all maximal chains of prime ideals of R are of the same length equal to $\dim R$.

3. Let R be a ring and let S consist of all $f \in R[X]$ such that the coefficients of f generate the unit ideal. Show that S is a multiplicatively closed set in $R[X]$ and $\dim R[X]_S = \dim R$.

4. If R is a Noetherian local ring, show that

$$\dim R[[X]] = 1 + \dim R$$

(**Hint:** If $d = \dim R$ and $I = (a_1, a_2, \ldots, a_d)$ is an $\mathfrak{m}$-primary ideal, then $(a_1, a_2, \ldots, a_d, X)$ is an $(\mathfrak{m}, X)$ primary ideal in $R[[X]]$).

5. Let $\mathfrak{m}'$ be a maximal ideal of $R[[X]]$ whose image under the natural map $\phi : R[[X]] \to R$ given by $\phi(X) = 0$, $\phi|_R = \mathrm{Id}$ is $\mathfrak{m}$. Show that $R[[X]]_{\mathfrak{m}'} \simeq R_{\mathfrak{m}}[[X]]$ (see Exercise 5, 2.1).

6. If R is a Noetherian ring show that $\dim R = \mathrm{Sup} \dim R_{\mathfrak{m}}$, where the supremum is taken over all the maximal ideals of R.

7. Deduce from Exercises 5 and 6 that if R is Noetherian, then

$$\dim R[[X_1, X_2, \ldots, X_n]] = n + \dim R.$$

8.4 Depth

Let R be a Noetherian local ring with maximal ideal $\mathfrak{m}$ and M a finitely generated R-module. We know from the Dimension Theorem that there exist $a_1, \ldots, a_n \in \mathfrak{m}$ ($n = \dim M$) such that the quotient module $M/(a_1, a_2, \ldots, a_n)M$ has finite length.

Definition 8.4.1 Let M be a finitely generated module over a Noetherian local ring R with maximal ideal $\mathfrak{m}$. A set of elements $a_1, a_2, \ldots, a_n$ of $\mathfrak{m}$ ($n = \dim M$) such that $M/(a_1, \ldots, a_n)M$ has finite length is called a *system of parameters* for M.

Example 8.4.2 If R is a Noetherian local ring of dimension d, and $a_1, \ldots, a_d$ is a generating set for an ideal of definition of R, then $a_1, \ldots, a_d$ is a system of parameters for R. In particular, if

$$R = k[[X_1, X_2, \ldots, X_n]],\ k \text{ a field},$$

then $X_1, X_2, \ldots, X_n$ is a system of parameters for R.

Lemma 8.4.3 *Let M be a finitely generated R-module over a Noetherian local ring R with maximal ideal $\mathfrak{m}$ and let $a \in \mathfrak{m}$, Then*

$$s(M/aM) \geq s(M) - 1,$$

i.e., $\dim(M/aM) \geq \dim(M) - 1$.

Proof Let $N = M/aM$ and $r = s(N)$. Then there exist $b_1, \ldots, b_r \in \mathfrak{m}$ such that $N/(\overline{b}_1, \ldots, \overline{b}_r)N$ has finite length (where $\overline{b}_i$ denotes the image of b_i in $\mathfrak{m}/aR \subset R/aR$). Thus $M/(a, b_1, b_2 \ldots, b_r)M$ has finite length so that $s(M) \leq r + 1$, i.e., $s(M/aM)$ $(= r) \geq s(M) - 1$. □

Theorem 8.4.4 *Let M be a finitely generated R-module of dimension n over a Noetherian local ring R with maximal ideal $\mathfrak{m}$. If $a_1, a_2, \ldots, a_t \in \mathfrak{m}$, then* $\dim \dfrac{M}{(a_1, \ldots, a_t)M} \geq n - t$ *and equality holds if and only if $a_1, \ldots, a_t$ is a part of a system of parameters for M.*

Proof Let I and J be ideals of R and $N = M/IM$. The composite of the maps $M \to M/IM = N \to N/JN$ is surjective and has kernel $(I + J)M$. Hence

$$\frac{M}{(I + J)M} \simeq \frac{N}{JN}.$$

In particular, if $I = (a_2, \ldots, a_i)$ and $J = (a_1)$, then

$$\frac{M}{(a_1, \ldots, a_i)M} \simeq \frac{N}{JN},$$

where $N = M/(a_2, \dots, a_i)M$. We now prove the proposition by induction on t. Let $t = 1$. Then by Lemma 8.4.3,

$$\dim\left(\frac{M}{a_1M}\right) = s\left(\frac{M}{a_1M}\right) \geq s(M) - 1 = \dim M - 1.$$

Now let $t > 1$ and $N = M/(a_1, \dots, a_t)M$. Note that

$$\frac{M}{(a_1, \dots, a_t)M} = \frac{N}{a_1N},$$

so that

$$\dim\frac{M}{(a_1, \dots, a_t)M} = \dim\frac{N}{a_1N}.$$

By induction, and by Lemma 8.4.3, $\dim N \geq \dim M - (t-1)$ and

$$\dim\frac{M}{(a_1, \dots, a_t)M} = \dim\frac{N}{a_1N} \geq \dim N - 1 \geq \dim M - t.$$

Suppose now that equality holds, so that

$$\dim\frac{M}{(a_1, \dots, a_t)M} = n - t.$$

Choose a system of parameters $a_{t+1}, a_{t+2}, \dots, a_n \in \mathfrak{m}$ for the module $N = M/(a_1, \dots, a_t)M$ so that

$$\frac{N}{(a_{t+1}, \dots, a_n)N} \simeq \frac{M}{(a_1, \dots, a_t, a_{t+1}, \dots, a_n)M}$$

has finite length. This implies that $a_1, \dots, a_n$ is a system of parameters for M.

Conversely, assume that $a_1, \dots, a_t$ is a part of a system of parameters $a_1, a_2, \dots, a_n$ for M. Then there exist $a_{t+1}, a_{t+2}, \dots, a_n$ in $\mathfrak{m}$ such that $N/(a_{t+1}, \dots, a_n)N$ has finite length, where $N = M/(a_1, \dots, a_t)M$. Hence $\dim N \leq n - t$ and therefore $\dim N = n - t$. □

Corollary 8.4.5 *Let $a_1, a_2, \dots, a_t \in \mathfrak{m}$ be a part of system of parameters for R such that $I = (a_1, \dots, a_t)$ has height t. Then*

$$\operatorname{ht}(I) + \operatorname{Coht}(I) = \dim R.$$

Proof From the above theorem, $\dim R/I = d - t$, where $d = \dim R$, so that $\operatorname{ht}(I) + \operatorname{Coht}(I) = \dim R$. □

The following example shows that if $a_1, \dots, a_t$ is a part of a system of parameters for R, then $I = (a_1, \dots, a_t)$ need not have height equal to t.

Example 8.4.6 Let $S = k[[X, Y, Z]]$, k a field and $R = \dfrac{S}{(XY, XZ)}$. Then we have seen in Example 8.2.22 that $\dim R = 2$. If $\overline{\mathfrak{m}} = (\overline{X}, \overline{Y}, \overline{Z})$ and $J = (\overline{Z}, \overline{X} + \overline{Y})$, then $\overline{\mathfrak{m}}^2 = (\overline{X}^2, \overline{Y}^2, \overline{Z}^2, \overline{YZ}) \subset J$. Hence J is an ideal of definition of R and $\{\overline{Z}, \overline{X} + \overline{Y}\}$ is a system of parameters for R. Thus $\{\overline{Z}\}$ is a part of a system of parameters for R. Since $(\overline{X}) \cap (\overline{Y}, \overline{Z}) = 0$, $(\overline{Y}, \overline{Z})$ has height 0, and hence $\mathrm{ht}(\overline{Z}) = 0$.

Remark 8.4.7 In the above Example, $\overline{Z}$ was a zero-divisor of R. However, if $a \in \mathfrak{m}$ is not a zero-divisor, then $\mathrm{ht}(a) = 1$ by Krull's Principal Ideal Theorem. Then $\dim R/(a) = \dim R - 1$ showing that $\{a\}$ will be a part of a system of parameters for R.

This leads us to the concept of M-sequences.

Definition 8.4.8 Let M be an R-module.
A set of elements $a_1, \dots, a_n$ of R is said to be an M-sequence if

(i) $(a_1, \dots, a_n)M \neq M$ and

(ii) a_i is not a zero-divisor of $\dfrac{M}{(a_1, \dots, a_{i-1})M}$, $1 \leq i \leq n$.

An M-sequence for $R = M$ is called an *R-sequence*. The condition (ii) for $i = 1$ means that a_1 is not a zero-divisor of M.

Example 8.4.9

(i) If $R = K[X_1, \dots, X_n]$ the polynomial ring over a field K, then $X_1, \dots, X_n$ is an R-sequence.

(ii) If $R = K[X, Y, Z]$, then $X, Y(1 - X), Z(1 - X)$ is an R-sequence.

Proposition 8.4.10 *Let M be an R-module. The following conditions are equivalent.*

(i) *$a_1, \dots, a_n$ is an M-sequence.*

(ii) *$a_1, \dots, a_i$ is an M-sequence and $a_{i+1}, \dots, a_n$ is an $\frac{M}{(a_1,\dots,a_i)M}$-sequence for all i $(1 \leq i \leq n)$.*

Proof If I and J are ideals of R and $N = M/IM$, then we have $M/(I + J)M \simeq N/JN$. The result now follows from the definition of M-sequence and by repeated use of the isomorphism for $I = (a_1, \dots, a_i)$ and $J = (a_{i+1})$, $(1 \leq i \leq n - 1)$. □

There are many similarities and differences between M-sequences and systems of parameters. We have seen in Example 8.4.6 that the members of a system of parameters can be zero-divisors. However, this cannot happen in an M-sequence over a Noetherian local ring. (It can happen if R is not a Noetherian local ring, *e.g.*, the sequence $\{\overline{X}, \overline{Z}(1-\overline{X})\}$ in the ring $R = k[X, Y, X]/Y(1-X)$.) We have also seen that if $a_1, \ldots, a_t$ is a part of a system of parameters for R, then $I = (a_1, \ldots, a_t)$ need not have height equal to t. However R-sequences are better behaved in this respect.

Proposition 8.4.11 *Let R be a Noetherian ring, $a_1, a_2, \ldots, a_t$ an R-sequence. If $I = (a_1 \ldots, a_t)$, then* $\operatorname{ht}(I) = t$.

Proof Clearly, $\operatorname{ht}(I) \leq t$ by Corollary 8.2.14. To show $\operatorname{ht}(I) \geq t$, consider a minimal $P \in \operatorname{Ass}(I)$. Then $P \supseteq P'$, $P' \in \operatorname{Ass}(a_1, \ldots, a_{t-1})$ and $P \supsetneq P'$, since a_t is not a zero-divisor of $R/(a_1, \ldots, a_{t-1})R$, i.e., $a_t \notin P'$, $a_t \in P$. Hence $\operatorname{ht} P \geq \operatorname{ht} P' + 1$. By induction $\operatorname{ht} P' \geq t - 1$. Hence $\operatorname{ht} P \geq t$, i.e., $\operatorname{ht} I \geq t$. □

Proposition 8.4.12 *Let M be a finitely generated module over a Noetherian local ring and $a_1, \ldots, a_t$, an M-sequence. Then $a_1, \ldots, a_t$ is a part of a system of parameters for M.*

Proof The proof is by induction on t. If $t = 1$, then a_1 is not a zero-divisor of M and $\dim M/aM = \dim M - 1$. By Theorem 8.4.4, $\{a_1\}$ is a part of system of parameters for M. If $t > 1$, by induction, $a_1, \ldots, a_{t-1}$ is a part of system of parameters for M and hence

$$\dim \frac{M}{(a_1, \ldots, a_{t-1})M} = n - (t-1),$$

where $n = \dim M$. Since a_t is not a zero-divisor of

$$N = \frac{M}{(a_1, \ldots, a_{t-1})M}, \quad \dim \frac{N}{a_t N} = \dim N - 1.$$

But $N/a_tN \simeq M/(a_1, \ldots, a_t)M$, so that

$$\dim \frac{M}{(a_1, \ldots, a_t)M} = n - (t-1) - 1 = n - t.$$

Hence, again by Theorem 8.4.4, $a_1, \ldots, a_t$ is a part of a system of parameters for M. □

Corollary 8.4.13 *If R is a Noetherian local ring, then any R-sequence is a part of a system of parameters for R.*

The following example shows that a permutation of an R-sequence need not be an R-sequence.

Example 8.4.14 Let $R = k[X, Y, Z]$, k a field. Then $X, Y(1-X)$, and $Z(1-X)$ is an R-sequence, but $Y(1-X), Z(1-X)$, and X is not an R-sequence.

However the following theorem shows that any permutation of an M-sequence contained in the Jacobson radical is an M-sequence.

Theorem 8.4.15 *Let R be a Noetherian ring, M a finitely generated R-module and $a_1, \ldots, a_n \in J(R)$ an M-sequence. Then any permutation of $a_1, \ldots, a_n$ is also an M-sequence.*

Proof It is sufficient to prove the result for a transposition of successive elements. In view of Proposition 8.4.10, it is sufficient to prove this for the transposition $(1, 2)$, i.e., to prove that

(i) a_2 is not a zero-divisor of M, and

(ii) a_1 is not a zero-divisor of M/a_2M.

To prove (ii), let $a_1\overline{x} = 0$ for some $\overline{x} \in M/a_2M$. Then $a_1x \in a_2M$, i.e., $a_1x = a_2y$, $y \in M$. Since a_2 is not a zero-divisor of M/a_1M, $y \in a_1M$, i.e., $y = a_1z$, $z \in M$. Hence $a_1x = a_2a_1z$ and $x = a_2z$, as a_1 is not a zero-divisor of M. Hence $\overline{x} = 0$.

To prove (i), let N be the submodule of M annihilated by a_2. If $x \in N$, then $a_2x = 0$. Since a_2 is not a zero-divisor of M/a_1M, $x \in a_1M$, i.e., $x = a_1y$ for some $y \in M$. Now, $a_2x = 0$ implies $a_2a_1y = 0$, i.e., $a_2y = 0$ as a_1 is not a zero-divisor of M.

Hence any $x \in N$ can be expressed as $x = a_1y$, $y \in N$, i.e., $N = a_1N$. Since $a_1 \in J(R)$, $N = 0$ by Nakayama Lemma. □

Corollary 8.4.16 *Let R be a Noetherian local ring and M a finitely generated R-module. Any permutation of an M-sequence is an M-sequence.*

Proof Since an M-sequence cannot contain units they are contained in $\mathfrak{m} = J(R)$. □

Theorem 8.4.17 *Let R be a Noetherian ring, M a finitely generated R-module, I an ideal of R with $IM \neq M$. Then any two maximal M-sequences contained in I have the same length.*

Proof We first show that if $a_1, \ldots, a_n$ is an M-sequence and N is an R-module annihilated by $(a_1, \ldots, a_n)$, then

$$\operatorname{Ext}_R^n(N, M) \simeq \operatorname{Hom}_R\left(N, \frac{M}{(a_1, \ldots, a_n)M}\right). \tag{8.5}$$

As $\operatorname{Ext}_R^0(N, M) = \operatorname{Hom}_R(N, R)$, equation (8.5) holds for $n = 0$. Consider the exact sequence $0 \longrightarrow M \stackrel{\lambda_{a_1}}{\rightarrow} M \rightarrow \dfrac{M}{a_1M} \rightarrow 0$, where λ_{a_1} is scalar multiplication by λ_{a_1}. This induces an exact sequence

$$\to \operatorname{Ext}_R^{n-1}(N, M) \to \operatorname{Ext}_R^{n-1}\left(N, \frac{M}{a_1 M}\right) \xrightarrow{\delta} \operatorname{Ext}_R^n(N, M) \xrightarrow{\lambda_{a_1}^*} \operatorname{Ext}_R^n(N, M) \quad (8.6)$$

where $\lambda_{a_1}^*$ is also multiplication by λ_{a_1} (Exercise 7, Section 7.2). Since $a_1 N = 0$, $\lambda_{a_1}^* = 0$ and hence δ is surjective. By induction,

$$\operatorname{Ext}_R^{n-1}(N, M) \simeq \operatorname{Hom}_R\left(N, \frac{M}{(a_1, \ldots, a_{n-1})M}\right) = 0,$$

and $\operatorname{Hom}_R(N, M/(a_1, \ldots, a_{n-1})M) = 0$ as $a_n N = 0$ and a_n is not a zero-divisor of $M/(a_1, \ldots, a_{n-1})M$. It follows then from (8.6) that $\operatorname{Ext}_R^{n-1}(N, M/a_1 M) \simeq \operatorname{Ext}_R^n(N, M)$. If $\overline{M} = M/a_1 M$, by induction

$$\begin{aligned} \operatorname{Ext}_R^{n-1}\left(N, \frac{M}{a_1 M}\right) &\simeq \operatorname{Hom}_R\left(N, \frac{\overline{M}}{(a_2 \ldots, a_{n-1})\overline{M}}\right) \\ &\simeq \operatorname{Hom}_R\left(N, \frac{M}{(a_1, \ldots, a_n)M}\right) \end{aligned}$$

and the proof of (8.5) is complete.

If we take $N = R/I$ in (8.5) with $a_1, a_2, \ldots, a_n \in I$, we have

$$\operatorname{Ext}_R^n(R/I, M) \simeq \operatorname{Hom}_R\left(R/I, \frac{M}{(a_1, \ldots, a_n)M}\right).$$

Now, $\operatorname{Hom}_R\left(R/I, \frac{M}{(a_1, \ldots, a_n)M}\right) \neq 0$ if and only if there is a non-zero element of $\overline{M} = \dfrac{M}{(a_1, \ldots, a_n)M}$ annihilated by I, i.e., $I \subseteq P$, $P \in \operatorname{Ass}(\overline{M})$. Hence

$$\begin{aligned} \operatorname{Ext}_R^n(R/I, M) = 0 \quad &\Leftrightarrow \quad I \not\subset \cup P, P \in \operatorname{Ass}(\overline{M}). \\ &\Leftrightarrow \quad a_1, \ldots, a_n \text{ is not a maximal} \\ &\qquad M\text{-sequence in } I, \text{ as } IM \neq M. \end{aligned}$$

Hence the smallest integer n for which $\operatorname{Ext}_R^n(R/I, M) \neq 0$ is the number of elements in any maximal M-sequence contained in I. This completes the proof. □

Definition 8.4.18 Let R be a Noetherian ring and M a finitely generated R-module, I an ideal of R with $IM \neq M$. Then $\operatorname{depth}_I(M)$ is the number of elements in a maximal M-sequence contained in I. Clearly, it is the smallest integer n for which $\operatorname{Ext}_R^n(R/I, M) \neq 0$.

If R is a local ring with maximal ideal $\mathfrak{m}$, then $\operatorname{depth}_{\mathfrak{m}}(M)$ is called the *depth* of M and is denoted by $\operatorname{depth}(M)$.

Example 8.4.19 Let R be a Noetherian local ring with maximal ideal $\mathfrak{m}$. Then $\operatorname{depth}(M) = 0$ if and only if $\operatorname{Hom}_R(R/\mathfrak{m}, M) \neq 0 \Leftrightarrow \mathfrak{m} \in \operatorname{Ass}(M)$.

Remark 8.4.20 Proposition 8.4.12 shows that if M is a finitely generated module over a Noetherian local ring, then $\operatorname{depth}(M) \leq \dim M$. Modules for which depth and dimension coincide form a special class and we will study them in the next section.

If I is an ideal in a Noetherian ring with $\operatorname{depth}_I(R) = n$, then we have $\operatorname{Ext}^n_R(R/I, R) \neq 0$, so that $\operatorname{pd}_R(R/I) \geq n$. We show that if $a_1, \ldots, a_n$ is an R-sequence and $I = (a_1, \ldots, a_n)$, then $\operatorname{pd}_R R/I = n$. We do this by constructing an explicit free resolution of R/I as an R-module. This is known as the Koszul resolution and we construct it more generally for an R-module M.

Let M be an R-module, $a_1, \ldots, a_n$ non-units of R and $I = (a_1, a_2, \ldots, a_n)$. Let F be a free module of rank n over R with basis $e_1, \ldots, e_n$. Let $\wedge^r F$ be the r-th homogeneous component of the exterior R-algebra $\wedge F$ of F. Define R-linear maps

$$\partial_r : \wedge^r F \to \wedge^{r-1} F \quad \text{by}$$

$$\partial_r(e_{i_1} \wedge e_{i_2} \wedge \cdots \wedge e_{i_r}) = \sum_{j=1}^{r} (-1)^{j+1} a_{i_j} e_{i_1} \wedge \cdots \wedge \widehat{e}_{i_j} \wedge \cdots \wedge e_{i_r},$$

$(1 \leq i_1 < \cdots < i_r \leq n)$, where $e_{i_1} \wedge \cdots \widehat{e}_{i_j} \cdots \wedge e_{i_r}$ is the term obtained by omitting e_{i_j} from $e_{i_1} \wedge e_{i_2} \wedge \cdots \wedge e_{i_r}$. Since $\partial_{r-1}\partial_r = 0$, $r \geq 1$, we have a complex of R-modules. Tensoring this complex with an R-module M, we have the Koszul complex $K(M, a_1, a_2, \ldots, a_n)$

$$0 \to X_n \xrightarrow{d_n} X_{n-1} \to \cdots X_r \to X_{r-1} \to \cdots \to X_1 \xrightarrow{d_1} X_0 \to 0,$$

where $X_r = M \otimes_R \wedge^r F$ and $d_r : X_r \to X_{r-1}$ $(r > 1)$ is given by

$$d_r(x \otimes e_{j_1} \wedge \cdots \wedge e_{j_r}) = \sum_{j=1}^{r} (-1)^{j+1} a_{i_j} x \otimes e_{i_1} \wedge \cdots \widehat{e}_{i_j} \cdots \wedge e_{i_r}.$$

Theorem 8.4.21 *Let M be an R-module and $a_1, \ldots, a_n$ an M-sequence. The homology of the Koszul complex*

$$X = K(M, a_1, a_2, \ldots, a_n)$$

is given by

$$H_i(X) = 0, \quad i \geq 1 \quad \text{and} \quad H_0(X) \simeq \frac{M}{(a_1, \ldots, a_n)M}.$$

Proof Let $I = (a_1, a_2, \ldots, a_n)$. It is sufficient to show that the augmented complex

$$\overline{X} : 0 \to X_n \overset{d_n}{\to} X_{n-1} \to \cdots \to X_r \to \cdots \to X_1 \to X_0 \overset{d_0}{\to} \frac{M}{IM} \to 0$$

is acyclic, where $d_0 : X_0 = M \to \frac{M}{IM}$ is the natural projection. We denote by $\overline{X}^{(k)}$ the augmented Koszul complex with respect to the elements $a_1, \ldots, a_k$. We prove by induction on k that $\overline{X}^{(k)}$ is acyclic for each k so that $\overline{X}^{(n)} = \overline{X}$ will be acyclic. For $k = 0$, $\overline{X}^{(0)}$ is the complex $0 \to M \overset{\text{Id}}{\to} M \to 0$ and is clearly acyclic. Assume now that the complex $\overline{X}^{(k-1)}$ is acyclic. Consider the truncated complex $Y^{(k)}$ obtained from $X^{(k)}$

$$Y^{(k)} : 0 \to X_k^{(k)} \to X_{k-1}^{(k)} \to \cdots \to X_1^{(k)} \to I_k M \to 0,$$

where $I_k = (a_1, \ldots, a_k)$. Clearly, the acyclicity of $\overline{X}^{(k)}$ is equivalent to the acyclicity of $Y^{(k)}$. Since by induction $Y^{(k-1)}$ is acyclic, it is sufficient to show that the quotient complex $Y^{(k)}/Y^{(k-1)}$ is acyclic (Exercise 2, Section 7.1). Consider the following commutative diagram

$$\begin{array}{ccccccccccc}
\overline{X}^{(k-1)} : & 0 \to & X_{k-1}^{(k-1)} & \to & X_{k-2}^{(k-1)} & \to \cdots \to & X_0^{(k-1)} & \to & \frac{M}{I_{k-1}M} & \to 0 \\
 & & \downarrow & & \downarrow & & \downarrow & & \downarrow & \\
\frac{Y^{(k)}}{Y^{(k-1)}} : & 0 \to & X_k^{(k)} & \to & \frac{X_{k-1}^{(k)}}{X_{k-1}^{(k-1)}} & \to \cdots \to & \frac{X_1^{(k)}}{X_1^{(k-1)}} & \to & \frac{I_k M}{I_{k-1}M} & \to 0
\end{array}$$

where the vertical maps are induced by multiplication by e_k and the last map $M/I_{k-1}M \to I_k M/I_{k-1}M$ is induced by multiplication by a_k. All the vertical maps are isomorphisms as $a_1, \ldots, a_k$ is an M-sequence. Since by induction the complex above is exact, the complex below is also exact. □

Corollary 8.4.22 *If I is an ideal of R generated by an R-sequence $a_1, \ldots, a_n$, then $\text{pd}_R(R/I) = n$.*

Proof The complex $\overline{X}$ for $M = R$ is given by

$$X : 0 \to X_n \to X_{n-1} \to \cdots \to X_r \to \cdots \to X_1 \overset{d_1}{\to} X_0 \overset{d_0}{\to} R/I \to 0. \tag{8.7}$$

It is a free resolution of R/I as an R-module, as each $X_r = \wedge^r F$ is a free R-module $(r \geq 0)$. Hence $\text{pd}_R(R/I) \leq n$. Using the projective resolution (8.7) to compute $\text{Ext}_R^n(R/I, R/I)$, we see that

$$\text{Ext}_R^n(R/I, R/I) \simeq R/I \neq 0.$$

Hence $\text{pd}_R(R/I) \geq n$, i.e., $\text{pd}_R(R/I) = n$.

Theorem 8.4.23 *Let R be a Noetherian local ring and $a_1, \ldots, a_r$ an R-sequence. If $I = (a_1, a_2, \ldots, a_r)$, then the natural map*

$$\phi : \frac{R}{I}[X_1, X_2, \ldots, X_r] \to \mathrm{gr}_I(R)$$

given by

$$\phi(X_i) = a_i + I^2 \ \ (1 \le i \le r), \ \ \phi|_{R/I} = \mathrm{Id}$$

can be extended to a graded algebra homomorphism and is an isomorphism.

We first prove the following lemma.

Lemma 8.4.24 *Let R be a Noetherian local ring and I an ideal of R generated by an R-sequence. If for some $c \in R$, $I : (c) = I$, then $I^d : (c) = I^d$ for all $d \ge 1$.*

Proof The proof is by induction on d. The result is true for $d = 1$ by assumption. Assume the result for d and all ideals generated by R-sequences. Let $I = (a_1, \ldots, a_r)$, where $a_1, a_2, \ldots, a_r$ is an R-sequence. If $b \in I^{d+1} : (c)$, then $bc \in I^{d+1} \subseteq I^d$ and by induction, $b \in I^d$. Write b as a homogeneous polynomial of degree d in $a_1, a_2, \ldots, a_r$ in the form

$$b = a_1u_1 + a_2u_2 + \cdots + a_qu_q, \ \ (q \le r), \ \ u_i \in (a_1, a_2, \ldots, a_i)^{d-1}.$$

It is sufficient to show that $u_i \in I^d$ $(1 \le i \le q)$. We prove this by induction on q.

Write $b = b' + a_qu_q$, where $b' = a_1u_1 + \cdots + a_{q-1}u_{q-1}$. Then

$$cb = cb' + ca_qu_q \in I^{d+1} = a_qI^d + K^{d+1},$$

where $K = (a_1, a_2, \ldots, a_{q-1}, a_{q+1}, \ldots, a_r)$. Then $cb = \lambda a_q + \mu$, $\lambda \in I^d$, $\mu \in K^{d+1} \subseteq K^d$ and $a_q(cu_q - \lambda) = \mu - b'c \in K^d$. By inductive assumption applied to K, we have

$$cu_q - \lambda \in K^d \subseteq I^d, \ \ i.e., \ \ cu_q \in I^d.$$

Since $I^d : (c) = I^d$, we have $u_q \in I^d$.

Now, $cb' = cb - ca_qu_q \in I^{d+1}$ and since $b' = a_1u_1 + \cdots + a_qu_{q-1}$, by induction on q, $u_i \in I^d$ $(1 \le i \le q-1)$. Hence $u_i \in I^d$ $(1 \le i \le q)$ and the proof is complete. □

Proof of Theorem 8.4.23 Clearly, the mapping ϕ is a surjective graded algebra homomorphism. It is sufficient to show that $\mathrm{Ker}\,\phi = 0$, i.e., if $F(X_1, \ldots, X_r)$ is a homogeneous polynomial of degree d with $F(a) = F(a_1, a_2, \ldots, a_r) \in I^{d+1}$, then all the coefficients of F are in I. We prove this by double induction on d and r.

The result is clear for $d = 0$. It is also true for $r = 1$, as a_1 is a non-zero-divisor of R. Consider now a homogeneous polynomial $F(X_1, X_2, \ldots, X_r)$ of degree d with $F(a) = F(a_1, a_2 \ldots, a_r) \in I^{d+1}$. Then there exist homogeneous polynomials $P_1, \ldots, P_r$, each of degree d, such that $F(a) = a_1P_1(a) + \cdots + a_rP_r(a)$. If

$$F(X_1, \ldots, X_r) = a_1P_1(X_1, \ldots, X_r) + \cdots + a_rP_r(X_1, \ldots, X_r),$$

then we are through. Otherwise, $\widetilde{F} = F - \Sigma_{i=1}^{r} a_i P_i$ is a homogeneous polynomial of degree d with $\widetilde{F}(a) = 0$. Replacing F by $\widetilde{F}$, we may assume without loss of generality that $F(a) = 0$.

Let $F(X_1, \ldots, X_r) = X_1 G(X_1, \ldots, X_r) + H(X_2, \ldots, X_r)$, where G and H are homogeneous polynomials of degrees $d - 1$ and d respectively. Then

$$0 = F(a) = a_1 G(a) + H(a). \tag{8.8}$$

This implies $a_1 G(a) \in (a_2, \ldots, a_r)^d$ and by Lemma 8.4.24, $G(a) \in (a_2, \ldots, a_r)^d \subset I^d$. Since $\deg G = d - 1$, by induction G has all its coefficients in I. Consider now the relation (8.8) modulo (a_1) and denote the cosets by putting bars. We have $0 = \overline{H}(\overline{a}_2, \ldots, \overline{a}_r)$. Since $\overline{a}_2, \ldots, \overline{a}_r$ is an $R/(a_1)$-sequence, we see by induction on r that the coefficients of $\overline{H}$ lie in $(\overline{a}_2, \ldots, \overline{a}_r)$. Hence the coefficients of H lie in $(a_1, a_2, \ldots, a_r) = I$. Thus the coefficients of F are in I and the proof is complete. □

Corollary 8.4.25 *Let R be a Noetherian local ring, and $a_1, \ldots, a_r$, an R-sequence. If $I = (a_1, a_2, \ldots, a_r)$, then I/I^2 is a free R/I-module of rank r. If $S(I/I^2)$ denotes the symmetric algebra of I/I^2 over R/I, then its n-th homogeneous component $S^n(I/I^2)$ is isomorphic to I^n/I^{n+1} ($n \geq 1$).*

Proof The result follows as ϕ is a graded algebra isomorphism. □

8.4. EXERCISES

1. Let R be a Noetherian local ring with maximal ideal $\mathfrak{m}$, $a_1, a_2, \ldots, a_n \in \mathfrak{m}$, M a finitely generated R-module, and $\widehat{M}$ its $\mathfrak{m}$-adic completion. Show that the following are equivalent:

   (i) $a_1, a_2, \ldots, a_n$ is a system of parameters for M.

   (ii) $a_1, a_2, \ldots, a_n$ is a system of parameters for $\widehat{M}$.

   (iii) $a_1, a_2, \ldots, a_n$ is a system of parameters for $R/\operatorname{Ann}(M)$.

2. Let R be a ring, S a multiplicatively closed set, and $a_1, \ldots, a_n$ an R-sequence. Show that $(a_1/1), \ldots, (a_n/1)$ is an R_S-sequence provided they generate a proper ideal in R_S.

3. Let M be an R-module, $a_1, \ldots, a_n \in R$ such that $a_i = b_i c_i$. Show that

   (i) If $a_1, \ldots, a_{i-1}, b_i, a_{i+1}, \ldots, a_n$ and $a_1, \ldots, a_{i-1}, c_i, a_{i+1}, \ldots, a_n$ are M- sequences, then so is $a_1, \ldots, a_i, \ldots, a_n$.

(ii) If $a_1, \ldots, a_n$ is an M-sequence and

$$(a_1, \ldots, a_{i-1}, b_i, a_{i+1}, \ldots, a_n)M \neq M,$$

then $(a_1, \ldots, a_{i-1}, b_i, a_{i+1}, \ldots, a_n)$ is an M-sequence.

(iii) Deduce that for arbitrary positive integers $k_1, k_2, \ldots, k_n$, $a_1^{k_1}, \ldots, a_n^{k_n}$ is an M-sequence if and only $a_1, \ldots, a_n$ is an M-sequence.

4. Let M be a finitely generated module over a Noetherian local ring and $a \in \mathfrak{m}$ a non-zero-divisor on M. Show that

$$\operatorname{depth}(M/aM) = \operatorname{depth}(M) - 1.$$

5. Let R be a Noetherian ring, M a finitely generated R-module, I an ideal of R. Show that

$$\operatorname{depth}(M) = \operatorname*{Inf}_{P \supseteq I} \operatorname{depth}_{I_P} M_P,$$

where P runs through the set of prime ideals of R containing I.

6. Let $f : R \to S$ be a local homomorphism of Noetherian local rings with maximal ideals $\mathfrak{m}$ and $\mathfrak{n}$, i.e., f is a ring homomorphism with $f(\mathfrak{m}) \subset \mathfrak{n}$, M a finitely generated S-module. Show that $\operatorname{depth}_R(M) = \operatorname{depth}_S(M)$, when S is a finitely generated R-module.

7. Let R be a local ring with maximal ideal $\mathfrak{m}$ and $I \subsetneq \mathfrak{m}$, $a \in \mathfrak{m} - I$, and $J = (I, a)$. Show that for an R-module M, $\operatorname{depth}_J(M) \leq 1 + \operatorname{depth}_I(M)$.

8. Let $0 \to M' \to M \to M'' \to 0$ be an exact sequence of finitely generated R-modules, I an ideal of R with $IM' \neq M'$, $IM \neq M$ and $IM'' \neq M''$. Show that

   (i) $\operatorname{depth}_I(M') = \operatorname{depth}_I(M)$ if $\operatorname{depth}_I(M) < \operatorname{depth}_I(M'')$,

   (ii) $\operatorname{depth}_I(M') = 1 + \operatorname{depth}_I(M'')$ if $\operatorname{depth}_I(M) > \operatorname{depth}_I(M'')$,

   (iii) $\operatorname{depth}_I(M') \geq \operatorname{depth}_I(M)$ if $\operatorname{depth}_I(M) = \operatorname{depth}_I(M'')$.

9. Let R be a Noetherian local ring with maximal ideal $\mathfrak{m}$ and $\mathfrak{m}$-adic completion $\widehat{R}$. If M is a finitely generated R-module, then show that any maximal M-sequence of R is a maximal $\widehat{M}$-sequence of $\widehat{R}$. Deduce that $\operatorname{depth}(M) = \operatorname{depth}(\widehat{M})$.

8.5 Cohen–Macaulay modules

We have seen that if R is a Noetherian ring and I an ideal generated by n elements, then any minimal prime ideal of I has ht $\leq n$ (Corollary 8.2.14). The following example shows that the different minimal prime ideals of I can have different heights.

Example 8.5.1 $R = k[X, Y]$, k a field and $I = (XY, X(X-1)) = (X) \cap (X-1, Y)$. I has minimal prime ideals (X) and $(X-1, Y)$ of heights 1 and 2 respectively.

However, if $\operatorname{ht} I = n$, all minimal prime ideals of I will have height equal to n, but all prime ideals associated to I need not be of the same height as the following example shows.

Example 8.5.2 $R = k[X.Y]$, k a field. $S = k[X^2, Y^2, XY^2, X^3]$. The $S \subset R$ is an integral extension and $\dim S = \dim R = 2$. If $I = (X^3)$ in S, then by Krull's Principal Ideal Theorem, $\operatorname{ht} I = 1$. However, the ideal $P = S \cap (X, Y)$ has height 2 and is associated with I as P is the annihilator of $\overline{X}^4 \in S/I$.

For some special types of rings, an ideal I generated by $\operatorname{ht} I$ elements has all its associated prime ideals of the same height. A classical theorem of Macaulay says that this is true in the polynomial ring $k[X_1, X_2, \ldots, X_n]$, over a field k. Rings with this property are called Cohen–Macaulay rings. We give an alternative description of these rings and show that it coincides with the classical definition.

Let R be a Noetherian local ring and M a finitely generated R-module.

Definition 8.5.3 M is called a *Cohen–Macaulay* (C.M.) *module* if $\operatorname{depth}(M) = \dim(M)$. R is called a Cohen–Macaulay (C.M.) ring if the R-module R is C.M.

Example 8.5.4 (i) If $\dim M = 0$, then $\operatorname{depth} M = 0$ as $\operatorname{depth} M \leq \dim M$. Hence M is C.M. In particular an Artinian local ring is C.M.

(ii) If $\dim R = 1$, then R is C.M. if $\operatorname{depth}(R) = 1$, i.e., if $\mathfrak{m}$ is not an associated prime of (0). Thus, any one-dimensional local domain is C.M.

(iii) If $\dim R = 1$ and all the elements of the maximal ideal $\mathfrak{m}$ are zero-divisors, then $\operatorname{depth}(R) = 0$ and R is not a C.M. ring. In particular, over a field k, the localisation of $\dfrac{k[X, Y]}{(X^2Y, XY^2)}$ at the maximal ideal (X, Y) is not C.M.

Theorem 8.5.5 *Let R be a Noetherian local ring and M a finitely generated R-module. Then $\operatorname{depth}(M) \leq \dim R/P$ for every $P \in \operatorname{Ass}(M)$*

Proof Let $\mathfrak{m}$ be the unique maximal ideal of R. We first show that if $a \in \mathfrak{m}$ and $P \in \text{Ass}(M)$, then there exists some

$$Q \in \text{Ass}(M/aM) \text{ with } Q \supseteq P + Ra.$$

Since $P \in \text{Ass}(M)$, $\text{Hom}_R(R/P, M) \neq 0$

Consider the exact sequence

$$0 \to M \xrightarrow{\lambda_a} M \to M' \to 0, \quad \text{where} \quad M' = M/aM$$

and λ_a is scalar multiplication by a. Then the sequence

$$0 \to \text{Hom}_R(R/P, M) \xrightarrow{\lambda_a^\star} \text{Hom}_R(R/P, M) \to \text{Hom}_R(R/P, M')$$

is exact. This implies $\text{Hom}_R(R/P, M') \neq 0$; for otherwise,

$$\text{Hom}_R(R/P, M) = 0$$

by Nakayama Lemma, a contradiction, as $P \in \text{Ass}(M)$. Hence

$$\text{Hom}_R\left(\frac{R}{P + Ra}, M'\right) = \text{Hom}_R(R/P, M') \neq 0.$$

This implies that there exists $Q \in \text{Ass}(M')$ with $Q \supseteq P + Ra = I$, for otherwise $I \nsubseteq \bigcup_{Q \in \text{Ass}(M)} Q$, i.e., for some $b \in I$, $\{b\}$ is an M'-sequence, showing that $\text{Hom}_R(R/I, M') = 0$. We now prove the proposition by induction on $\dim R/P$.

If $\dim R/P = 0$, $\mathfrak{m} = P \in \text{Ass}(M)$ and $\text{depth}(M) = 0$. Assume $\dim R/P > 0$. If $\text{depth}(M) = 0$ we are through. Otherwise choose $a \in \mathfrak{m}$ such that a is not a zero-divisor of M. If $N = M/aM$, there exists $Q \in \text{Ass}(N)$, with $Q \supseteq P + Ra$. By induction,

$$\text{depth}(N) \leq \dim R/Q < \dim R/P.$$

By Exercise 4, 8.4,

$$\text{depth}(N) = \text{depth}(M) - 1 < \dim R/P.$$

Hence $\text{depth}(M) \leq \dim R/P$. □

Corollary 8.5.6 *If M is a C.M. module, then*

$$\text{depth}(M) = \dim R/P = \dim M$$

for every $P \in \text{Ass}(M)$.

Proof $\operatorname{depth}(M) \leq \dim R/P \leq \sup_{Q \in \operatorname{Ass} M} \dim(R/Q) = \dim M$. Since M is C.M., $\operatorname{depth}(M) = \dim M$ and the result follows. □

Corollary 8.5.7 *If M is a C.M. module, then every $P \in \operatorname{Ass}(M)$ is minimal.*

Proof If $P \supseteq P' \in \operatorname{Ass}(M)$, then $\dim R/P = \dim R/P' = \dim M$. Hence $P = P'$. □

Proposition 8.5.8 *Let M be a C.M. module and $a \in \mathfrak{m}$. The following conditions are equivalent.*

(i) $\{a\}$ *is an M-sequence.*

(ii) $\dim(M/aM) = \dim M - 1$.

If any of the above two conditions are satisfied, then M/aM is C.M.

Proof Clearly, $(i) \Rightarrow (ii)$ by Exercise 5, 8.2.

Assume now condition (ii). By Theorem 8.5.5, for every $P \in \operatorname{Ass}(M)$, $\dim(M) = \dim(R/P)$. If $P \in \operatorname{Supp}(M/aM)$, then

$$\dim M/aM \geq \operatorname{Coht}(P) = \dim M$$

contradicting (ii). Hence $P \notin \operatorname{Supp}(M/aM)$. Now

$$M/aM \simeq M \otimes_R R/(a)$$

and

$$\operatorname{Supp}(M/aM) = \operatorname{Supp}(M) \cap \operatorname{Supp}(R/(a)).$$

Since $P \in \operatorname{Supp}(M)$ we have $P \notin \operatorname{Supp}(R/a)$, *i.e.* $a \notin P$, for all $P \in \operatorname{Ass}(M)$. Thus a is not a zero-divisor of M and $\{a\}$ is an M-sequence.

Assume now that the condition (i) or (ii) is satisfied. By Exercise 4, 8.4,

$$\operatorname{depth}(M/aM) = \operatorname{depth}(M) - 1 = \dim M - 1 = \dim(M/aM)$$

so that M/aM is C.M. □

Proposition 8.5.9 *If M is a C.M. R-module, then every system of parameters of M is an M-sequence. Conversely, if every system of parameters of M is an M-sequence, then M is C.M.*

Proof The converse is trivially true, for, if $a_1, \ldots, a_n$ is a system of parameters of M ($n = \dim M$), then it is an M–sequence, so that $\operatorname{depth}(M) \geq n = \dim M$. Hence $\operatorname{depth}(M) = \dim M$, i.e., M is C.M.

To prove the first part assume that M is C.M. and let $a_1, \ldots, a_n$ be a system of parameters for M. We prove, by induction on k, that $a_1, \ldots, a_k$ is an M-sequence and that $\dfrac{M}{(a_1, \ldots, a_k)M}$ is C.M. ($0 \leq k \leq n$). This is clearly true for $k = 0$, as M is C.M. Assume now that $a_1, \ldots, a_{k-1}$ is an M-sequence and $\dfrac{M}{(a_1, \ldots, a_{k-1})M} = N$ is C.M. Now

$$\frac{N}{a_k N} \simeq \frac{M}{(a_1, \ldots, a_k)M}$$

and by Theorem 8.4.4,

$$\dim \frac{N}{a_k N} = \dim(M) - k \text{ and } \dim(N) = \dim(M) - (k-1).$$

Hence $\dim \dfrac{N}{a_k N} = \dim N - 1$ and, by Proposition 8.5.8, $\{a_k\}$ is not a zero-divisor of $\dfrac{M}{(a_1, \ldots, a_{k-1})M} = N$, i.e., $a_1, \ldots, a_k$ is an M–sequence, and $N/a_k N \simeq M/(a_1, \ldots, a_k)M$ is C.M. □

Corollary 8.5.10 *If M is a C.M. module of dimension n and if $a_1, \ldots, a_r$ is a part of a system of parameters for M, then $\frac{M}{(a_1,\ldots,a_r)M}$ is C.M. of dimension equal to $n - r$.*

Corollary 8.5.11 *If M is a C.M. module, then every maximal M-sequence is a system of parameters for M.*

Proof By Proposition 8.4.12, every M-sequence is a part of a system of parameters and since $\operatorname{depth}(M) = \dim M$, the result follows. □

Theorem 8.5.12 *Let M be a C.M. R-module and P a prime ideal of R. Then M_P is a C.M. R_P-module.*

Proof If $P \not\supseteq \operatorname{Ann}(M)$, then $M_P = 0$. We therefore assume that $P \supseteq \operatorname{Ann}(M)$. Then

$$\dim M_P = \operatorname{ht} \frac{P}{\operatorname{Ann}(M)}$$

and $\operatorname{depth} M_P = \operatorname{depth}_P(M)$.

The proof is by induction on $\operatorname{depth}_P(M)$. If $\operatorname{depth}_P(M) = 0$, then P is contained in some $Q \in \operatorname{Ass}(M)$ and since M is C.M., Q is a minimal element of $\operatorname{Ass}(M)$ so that $P = Q \in \operatorname{Ass}(M)$. Hence $\dim M_P = \operatorname{ht} \frac{P}{\operatorname{Ann}(M)} = 0$.

Let $\operatorname{depth}_P(M) > 0$ so that there exists $a \in P$ which is a non-zero-divisor of M. If $N = M/aM$, then $\dim N_P = \dim M_P/aM_P = \dim M_P - 1$ and

$$\operatorname{depth} N_P = \operatorname{depth} M_P - 1 < \operatorname{depth} M_P.$$

By Proposition 8.5.8, N is C.M. Hence by induction, N_P is C.M., i.e., $\dim N_P = \operatorname{depth} N_P$. Therefore, $\dim M_P = \operatorname{depth} M_P$, i.e., M_P is a C.M. module. □

Corollary 8.5.13 *If R is a C.M. local ring and P a prime ideal of R, then R_P is a C.M. ring.*

Corollary 8.5.14 *Let R be a C.M. local ring and I a proper ideal of R. Then* $\operatorname{ht}(I) + \operatorname{Coht}(I) = \dim R$.

Proof If the result is true for prime ideals, then it is true for any ideal I; for if $\operatorname{ht} I = r$, choose a prime ideal $P \supset I$ with $\operatorname{ht} P = r$. Then

$$\operatorname{ht} I + \operatorname{Coht} I = \operatorname{ht} P + \operatorname{Coht} I \geq \operatorname{ht} P + \operatorname{Coht} P = \dim R.$$

It is therefore sufficient to prove the result when I is a prime ideal. Let $I = P$, a prime ideal, and $\operatorname{ht} P = r$. By Corollary 8.5.13 above, R_P is a C.M. of dimension r, i.e., $\operatorname{depth} R_P = \operatorname{depth}_P R = r$. Hence there exist $a_1, \ldots, a_r \in P$ which is an R-sequence. If $I = (a_1, \ldots, a_r)$, then R/I is C.M. and has dimension $n - r$, where $n = \dim R$. Since P is of height r containing an R-sequence $a_1, \ldots, a_r$, P is a minimal prime ideal of $I = (a_1, \ldots, a_r)$. Hence, by Corollary 8.5.6,

$$\dim R/P = \operatorname{depth} R/I = \dim R/I = n - r.$$

Hence $\qquad \operatorname{ht} P + \dim R/P = n = \dim R.$ □

Definition 8.5.15 A chain of prime ideals $P_0 \subsetneqq P_1 \subsetneqq \cdots \subsetneqq P_n$ is said to be *saturated* if there exists no prime ideal strictly lying between P_i, and P_{i+1} $(0 \leq i \leq n-1)$.

The following example shows that for a prime ideal P of R, the length of a saturated chain contained in P need not be equal to $\operatorname{ht} P$.

Example 8.5.16 Consider Example 8.2.22 where

$$R = \frac{k[[X, Y, Z]]}{(XY, XZ)}, \ k \text{ a field}, \ \mathfrak{m} = (\overline{X}, \overline{Y}, \overline{Z}).$$

We have seen that $Q = (\overline{Y}, \overline{Z})$ is a minimal prime ideal of R and the chain $Q \subset \mathfrak{m} = (\overline{X}, \overline{Y}, \overline{Z})$ is saturated, but $\operatorname{ht} \mathfrak{m} = 2$.

However, this will not happen in a C.M. local ring.

Corollary 8.5.17 *Let R be a C.M. local ring and $Q \subset P$ be prime ideals of R. Then all saturated chains of prime ideals from Q to P are of the same length, which is equal to* $\operatorname{ht}(P) - \operatorname{ht}(Q)$.

Proof It is sufficient to show that if $Q \subsetneq P$ are such that there is no prime ideal strictly lying between Q and P, then $\operatorname{ht} P - \operatorname{ht} Q = 1$. Clearly, by assumption, $\dim(R_P/QR_P) = 1$. Since R_P is C.M., we have by Corollary 8.5.14, $\operatorname{ht}(QR_P) + \dim(R_P/QR_P) = \dim R_P$. Since $\operatorname{ht}(QR_P) = \operatorname{ht}(Q)$, we have $\operatorname{ht}(P) - \operatorname{ht}(Q) = 1$. □

Corollary 8.5.18 *Let S be a local domain which is the quotient of a C.M. ring and $\overline{Q}$ a prime ideal of S. Then*

$$\operatorname{ht}(\overline{Q}) + \operatorname{Coht}(\overline{Q}) = \dim S.$$

Proof Write $S = R/P$, where R is a C.M. local ring and P a prime ideal of R. If $\overline{Q}_1 = \frac{Q_1}{P} \subset \frac{Q_2}{P} = \overline{Q}_2$ are prime ideals of S, then all saturated chains of prime ideals between them have the same length equal to

$$\operatorname{ht} Q_2 - \operatorname{ht} Q_1 = \operatorname{ht}(\overline{Q}_2) - \operatorname{ht}(\overline{Q}_1)$$

as R is C.M. Let $\overline{Q} = Q/P$ be any prime ideal of S. Choose maximal chains of prime ideals of S

$$\overline{Q}_0 \subset \overline{Q}_1 \subset \cdots \subset \overline{Q}_h = \overline{Q} \subset \overline{Q}'_1 \subset \cdots \subset \overline{Q}'_t,$$

where $h = \operatorname{ht}(\overline{Q})$ and $t = \operatorname{Coht}(\overline{Q})$. Since the above chain is saturated,

$$h + t = \operatorname{ht}(\overline{Q}'_t) - \operatorname{ht}(\overline{Q}_0) = \dim S$$

as $\overline{Q}'_t$ is maximal and $\overline{Q}_0$ a minimal prime ideal of S. □

Definition 8.5.19 A Noetherian ring R is called a *Coehn–Macaulay* (C.M.) ring if R_P is C.M. for every prime ideal P of R.

In view of Corollary 8.5.13, this definition coincides with the one already given for local rings.

Example 8.5.20 The rings

(i) $k[X]$, k a field,

(ii) Dedekind domain and

(iii) $\dfrac{k[X, Y]}{(X^2 - Y^3)}$, k a field

are C.M. rings as their localisations are one dimensional local domains.

We now show that Cohen–Macaulay rings are precisely the class of rings for which the following classical unmixedness property holds.

Definition 8.5.21 An ideal I of a Noetherian ring is said to be *height unmixed* if $\operatorname{ht}(I) = \operatorname{ht}(P_i)$ for all $P_i \in \operatorname{Ass}(R/I)$. A Noetherian ring has unmixedness property if every ideal of height r generated by r elements (and the zero ideal) is height unmixed.

Theorem 8.5.22 *Let R be a Noetherian ring. The following conditions are equivalent.*

(i) *R has the unmixedness property.*

(ii) *R_P is C.M. for every prime ideal P of R.*

(iii) *$R_{\mathfrak{m}}$ is C.M. for every maximal ideal $\mathfrak{m}$ of R.*

Proof We first prove the following. If P is a prime ideal of ht r, then there exist $a_1, \ldots, a_r \in P$ such that $\operatorname{ht}(a_1, \ldots, a_i) = i$, $1 \leq i \leq r$. We do this by induction on r.

For $r = 1$, ht $P = 1$. Since $P \nsubseteq \underset{P_i}{\cup} P_i$ (union over minimal primes), there exists $a_1 \in P - \underset{P_i}{\cup} P_i$ (union over minimal primes). Now $\operatorname{ht}(a_1) \leq 1$ by Krull's Principle Ideal Theorem and $\operatorname{ht}(a_1) \neq 0$ as $a_1 \notin P_j$ for any minimal prime P_j. Therefore, $\operatorname{ht}(a_1) = 1$.

Assume that $a_1, \ldots, a_{r-1} \in P$ have been constructed with the property

$$\operatorname{ht}(a_1, \ldots, a_i) = i, \quad 1 \leq i \leq r - 1.$$

Let $Q_1, Q_2, \ldots, Q_s$ be the minimal prime ideals of $(a_1, \ldots, a_{r-1})$. Then these are the only prime ideals of height $(r - 1)$ containing $(a_1, \ldots, a_{r-1})$. Now, $P \not\subset \cup_i Q_i$, as $\operatorname{ht}(P) = r > \operatorname{ht} Q_i$ $(1 \leq i \leq s)$. Choose $a_r \in P$, $a_r \notin Q_i$, $(1 \leq i \leq s)$. It is easy to see that

$$\operatorname{ht}(a_1, \ldots, a_r) = r.$$

$(i) \Rightarrow (ii)$. By (i), the ideal $(a_1, \ldots, a_i)$ is unmixed and hence $a_{i+1} \notin Q$ for any $Q \in \operatorname{Ass}\left(\dfrac{R}{(a_1, \ldots, a_i)}\right)$. Hence $a_1, \ldots, a_r$ is an R-sequence in P. This implies depth $R_P \geq r = \dim R_P$, i.e., R_P is C.M..

$(ii) \Rightarrow (iii)$. is clear.

$(iii) \Rightarrow (i)$. Since unmixedness property is a local property it is sufficient to show that $R_{\mathfrak{m}}$ has unmixedness property for all $\mathfrak{m}$, i.e., if R is a C.M. local ring, then R has unmixedness property. By Corollary 8.5.7, the zero ideal is unmixed. Let $I = (a_1, \ldots, a_r)$ be an ideal of ht r. Then $\dim R/I = \dim R - r$ and hence $a_1, \ldots, a_r$ is a part of a system of parameters (Theorem 8.4.4). Since R is C.M., $a_1, \ldots, a_r$ is an R-sequence. Hence $R/(a_1, \ldots, a_r)$ is C.M., i.e., $(a_1, \ldots, a_r)$ has unmixedness property. □

Theorem 8.5.23 *If a Noetherian ring R is* C.M., *then $R[X_1, \ldots, X_n]$ is C.M.*

Proof It is sufficient to prove this for the case $n = 1$. We first observe that if $\mathfrak{m}$ is a maximal ideal of $R[X]$, then $\mathfrak{m}$ contains a non-zero-divisor; for if $\mathfrak{m}$ consists entirely of zero-divisors, then $X \notin \mathfrak{m}$ and hence $(\mathfrak{m}, X) = R[X]$, i.e., $1 = Xf(X) + g(X)$, for some $g(X) \in \mathfrak{m}$. But $g(X) = 1 - Xf(X)$ has constant term 1. This implies that $g(X)$ cannot be a zero-divisor, which is a contradiction as $g \in \mathfrak{m}$.

Let $\mathfrak{m}$ be a maximal ideal of $R[X]$ and $P = \mathfrak{m} \cap R$. Since R is C.M., so is R_P, i.e., $\operatorname{ht} P = \operatorname{depth}_P(R) = r$, say. By Propositions 8.3.1 and 8.3.3, it is easy to see that $PR[X] \neq \mathfrak{m}$ and $\operatorname{ht} \mathfrak{m} = r + 1$.

If we show that $\operatorname{depth}_{\mathfrak{m}} R[X] = r + 1$, then $R[X]_{\mathfrak{m}}$ will be C.M. for every maximal ideal $\mathfrak{m}$ of $R[X]$ and $R[X]$ will be C.M..

Let $a_1, \ldots, a_r$ be a maximal R-sequence in P. Then it is also an $R[X]$-sequence in $PR[X] \subseteq \mathfrak{m}$. Let $I = (a_1, \ldots, a_r)$ and $\overline{\mathfrak{m}}$ the image of $\mathfrak{m}$ in

$$R[X]/IR[X] \simeq (R/I)[X].$$

Then $\overline{\mathfrak{m}}$ is a maximal ideal in $(R/I)[X]$ and hence $\overline{\mathfrak{m}}$ must contain a non-zero-divisor $\overline{a}_{r+1} \in R[X]/IR[X]$. Then $a_1, \ldots, a_r, a_{r+1}$ is an $R[X]$-sequence in $\mathfrak{m}$, i.e.,

$$\operatorname{depth}_{\mathfrak{m}} R[X] \geq r + 1.$$

But

$$\operatorname{depth}_{\mathfrak{m}} R[X] \leq \dim R[X]_{\mathfrak{m}} = r + 1$$

and $\operatorname{depth}_{\mathfrak{m}} R[X] = r + 1$. □

Corollary 8.5.24 (**Macaulay's Theorem**)

If K is a field, then $K[X_1, X_2, \ldots, X_n]$ is a C.M. ring.

8.5. EXERCISES

1. Show that if R is an integrally closed Noetherian local domain of dimension 2, then R is C.M. (**Hint:** If R is a Noetherian integrally closed domain, then principal ideals of R have no embedded components.)

2. Let M be an R-module over a Noetherian local ring with maximal ideal $\mathfrak{m}$. Show that M is C.M. over R if and only if $\widehat{M}$ is C.M. over $\widehat{R}$.

3. Let $f : R \to S$ be a homomorphism of Noetherian local rings with maximal ideals $\mathfrak{m}$ and $\mathfrak{n}$ such that $f(\mathfrak{m}) \subseteq \mathfrak{n}$. Assume that S is a finitely generated R-module. Show that a finitely generated S-module M is C.M. over S if and only if it is C.M. over R.

4. Let M be a finitely generated module of dimension n over a Noetherian local ring R with maximal ideal $\mathfrak{m}$. Suppose for every $a_1, \ldots, a_r \in \mathfrak{m}$ with $\dim \dfrac{M}{(a_1, \ldots, a_r)M} = n - r$, we have $\dim R/P = n - r$ for every $P \in \operatorname{Ass}\left(\dfrac{M}{(a_1, \ldots, a_r)M}\right)$. Show that M is C.M.

5. Let S be a C.M. local ring, I an ideal in S and $R = S/I$. Show that R is C.M. if I is generated by $\dim S - \dim R$ elements.

6. Let R be a Noetherian local ring having a system of parameters $a_1, \ldots, a_r$ and $I = (a_1, \ldots, a_r)$. Assume that the mapping $\phi : (R/I)[X_1, X_2, \ldots, X_r] \to \operatorname{gr}_I(R)$, given by $\phi(X_i) = a_i + I^2$, $(1 \leq i \leq r)$, $\phi|_{R/I} = \mathrm{Id}$, is an isomorphism. Show that R is C.M.

7. If R is a Noetherian local ring, M a finitely generated R-module with proj. $\dim_R M < \infty$, then

$$\mathrm{pd}_R M + \mathrm{depth}_R M = \mathrm{depth}(R).$$

CHAPTER

9

REGULAR LOCAL RINGS

The local ring of an affine variety at a point P reflects the local properties of the variety at P. The point P will be non-singular if and only if the local ring at P is a regular local ring. We define the concept of regularity for any Noetherian local ring R and obtain equivalent characterisations of regularity including the homological characterisation, *viz.* $\operatorname{gl\,dim} R$ is finite. As a consequence, R_P will be a regular local ring when R is regular local and P a prime ideal of R. We also show that a regular local ring has UFD property and is a Cohen–Macaulay ring.

We then investigate two conditions concerning depth and regularity *viz.* (R_k) and (S_k) $(k \geq 0)$ for a Noetherian ring R. The ring R will have no nilpotent elements if and only if it satisfies the conditions R_0 and S_1. The conditions R_1 and S_2 are equivalent to the normality of the ring, *i.e.* the localisation at every prime ideal is an integrally closed domain. In the last section we study complete local rings. We prove the Cohen Structure Theorem on the existence of a coefficient ring for any complete local ring and deduce the structure of complete regular local rings.

9.1 Regular local rings

Let R be a Noetherian local ring with maximal ideal $\mathfrak{m}$ and dimension d and let $k = R/\mathfrak{m}$. From the Dimension Theorem we know that any generating set for $\mathfrak{m}$ has at least d elements.

Definition 9.1.1 Let R be a Noetherian local ring with maximal ideal $\mathfrak{m}$ and dimension d. R is called *regular* if $\mathfrak{m}$ has a generating set of d elements. A generating set of d elements for $\mathfrak{m}$ is called a *regular system of parameters* of R.

Example 9.1.2

1. If $\dim R = 0$, then R is regular if and only if R is a field.
2. If $\dim R = 1$, then R is regular if and only if R is a DVR.
3. $R = k[[X_1, X_2, \ldots, X_d]]$; k, a field, is regular of dimension d and $X_1, X_2, \ldots, X_d$ is a regular system of parameters.
4. Let R be the localisation of $\dfrac{k[X,Y]}{(X^3 - Y^2)}$ at the maximal ideal $\mathfrak{m} = (\overline{X}, \overline{Y})$, k a field. Then R is not regular. We have $\dim R = \operatorname{ht} \overline{\mathfrak{m}} = 1$, as $\operatorname{ht}(X, Y) = 2$ and $X^3 - Y^2$ is a non-zero-divisor of $k[X, Y]$. But $(\overline{X}, \overline{Y})$ is a minimal generating set for $\overline{\mathfrak{m}}$.
5. Let $R = \left(\dfrac{k[X,Y]}{(X^3 - Y^2)}\right)_{\mathfrak{m}}$, k a field, where $\mathfrak{m} = (X - 1, Y - 1)$. Then $\dim R = 1$ and R is regular. In this case $(\overline{X} - 1, \overline{Y} - 1)$ is not a minimal generating set for $\overline{\mathfrak{m}}$, as $(\overline{Y} + 1)(\overline{Y} - 1) = (\overline{X}^2 + \overline{X} + 1)(\overline{X} - 1)$ and $\overline{Y} + 1$, $\overline{X}^2 + \overline{X} + 1$ are units in R.

Theorem 9.1.3 *Let R be a Noetherian local ring of dimension d, maximal ideal $\mathfrak{m}$ and $k = R/\mathfrak{m}$. The following condition are equivalent.*

(i) *R is regular.*

(ii) *$\operatorname{gr}_{\mathfrak{m}}(R)$ is isomorphic to $k[X_1, X_2, \ldots, X_d]$ as a graded k-algebra.*

(iii) $\dim R = \dim_k(\mathfrak{m}/\mathfrak{m}^2)$.

Proof $(i) \to (ii)$. Let $a_1, a_2, \ldots, a_d$ be a regular system of parameters for R. The mapping $\phi : k[X_1, \ldots, X_d] \to \operatorname{gr}_{\mathfrak{m}}(R)$ given by $\phi(X_i) = a_i + \mathfrak{m}^2$ $(1 \le i \le d)$, can be extended to a graded k-algebra homomorphism. By Corollary 8.1.16, ϕ is an isomorphism as $\deg P_{\mathfrak{m}}(R, n) = d = \dim R$.

$(ii) \Rightarrow (iii)$. Let $\phi : k[X_1, \ldots, X_d] \to \operatorname{gr}_{\mathfrak{m}}(R)$ be a graded k-algebra isomorphism. Then the first homogeneous components are isomorphic, as k-spaces. In particular, $\dim_k(\mathfrak{m}/\mathfrak{m}^2) = d$.

$(iii) \Rightarrow (i)$. If $\dim_k(\mathfrak{m}/\mathfrak{m}^2) = d = \dim R$, then any basis of d elements of $\mathfrak{m}/\mathfrak{m}^2$ can be lifted to a generating set of $\mathfrak{m}$. Hence R is regular. □

Corollary 9.1.4 *A regular local ring is a domain.*

Proof Since $\operatorname{gr}_{\mathfrak{m}}(R) \simeq k[X_1, \ldots, X_d]$ is a domain, R is also a domain by Proposition 6.4.7.

□

We will show in the next section that a regular local ring is a UFD.

Definition 9.1.5 Let R be a local ring and $\widehat{R}$ its $\mathfrak{m}$-adic completion. Then R is regular if and only if $\widehat{R}$ is regular.

Proof By Proposition 6.4.1, $\mathrm{gr}_{\mathfrak{m}}(R) \simeq \mathrm{gr}_{\widehat{\mathfrak{m}}}(\widehat{R})$, as graded rings. The result now follows from condition (ii) of Theorem 9.1.3. □

If R is a regular local ring and P a prime ideal of R, R/P need not be a regular local ring in general, as is shown by Example (iv). The following Proposition shows conditions under which R/P will be regular.

Proposition 9.1.6 *Let R be a regular local ring of dimension d and $a_1, \ldots, a_t \in \mathfrak{m}$ $(1 \leq t \leq d)$. The following conditions are equivalent.*

(i) *$a_1, a_2, \ldots, a_t$ form a part of a regular system of parameters for R.*
(ii) *The images $\overline{a}_1, \overline{a}_2, \ldots, \overline{a}_t$ of $a_1, \ldots, a_t$ modulo $\mathfrak{m}^2$ are linearly independent over k.*
(iii) *$R/(a_1, \ldots, a_t)$ is a regular local ring of dimension $d - t$.*

Proof $(i) \Leftrightarrow (ii)$. The elements $a_1, \ldots, a_t$ form a part of a regular system of parameters $a_1, \ldots, a_d$ if and only if $\overline{a}_1, \overline{a}_2, \ldots, \overline{a}_t$ form a part of $\overline{a}_1, \overline{a}_2, \ldots, \overline{a}_d$, a k-basis of $\mathfrak{m}/\mathfrak{m}^2$, i.e., if and only if $\overline{a}_1, \overline{a}_2, \ldots, \overline{a}_t$ are k-linearly independent.
$(ii) \Rightarrow (iii)$. Let $\overline{R} = R/(a_1, \ldots, a_t)$ and $\overline{\mathfrak{m}} = \mathfrak{m}/(a_1, \ldots, a_t)$. Let $a_1, a_2, \ldots, a_t, a_{t+1}, \ldots, a_d$ be a regular system of parameters for R. Their images generate $\overline{\mathfrak{m}}$. By Theorem 8.4.4, $\dim \overline{R}$ as an R-module is $d - t$. This implies $\dim \overline{R} = d - t$. Since $\overline{\mathfrak{m}}$ is generated by $d - t$ elements, $\overline{R}$ is regular.
$(iii) \Rightarrow (i)$. Let $a_{t+1}, \ldots, a_d \in \mathfrak{m}$ be such that their images in $\overline{\mathfrak{m}} = \frac{\mathfrak{m}}{(a_1,\ldots,a_t)}$ form a regular system of parameters. Then $\mathfrak{m}$ is generated by $a_1, \ldots, a_t, a_{t+1}, \ldots, a_d$. Hence R is regular and $a_1, \ldots, a_t$ is a part of a regular system of parameters. □

Corollary 9.1.7 *Let R be a Noetherian local ring with maximal ideal $\mathfrak{m}$. Then R is regular $\Leftrightarrow$ $\mathfrak{m}$ is generated by an R-sequence.*

Proof Let R be a regular ring and $a_1, \ldots, a_d$ a regular system of parameters. For each t $(0 \leq t \leq d)$, $\overline{R} = R/(a_1, \ldots, a_t)$ is regular and $\bar{a}_{t+1}, \ldots, \bar{a}_d$ generate $\overline{\mathfrak{m}} = \frac{\mathfrak{m}}{(a_1,\ldots,a_t)}$, *i.e.*, it is a regular system of parameters for $\overline{R}$. Since $\overline{R}$ is a domain, $\overline{a}_{t+1} \neq 0$ is not a zero-divisor of $\overline{R}$, *i.e.*, a_{t+1} is not a zero-divisor of $R/(a_1, \ldots, a_t)$, $(0 \leq t \leq d)$. Hence $a_1, a_2, \ldots, a_d$ is an R-sequence.

Conversely, let $\mathfrak{m}$ be generated by an R-sequence $a_1, \ldots, a_d$, so that $d \leq \operatorname{depth} R \leq \dim R \leq d$. Hence $\dim R = d$ and R is regular. □

Corollary 9.1.8 *A regular local ring is C.M.*

Proof If $\mathfrak{m}$ is generated by an R-sequence $a_1, a_2, \ldots, a_d$, then depth $R = \dim R = d$. $\square$

Corollary 9.1.9 *Let R be a regular local ring and P a prime ideal of R. The following conditions are equivalent.*

(i) R/P is a regular local ring.
(ii) P is generated by a part of a regular system of parameters.

Proof $(ii) \Rightarrow (i)$ follows from Proposition 9.1.6.

$(i) \Rightarrow (ii)$ Let R/P be regular of dimension $d-t$. If $\overline{\mathfrak{m}} = \mathfrak{m}/P$, then $\dim_k(\overline{\mathfrak{m}}/\overline{\mathfrak{m}}^2) = d-t$. Since

$$\frac{\overline{\mathfrak{m}}}{\overline{\mathfrak{m}}^2} \simeq \frac{\mathfrak{m}}{\mathfrak{m}^2 + P}$$

from the exact sequence

$$0 \longrightarrow \frac{(\mathfrak{m}^2 + P)}{\mathfrak{m}^2} \longrightarrow \frac{\mathfrak{m}}{\mathfrak{m}^2} \longrightarrow \frac{\mathfrak{m}}{(\mathfrak{m}^2 + P)} \longrightarrow 0,$$

we have

$$\dim_k\left(\frac{\mathfrak{m}^2 + P}{\mathfrak{m}^2}\right) = \dim_k\left(\frac{\mathfrak{m}}{\mathfrak{m}^2}\right) - \dim_k\left(\frac{\mathfrak{m}}{\mathfrak{m}^2 + P}\right) = d - (d-t) = t.$$

Choose $a_1, \ldots, a_t \in P$ such that their images modulo $\mathfrak{m}^2$ span $(\mathfrak{m}^2 + P)/\mathfrak{m}^2$ over k. Choose $a_{t+1}, \ldots, a_d \in \mathfrak{m}$ which together with $a_1, \ldots, a_t$ span $\mathfrak{m}$ modulo $\mathfrak{m}^2$. Then $a_1, \ldots, a_d$ is a regular system of parameters for R. By Corollary 9.1.8, it is an R-sequence. Hence $P' = (a_1, \ldots, a_t)$ is a prime ideal of height t contained in P. Since $P' \subset P$ and $\dim R/P = d - t$, we have $P' = P$. $\square$

The following Theorem is the algebraic analogue of the geometric result which says that P is a non-singular point of a variety if and only if the local ring at P is a regular local ring.

Theorem 9.1.10 *Let $R = k[X_1, \ldots, X_n]$; k a field, $I = (f_1, \ldots, f_t)$ an ideal of R and $\mathfrak{m} = (X_1 - a_1, \ldots, X_n - a_n)$ a maximal ideal of R containing I. If $S = (R/I)_{\mathfrak{m}/I}$, then S is a regular local ring if and only if $\dim S = n - r$, where r is the rank of the matrix $\left(\frac{\partial f_i}{\partial x_j}\right)_{(a)}$, $(a) = (a_1, a_2, \ldots, a_n)$.*

Proof Consider the k-linear map $\lambda : R \to k^n$ defined by

$$\lambda(f) = \left(\left(\frac{\partial f}{\partial x_1}\right)_a, \ldots, \left(\frac{\partial f}{\partial x_n}\right)_a\right), \quad f \in R.$$

Then λ maps $\mathfrak{m}$ onto k^n and $\mathfrak{m}^2$ onto zero, so that it induces a k-isomorphism $\overline{\lambda} : \mathfrak{m}/\mathfrak{m}^2 \simeq k^n$. If r = rank $\left(\frac{\partial f_i}{\partial x_j}\right)_{(a)}$, then

$$r = \dim_k \overline{\lambda}(I) = \dim_k \frac{(I + \mathfrak{m}^2)}{\mathfrak{m}^2}.$$

From the exact sequence

$$0 \to \frac{I + \mathfrak{m}^2}{\mathfrak{m}^2} \to \frac{\mathfrak{m}}{\mathfrak{m}^2} \to \frac{\mathfrak{m}}{I + \mathfrak{m}^2} \to 0$$

we have

$$\dim_k \left(\frac{I + \mathfrak{m}^2}{\mathfrak{m}^2}\right) = \dim_k \left(\frac{\mathfrak{m}}{\mathfrak{m}^2}\right) - \dim_k \left(\frac{\mathfrak{m}}{I + \mathfrak{m}^2}\right).$$

Thus $\dim_k \left(\frac{\mathfrak{m}}{I+\mathfrak{m}^2}\right) = n - r$. Let $\overline{\mathfrak{m}}$ be the unique maximal ideal of S, so that

$$\frac{\overline{\mathfrak{m}}}{\overline{\mathfrak{m}}^2} \simeq \frac{\mathfrak{m}}{I + \mathfrak{m}^2}$$

as k-spaces. Thus S is regular $\Leftrightarrow \dim S = \dim_k \overline{\mathfrak{m}}/\overline{\mathfrak{m}}^2 = n - r$. □

9.1. EXERCISES

1. Give an example of a Cohen–Macaulay local ring which is not a regular local ring.
2. Give examples of local rings $R \subset S$, where either may be regular without the other being so.
3. Let R be a Noetherian ring, I an ideal of R and $S = R/I$. S is said to be regularly imbedded in R if I is generated by an R-sequence. Show that if R and S are regular local rings, then S is regularly imbedded in R.
4. Let R be a Noetherian local ring with maximal ideal $\mathfrak{m}$ and $a \in \mathfrak{m} - \mathfrak{m}^2$. Show that R is regular implies $R/(a)$ is regular and that the converse holds if a does not belong to any minimal prime ideal of R.
5. Let R be a local ring. Show that R is regular if and only if $R[[X]]$ is regular.

9.2 Homological characterisation

Let R be a local ring. We obtain a homological characterisation for the regularity of R, *viz.* R is regular if and only if $\text{gl}.\dim R < \infty$.

Theorem 9.2.1 *Let R be a regular local ring. Then*

(i) $\operatorname{gl.dim} R < \infty$ *and is equal to* $\dim R$.
(ii) $\operatorname{pd}_R M + \operatorname{depth}_R M = \dim R$ *for any finitely generated R-module M.*

Proof (i) Let $a_1, \dots, a_d$ be a regular system of parameters for R, so that it is also an R-sequence (Corollary 9.1.8). Since

$$R/(a_1, \dots, a_d) \simeq k,$$

we have by Corollary 8.4.22, $\operatorname{pd}_R k = d$. But by Corollary 7.3.30, $\operatorname{gl.dim} R = \operatorname{pd}_R k = d = \dim R$.

(ii) This is proved by induction on $\operatorname{depth}_R M = t$. If $t = 0$, $\mathfrak{m} \in \operatorname{Ass}(M)$, so that there exists an injection $0 \to k \to M$. Since

$$\operatorname{pd}_R k = d, \quad \operatorname{Tor}_{d+1}^R\left(\frac{M}{k}, k\right) = 0,$$

so that the exact sequence

$$0 \to k \to M \to M/k \to 0$$

induces an exact sequence

$$0 = \operatorname{Tor}_{d+l}^R\left(\frac{M}{k}, k\right) \to \operatorname{Tor}_d^R(k, k) \to \operatorname{Tor}_d^R(M, k).$$

Since $\operatorname{Tor}_d^R(k, k) \neq 0$ (Theorem 7.3.29), $\operatorname{Tor}_d^R(M, k) \neq 0$. Hence $\operatorname{pd}_R M = d$ and the result is proved. Assume therefore that

$$\operatorname{depth}_R M = t > 0.$$

Then there exists some $a \in \mathfrak{m}$ which is not a zero-divisor of M. Let $M_1 = M/aM$ and consider the exact sequence $0 \to M \xrightarrow{\lambda_a} M \to M_1 \to 0$, where λ_a is scalar multiplication by a. Since $\operatorname{depth}_R M_1 = t-1$, we have by induction $\operatorname{pd}_R M_1 + \operatorname{depth}_R M_1 = d$. Consider the exact sequence

$$\to \operatorname{Tor}_i^R(M, k) \xrightarrow{\lambda_a^*} \operatorname{Tor}_i^R(M, k) \to \operatorname{Tor}_i^R(M_1, k) \xrightarrow{\delta}$$
$$\operatorname{Tor}_{i-1}^R(M, k) \xrightarrow{\lambda_a^*} \operatorname{Tor}_{i-1}^R(M, k)$$

where λ_a^* is multiplication by a. Since a annihilates k, we have an exact sequence

$$0 \to \operatorname{Tor}_i^R(M, k) \to \operatorname{Tor}_i^R(M_1, k) \to \operatorname{Tor}_{i-1}^R(M, k) \to 0.$$

Applying Corollary 7.3.28, we get $\operatorname{pd}_R M_1 = 1 + \operatorname{pd}_R M$. Hence $\operatorname{pd}_R M + \operatorname{depth}_R M = d$. □

Corollary 9.2.2 *Let M be a finitely generated R-module over a regular local ring R. Then* $\operatorname{pd}_R M = \dim R \Leftrightarrow \mathfrak{m} \in \operatorname{Ass}(M)$.

Proof $\operatorname{depth}_R M = 0 \Leftrightarrow \mathfrak{m} \in \operatorname{Ass}(M)$. □

Corollary 9.2.3 *Let R be a regular local ring and M a finitely generated R-module with* $\dim M = d = \dim R$. *Then M is C.M. $\Leftrightarrow$ M is free.*

Proof M is free $\Leftrightarrow \operatorname{pd}_R M = 0 \Leftrightarrow \operatorname{depth}_R M = \dim R = \dim M \Leftrightarrow M$ is C.M.. □

Corollary 9.2.4 *Let R be a regular local ring,* $\dim R = d$, *and S a local ring of dimension d which is an R-algebra of finite type, i.e., finitely generated as an R-module. Then S is C.M. $\Leftrightarrow$ S is free as an R-module.*

Proof By Exercise 3, 8.5, S is C.M. over $S \Leftrightarrow S$ is C.M. over R. By Corollary 9.2.3, this is equivalent to S being free over R.

Corollary 9.2.5 *Let R be a regular local ring,* $\dim R = d$, *and S a regular local ring of dimension d which is an R-algebra of finite type. Then S is free over R.*

Proof This follows from Corollaries 9.1.8 and 9.2.4. □

Proposition 9.2.6 *Let M be an R-module, $a \in R$ a non-zero-divisor of R as well as M. If* $\operatorname{pd}_R M < \infty$, *then* $\operatorname{pd}_{R/(a)}(M/aM) < \infty$.

Proof The proof is by induction on $n = \operatorname{pd}_R M$. If $n = 0$, M is R-projective and hence $M/aM = M \otimes R/(a)$ is $R/(a)$-projective. Assume $n > 0$ and consider the exact sequence

$$0 \to R \stackrel{\lambda_a}{\to} R \to R/(a) \to 0,$$

where λ_a is scalar multiplication by a. This induces an exact sequence

$$0 \to \operatorname{Tor}_1^R(M, R/(a)) \to M \stackrel{\lambda_a^*}{\to} M \to M/aM \to 0.$$

Since a is not a zero-divisor of M, $\operatorname{Tor}_1^R(M, R/(a)) = 0$. Consider an exact sequence

$$0 \to K \to F \to M \to 0,$$

with F free. Then the sequence

$$0 = \operatorname{Tor}_1^R(M, R/(a)) \to K/aK \to F/aF \to M/aM \to 0$$

is exact. By Exercise 10, 7.3, $\operatorname{pd}_R K = \operatorname{pd}_R M - 1$, so that induction applies to K because a is a non-zero-divisor on $K \subset F$. Since $\operatorname{pd}_R M < \infty$, we have $\operatorname{pd}_R K < \infty$. This implies

$$\operatorname{pd}_{R/(a)}(K/aK) < \infty \quad \text{and hence} \quad \operatorname{pd}_{R/(a)}(M/aM) < \infty.$$

□

Proposition 9.2.7 *Let R be a Noetherian local ring with maximal ideal $\mathfrak{m}$.*

(i) *For any $a \in \mathfrak{m} - \mathfrak{m}^2$, the exact sequence of $R/(a)$ modules,*

$$0 \to \frac{Ra}{\mathfrak{m}a} \to \frac{\mathfrak{m}}{\mathfrak{m}a} \to \frac{\mathfrak{m}}{Ra} \to 0 \quad splits.$$

(ii) *If $\mathfrak{m} \neq \mathfrak{m}^2$ and every $a \in \mathfrak{m} - \mathfrak{m}^2$ is a zero-divisor, any finitely generated R-module M with $\operatorname{pd}_R M < \infty$ is free.*

Proof (i) Let $d = \dim_k \mathfrak{m}/\mathfrak{m}^2$. Since $a \notin \mathfrak{m}^2$, there exist $a_1, a_2, \ldots, a_{d-1} \in \mathfrak{m}$ such that $(a, a_1, \ldots, a_{d-1})$ is a minimal generating set for $\mathfrak{m}$. If $I = (a_1, \ldots, a_{d-1})$, then $I \cap Ra = I \cap \mathfrak{m}a$; for if $b = ca \in I, c \in R$, c cannot be a unit, for otherwise $a \in I$, which is impossible as $a, a_1, \ldots, a_{d-1}$ is a minimal generating set. Hence $b \in I \cap \mathfrak{m}a$. Now,

$$\frac{\mathfrak{m}}{Ra} = \frac{I + Ra}{Ra} \simeq \frac{I}{I \cap Ra} = \frac{I}{I \cap \mathfrak{m}a} \simeq \frac{I + \mathfrak{m}a}{\mathfrak{m}a}.$$

Hence the natural map $\dfrac{\mathfrak{m}}{\mathfrak{m}a} \to \dfrac{\mathfrak{m}}{Ra}$ maps $\dfrac{I + \mathfrak{m}a}{\mathfrak{m}a}$ isomorphically onto $\dfrac{\mathfrak{m}}{Ra}$, whose inverse gives the required splitting $\dfrac{\mathfrak{m}}{Ra} \to \dfrac{\mathfrak{m}}{\mathfrak{m}a}$.

(ii) Assume now that each $a \in \mathfrak{m} - \mathfrak{m}^2$ is a zero-divisor. Then $\mathfrak{m} - \mathfrak{m}^2 \subset \bigcup_P P$ where the union is taken over all the associated prime ideals of R, *i.e.*, $\mathfrak{m} \subset (\bigcup_P P) \cup \mathfrak{m}^2$. Since $\mathfrak{m} \neq \mathfrak{m}^2$, we have $\mathfrak{m} \subset \bigcup_P P$ and hence $\mathfrak{m} = P$. This implies that there exists an injection of $k = R/\mathfrak{m}$ in R. If M is a finitely generated R-module with $\operatorname{pd}_R M = n$, then the exact sequence $0 \to k \to R \to R/k \to 0$ induces an exact sequence

$$0 \to \operatorname{Tor}_{n+1}^R(M, R/k) \to \operatorname{Tor}_n^R(M, k) \to \operatorname{Tor}_n^R(M, R).$$

Since $\operatorname{Tor}_n^R(M, k) \neq 0$, we have $\operatorname{Tor}_n^R(M, R) \neq 0$. This is possible only if $n = 0$. Hence M is projective and hence free. □

Corollary 9.2.8 *Let R be a Noetherian local ring with $\operatorname{gl.dim} R < \infty$ and $a \in \mathfrak{m} - \mathfrak{m}^2$, a non-zero-divisor of R. Then $\operatorname{gl.dim} R/(a) < \infty$.*

Proof Consider the exact sequence of $R/(a)$-modules

$$0 \to \mathfrak{m}/Ra \to R/Ra \to R/\mathfrak{m} \to 0.$$

Since $\operatorname{gl.dim} R/(a) = \operatorname{pd}_{R/(a)} R/\mathfrak{m}$, it is sufficient to show that $\operatorname{pd}_{R/(a)} \mathfrak{m}/Ra < \infty$. Since $\operatorname{gl.dim} R < \infty$, $\operatorname{pd}_R \mathfrak{m} < \infty$ and we have by Proposition 9.2.6 that $\operatorname{pd}_{R/(a)} \mathfrak{m}/\mathfrak{m}a < \infty$.

By Proposition 9.2.7, $\mathfrak{m}/Ra$ is an $R/(a)$-direct summand of $\mathfrak{m}/\mathfrak{m}a$, so that $\operatorname{pd}_{R/(a)} \mathfrak{m}/Ra < \infty$. □

Theorem 9.2.9 *Let R be a Noetherian local ring with* $\operatorname{gl.dim} R < \infty$. *Then R is regular.*

Proof The proof is by induction on $t = \dim_k \mathfrak{m}/\mathfrak{m}^2$. If $t = 0$, $\mathfrak{m} = \mathfrak{m}^2$, so that by Nakayama Lemma, $\mathfrak{m} = 0$ and R is regular. Assume $t > 0$. If every $a \in \mathfrak{m} - \mathfrak{m}^2$ is a zero-divisor, then $\operatorname{gl.dim} R = \operatorname{pd}_R k < \infty$ would imply (Proposition 9.2.7) that k is R-free, *i.e.*, $\mathfrak{m} = 0$, a contradiction. Hence there exists some $a \in \mathfrak{m} - \mathfrak{m}^2$ which is not a zero-divisor. By Corollary 9.2.8, we have $\operatorname{gl.dim} R/(a) < \infty$. If $\overline{\mathfrak{m}} = \mathfrak{m}/Ra$, $\overline{\mathfrak{m}}/\overline{\mathfrak{m}}^2$ is a k-vector space of dimension $t-1$, so that by induction $R/(a)$ is regular, *i.e.*, $\overline{\mathfrak{m}}$ is generated by an $R/(a)$-sequence $\overline{a}_1, \ldots, \overline{a}_{t-1}$. Then $a, a_1, \ldots, a_t$, is an R-sequence which generates $\mathfrak{m}$ and R is regular.

Corollary 9.2.10 *Let R be a regular local ring and P any prime ideal of R. Then R_P is a regular local ring.*

Proof Let $\operatorname{gl.dim} R = n$, so that $\operatorname{pd}_R R/P \leq n$. Hence it has a free resolution of length $t \leq n$, say

$$0 \to F_t \to F_{t-1} \to \cdots \to F_0 \to R/P \to 0.$$

Since R_P is R-flat, the sequence

$$0 \to R_P \otimes_R F_t \to R_P \otimes_R F_{t-1} \to \cdots \to R_P \otimes_R F_0 \to R_P \otimes_R R/P \to 0$$

is exact and each $R_P \otimes F_i$ $(0 \leq i \leq t)$ is R_P-free. Hence

$$\operatorname{pd}_{R_P} (R_P \otimes R/P) \leq n,$$

i.e., $\operatorname{pd}_{R_P} \left(\dfrac{R_P}{PR_P} \right) \leq n$. Hence $\operatorname{gl.dim} R_P \leq n$, *i.e.*, R_P is a regular local ring. □

Definition 9.2.11 A ring R is called *regular* if R_P is a regular local ring for every prime ideal P of R.

In view of the above Corollary 9.2.10, a regular local ring will also be regular in the sense of the above definition.

Theorem 9.2.12 *If R is a regular ring, then the polynomial ring $R[X_1, X_2, \ldots, X_n]$ is also regular.*

Proof It is sufficient to prove the result for $n = 1$. Let Q be a prime ideal of $S = R[X]$ lying above $P = R \cap Q$. Since S_Q is a localisation of $R_P[X]$ and R_P is a regular local ring, we may assume without loss of generality that R is a regular local ring with maximal ideal $\mathfrak{m}$. We have $S/\mathfrak{m}S \simeq R[X] \otimes_R k \simeq k[X]$, where $k = R/\mathfrak{m}$. Since $Q \supseteq \mathfrak{m}S$, we have either $Q = \mathfrak{m}S$ or $Q = \mathfrak{m}S + f(X)S$ for some monic polynomial $f(X) \in S(= R[X])$.

Since R is regular local, $\mathfrak{m}$ is generated by $d = \dim R$ elements $a_1, a_2, \ldots, a_d$. Hence Q is generated over S by d or $d+1$ elements, according as $Q = \mathfrak{m}S$ or $Q = \mathfrak{m}S + (f)$. Since $\mathrm{ht}(Q) \geq d$, we have $\mathrm{ht}(Q) = d$ if $Q = \mathfrak{m}S$ or $\mathrm{ht}(Q) = d+1$ if $Q = \mathfrak{m}S + (f)$. Hence S_Q is a regular local ring. □

Corollary 9.2.13 *If k is a field, then the polynomial ring $k[X_1, \ldots, X_n]$ is regular.*

Corollary 9.2.14 *If R is Noetherian regular, then* $\mathrm{gl.dim}\, R = \dim R$.

Proof

$$\begin{aligned} \mathrm{gl.dim}\, R &= \mathrm{Sup}\ \ \mathrm{gl.dim}\, R_{\mathfrak{m}} && \text{(Exercise 5, 7.3)} \\ &= \mathrm{Sup}\ \ \dim R_{\mathfrak{m}} && \text{(Theorem 9.2.1)} \\ &= \dim R && \text{(Exercise 6, 8.3).} \end{aligned}$$

□

Corollary 9.2.15 $\mathrm{gl.dim}\, k[X_1, X_2, \ldots, X_n] = n$, *$k$ field.*

□

Proof $\dim k[X_1, X_2, \ldots, X_n] = n$ by Corollary 8.3.5.
We have already proved that a regular local ring is a domain (Corollary 9.1.4). We now show that a regular local ring is a UFD.

Proposition 9.2.16 *Let R be a Noetherian domain. Then R is a UFD if and only if every ht 1 prime ideal of R is principal.*

Proof Assume R is a UFD and let P be a prime ideal of ht 1. Let $a \in P$, $a \neq 0$, and p be an irreducible factor of a such that $p \in P$. Then $Rp \subseteq P$ and since $\mathrm{ht}(P) = 1$ and Rp is prime, $Rp = P$, *i.e.*, P is principal.

Conversely, assume that every ht 1 prime ideal is principal. Since R is Noetherian, each $a \in R$ which is a non-unit can be expressed as a product of irreducible elements. It is sufficient to show that every irreducible element is prime. Let $a \in R$ be irreducible and P a minimal prime ideal of (a). Then $\mathrm{ht}(P) = 1$ and by assumption P is principal, *i.e.*, $P = Rp$, $p \in R$, p prime. Then $a = up$ for some $u \in R$. As p is prime dividing the irreducible element a, we have u is a unit and hence a is a prime element. Thus the proof is complete. □

Proposition 9.2.17 *Let R be a ring and M a projective R-module.*

(i) If M has a finite free resolution

$$0 \to F_n \to F_{n-1} \to \cdots \to F_0 \to M \to 0$$

where each F_i is free of finite rank ($0 \leq i \leq n$), then there exists a finitely generated free R-module F with $M \oplus F$ free of finite rank.

(*ii*) *If* $M \oplus R^n \simeq R^{n+1}$, *then* $M \simeq R$.

Proof (i) The proof is by induction on n. If $n = 0$, $M \simeq F_0$ is finitely generated free. Let $n > 0$ and let $K = \text{Ker}(F_0 \to M)$. Since M is projective, $F_0 \simeq M \oplus K$ and K has a free resolution of length $n-1$. By induction there exists a finitely generated free module F' with $K \oplus F' = F$, finitely generated free. Then

$$M \oplus F = M \oplus K \oplus F' \simeq F_0 \oplus F'$$

is finitely generated free.

(ii) Let M be such that $M \oplus R^n \simeq R^{n+1}$. Then $M_P \oplus R_P^n \simeq R_P^{n+1}$ for every prime ideal P so that M is projective of constant rank one. Moreover

$$(\wedge_R^i M)_P \simeq R_P \otimes_R \wedge_R^i M \simeq \wedge_{R_P}^i (R_P \otimes_R M) \simeq \wedge_{R_P}^i M_P = 0 \;\; (i > 1).$$

As this is true for every prime ideal P, $\wedge^i M = 0$, for $i > 1$.

$$\begin{aligned} \text{Now,} \quad R = \wedge^{n+1} R^{n+1} &\simeq \wedge^{n+1}(M \oplus R_n) \\ &\simeq \oplus \Sigma_{i+j=n+1} \wedge^i M \otimes \wedge^j R^n \\ &\simeq \wedge^0 M \otimes \wedge^{n+1} R^n \oplus \wedge^1 M \otimes \wedge^n R^n \\ &\simeq M \otimes_R R \simeq M. \end{aligned}$$

□

Corollary 9.2.18 *Let I be a non-zero projective ideal of R having a finite free resolution. Then I is principal.*

Proof By Proposition 9.2.17 (i), there exists a finitely generated free module F such that $I \oplus F = F'$, where F' is free of finite rank. Since $I_P \oplus F_P \simeq F'_P$, I is of constant rank equal to $\text{rank } F' - \text{rank } F$. Since I is a non-zero projective ideal, $\text{rank } I = 1$. Hence by Proposition 9.2.17 (ii), $I \simeq R$. □

Theorem 9.2.19 *If R be a regular local ring, then R is a UFD.*

Proof The proof is by induction on $d = \dim R$. If $d = 0$, R is a field. Assume $d > 0$. Since R is a domain, it is sufficient to show that every $\text{ht}\, 1$ prime ideal is principal. Let P be a prime ideal of $\text{ht}\, 1$. Since $d \geq 1$, $\mathfrak{m} \neq \mathfrak{m}^2$ and we can choose $a \in \mathfrak{m} - \mathfrak{m}^2$. Since $\{a\}$ is a part of a regular system of parameters (Exercise 4, 9.1) the ideal Ra is a prime ideal, *i.e.*, a is a prime element. If $a \in P$, then $Ra \subset P$ and since P is of $\text{ht}\, 1$, $P = Ra$ is principal. Assume therefore that $a \notin P$ and consider the multiplicatively closed set $S = \{a^n | n \geq 0\}$.

If $R' = R_S$, then $P' = PR'$ is a prime ideal of R' of ht 1. We show that P' is principal. Since $\text{gl}.\dim R < \infty$, P admits a finite free R-resolution and hence $P' = PR'$ admits a finite free R'-resolution. We show that P' is a projective ideal in R'. Let $Q' = QR'$ be any prime ideal of R' where Q is a prime ideal of R. Then $Q \neq \mathfrak{m}$ as $a \notin Q$. Then $R'_{Q'} \simeq R_Q$ is a regular local ring of dimension less than $\dim R$. Hence by induction, $R'_{Q'}$ is a UFD. Since $P'R'_{Q'}$ is a prime ideal of ht 1, it is principal. Hence $\text{pd}_{R'} P' = \text{Sup}\, \text{pd}_{R'_{Q'}} P'R'_{Q'} = 0$.

Hence P' is R'-projective. Now, by Corollary 9.2.18, P' is a principal ideal in R'. Write $P' = R'p$, $p \in P$. By removing the highest power of a dividing p, we may assume without loss of generality that a does not divide p. We claim that $P = Rp$. Clearly $Rp \subseteq P$. Let $x \in P = P' \cap R$. Write $x = bp/a^m$, *i.e.*, $a^m x = bp$. Since a is a prime element and a does not divide p, we have $a^m | b$, *i.e.*, $x \in Rp$ showing $P = Rp$. □

9.2. EXERCISES

1. Let R be a regular local ring and I an ideal of R. Then R/I is Cohen–Macaulay if and only if the ideal I satisfies the condition $\text{ht}(I) = \text{pd}_R(R/I)$.

2. Let M be a finitely generated module over a Noetherian ring R and N a submodule of M. Show that if N is expressed as a finite intersection of irreducible submodules $N = \cap_i N_i$ with $\cap_{k\neq j} N_k \not\subset N_j$ then the number of factors in the decomposition, depends only on N (called the *index of irreducibility* (I.I.D.) of N in M).

3. Let R be a C.M. local ring with $\dim R = d$. If $a_1, \ldots, a_d$ and $b_1, \ldots, b_d$ are two systems of parameters for R, show that the ideals $(a_1, \ldots, a_d)$ and $(b_1, b_2, \ldots, b_d)$ have the same I.I.D. (called the *index of irreducibility* (I.I.D.) of R).

4. Let R be a regular local ring of dimension d and $a_1, \ldots, a_d$, a system of parameters. Show that the ideal $(a_1 \ldots, a_d)$ is irreducible.

5. Let R be a local ring such that any ideal generated by a system of parameters is irreducible. Show that R is C.M.

6. Let R be a C.M. ring whose index of irreducibility is one. Then R is called a *Gorenstein ring*. Show that a local ring R with $\dim R = n$ is a Gorenstein ring $\Leftrightarrow id_R R = n$. (H. Bass: On the ubiquity of Gorenstein Rings: Math. Zeitschrift. Vol. 82 (1963)).

9.3 Normality conditions

Let R be a Noetherian ring. Consider the following conditions.

$(k \geq 0)$

(S_k): $\operatorname{depth} R_P \geq \operatorname{Min}(k, \dim R_P)$

(R_k): $\dim R_P \leq k$ implies R_P is regular.

Clearly (S_0) is always true.

The following Proposition shows that (S_1) is equivalent to the condition that R has no embedded primes.

Proposition 9.3.1 *R satisfies (S_1) if and only if R has no embedded primes.*

Proof Assume that R satisfies (S_1) and suppose P is a prime ideal of R which is not minimal. Then $\dim R_P \geq 1$ and hence by (S_1), $\operatorname{depth} R_P \geq 1$. This implies $P \notin \operatorname{Ass}(R)$; for if $P \in \operatorname{Ass}(R)$, then $PR_P \in \operatorname{Ass}(R_P)$, a contradiction as $\operatorname{depth} R_P \geq 1$. Hence every $P \in \operatorname{Ass}(R)$ is minimal.

Conversely, assume that every $P \in \operatorname{Ass}(R)$ is minimal. If $P \in \operatorname{Ass}(R)$, then $\operatorname{depth} R_P \geq 0 = \min(1, \dim R_P)$. If $P \notin \operatorname{Ass}(R)$, $\min(1, \dim R_P) = 1$ and $\operatorname{depth} R_P \geq 1$; for if $\operatorname{depth} R_P = 0$, then $PR_P \in \operatorname{Ass}(R_P)$, *i.e.*, $P \in \operatorname{Ass}(R)$, a contradiction. Hence (S_1) is satisfied. □

Proposition 9.3.2 *R is reduced, i.e., the nil radical $N(R) = 0$ if and only if R satisfies (S_1) and (R_0).*

Proof Clearly (R_0) is equivalent to the condition that R_P is a field for all minimal prime ideals P of R.

Assume that R is reduced. If P is a minimal prime ideal of R, R_P is an Artinian local ring and $N(R_P) = N(R)_P = 0$, so that R_P is a field, *i.e.*, R satisfies (R_0).

Suppose R does not satisfy (S_1). By Proposition 9.3.1, there exists $P \in \operatorname{Ass}(R)$ which is not minimal. Let $P_1, P_2, \ldots, P_r$ be the minimal prime ideals of R.

Then $P \not\subset \bigcup_i P_i$. Choose $a \in P$ such that $a \notin P_i$ $(1 \leq i \leq r)$. Since a is a zero-divisor there exists $b (\neq 0) \in R$ such that $ba = 0$. Then $(ba/1) = 0$ in R_{P_i} for each i $(1 \leq i \leq r)$. Since $(a/1)$ is a unit in R_{P_i}, we have $(b/1) = 0$ in R_{P_i}, i.e., $s_i b = 0$ in R for some $s_i \in R - P_i$ $(1 \leq i \leq r)$. This implies $b \in \bigcap_i P_i = 0$, a contradiction. Thus R satisfies (S_1).

Conversely, assume that R satisfies (S_1) and (R_0). Let $P_1, P_2, \ldots, P_r$ be the minimal prime ideals of R. If $\cap_i P_i \neq 0$, then there exists some $a \in \cap_i P_i$, $a \neq 0$. Since R_{P_i} is a field $(1 \leq i \leq r)$, $P_i R_{P_i} = 0$, *i.e.*, $s_i a = 0$ for some $s_i \notin P_i$ $(1 \leq i \leq r)$. Thus $\operatorname{Ann}(a) \not\subset P_i$ $(1 \leq i \leq r)$ and hence $\operatorname{Ann}(a) \not\subset \cup_{i=1}^r P_i$. By (S_1), R has no embedded primes, so that $\cup P_i = \bigcup_{P \in Ass\, R} P$. This implies $\operatorname{Ann}(a) \not\subset \bigcup_{P \in Ass\, R} P$, a contradiction. Hence $\cap_i P_i = 0$. □

Definition 9.3.3 A ring R is called *normal* if R_P is an integrally closed domain for every prime ideal P of R.

Theorem 9.3.4 *Let R be a ring and $Q(R)$ its total quotient ring. The following conditions are equivalent.*

(i) *R is normal.*

(ii) *R is reduced and integrally closed in $Q(R)$.*

(iii) *R is a finite product of integrally closed domains.*

Proof $(i) \Rightarrow (ii)$. Let R be normal. If $a \in R$ is nilpotent, $a^n = 0$ in R_P, for every prime ideal P and since R_P is a domain, $a = 0$ in R_P for every P, *i.e.*, $a = 0$. Hence R is reduced. Let $\alpha = a/b \in Q(R)$ be integral over R. For any maximal ideal $\mathfrak{m}$ of R, $\alpha \in Q(R_{\mathfrak{m}})$ is integral over $R_{\mathfrak{m}}$ and since $R_{\mathfrak{m}}$ is integrally closed, $\alpha \in R_{\mathfrak{m}}$. This implies $ca \in Rb$ for some $c \notin \mathfrak{m}$. Let $I = \{\lambda \in R \mid \lambda a \in Rb\}$. If I were a proper ideal, then I would have been contained in $\mathfrak{m}_0$, for some maximal ideal $\mathfrak{m}_0$. Now, for some $c_0 \notin \mathfrak{m}_0$, $c_0 a \in Rb$, a contradiction. Hence $1 \in I$ and $\alpha = a/b \in R$.

$(ii) \Rightarrow (iii)$. Since R is reduced, by Proposition 9.3.2, R satisfies S_1. Let $P_1, P_2, \ldots P_r$ be the minimal prime ideals of R. Then $\cap P_i = 0$ and by (S_1) these are all the prime ideals of $\text{Ass}(R)$. If T is the set of non-zero-divisors of R, $R_T = Q(R)$. Now, $P_i R_T$ $(1 \le i \le r)$ are mutually coprime and $\cap P_i R_T = 0$. By the Chinese Remainder Theorem

$$Q(R) = R_T = \prod_{i=1}^{r} \frac{R_T}{P_i R_T} \simeq \prod_i K_i,$$

where K_i is the quotient field of R/P_i, $1 \le i \le r$.

Consider the composite of injections

$$R \to \prod_{i=1}^{r} R/P_i \to \prod_{i=1}^{r} K_i \simeq Q(R).$$

By (ii), R is integrally closed in $Q(R)$.

If $e_i = (0, 0, \ldots, 1, 0, \ldots, 0)$, with 1 in the i-th place, then $e_i^2 = e_i$, and $e_i e_j = 0, i \neq j$. Since e_i is integral over R, $e_i \in R$ and hence $R \simeq \Pi_{i=1}^{r} Re_i$. Each Re_i is integrally closed as R is integrally closed in $Q(R)$. Hence R is a finite product of integrally closed domains.
$(iii) \Rightarrow (i)$. Let $R = \Pi_1^r R_i$ be a finite product of integrally closed domains R_i $(1 \le i \le r)$. It is easy to see that any prime ideal P of R is of the type $P = \Pi P_i$, where P_j is a prime ideal of R_j for some j and $P_i = R_i$ for some $i \neq j$ Then $R_P \simeq (R_j)_{P_j}$ is an integrally closed domain. Hence R is normal. □

Theorem 9.3.5 *Let R be a Noetherian ring. Then R satisfies (R_1) and (S_2) if and only if R is normal.*

Proof Assume R is normal and let P be a prime ideal of R. Then R_P is an integrally closed domain and if $\dim R_P \leq 1$, then R_P is either a field or a DVR and hence R_P is regular, *i.e.*, (R_1) is satisfied.

To prove (S_2) we have to prove, in addition, depth $R_P \geq 2$ whenever $\dim R_P \geq 2$.

Let $a \in PR_P$, $a \neq 0$. Since R_P is a domain, a is non-zero-divisor, *i.e.*, $\{a\}$ is an R_P-sequence. Since R_P is integrally closed, the ideal (a) has no embedded components (Exercise 1, 8.5). Thus, if $Q \in \text{Ass}(R_P/aR_P)$, by Krull's Principal Ideal Theorem, $Q \neq PR_P$ as $\text{ht}\, PR_P \geq 2$. Hence there exists $b \in PR_P$ with $b \notin \cup Q$, where Q runs over the set $\text{Ass}(R_P/aR_P)$. Thus $\{a, b\}$ is an R_P-sequence and depth $R_P \geq 2$.

Conversely, assume that R satisfies (R_1) and (S_2). This implies that R satisfies (R_0) and (S_1), *i.e.*, R is reduced by Proposition 9.3.2. Let $P_1, P_2, \ldots, P_r$ be the minimal prime ideals of R.

Then by Theorem 9.3.4, the total quotient ring of R is $Q(R) \simeq \Pi_i K_i$, where K_i is the quotient field of R/P_i $(1 \leq i \leq r)$. Consider the natural injection $R \to Q(R)$ given by $a \to (a + P_i)$ and identify R as a subring of $Q(R)$. We will show that R is integrally closed in $Q(R)$. Let $\alpha \in Q(R)$ be expressed as $\alpha = a/b$, $a \in R$, $b \in T$. Suppose α satisfies an integral relation of the type

$$\alpha^n + c_1\alpha^{n-1} + \cdots + c_n = 0, \quad c_i \in R \ (1 \leq i \leq n).$$

Then $a^n + \Sigma\, c_i a^{n-i} b^i = 0$.

Let P be a prime ideal of ht 1. By (R_1), R_P is regular and hence integrally closed. If $\overline{a} = (a/1)$ and $\overline{b} = (b/1)$ in R_P, the relation $\overline{a}^n + \Sigma_i \overline{c}_i \overline{a}^{n-i} \overline{b}^i = 0$ shows that $\overline{a}/\overline{b}$ is integral over R_P and hence belongs to R_P, *i.e.*, $\overline{a} \in \overline{b}R_P$ for every prime ideal P of ht 1. Since R satisfies (S_2) and $b \in R$ is a non-zero-divisor, R/bR satisfies (S_1) and by Proposition 9.3.1, R/bR has no embedded primes. Thus (bR) is unmixed of ht 1 and since $\overline{a} \in \overline{b}R_P$ for every prime P of ht 1, $a \in bR$, *i.e.*, $\alpha = a/b \in R$.

Since R is reduced and integrally closed in $Q(R)$, R is normal by Theorem 9.3.4. □

9.3. EXERCISES

1. Show that if for every prime ideal P of R, depth $R_P \geq k$ and $\dim R_P \leq k$ implies R_P is C.M., then R satisfies (S_k). State and prove the converse.

2. Show that R satisfies (S_k) if and only if for every $t \in R$ and R_t regular sequence $a_1, a_2, \ldots, a_r$ $(r < k)$, the R_t-module $\dfrac{R_t}{(a_1, a_2, \ldots, a_r)R_t}$ has no embedded primes.

3. A normal domain R is called a Krull ring if (i) for each $a \in R$, the number of ht 1 prime ideals containing (a) is finite and (ii) $R = \bigcap_{\text{ht}\, P = 1} R_P$. Show that a Noetherian normal ring is a Krull ring but not conversely.

4. Show that if R is a Noetherian domain, the integral closure of R in its quotient field is a Krull ring.

9.4 Complete local rings

Let R be a local ring with maximal ideal $\mathfrak{m}$ and residue field $R/\mathfrak{m} = k$.

Definition 9.4.1 R is said to be of *equicharacteristic* type if $\operatorname{ch} R = \operatorname{ch} k$. Otherwise it is said to be of *inequicharacteristic* type.

If R is of inequicharacteristic type, then $\operatorname{ch} R = 0$, $\operatorname{ch} k = p > 0$ with $p \in \mathfrak{m}$.

Definition 9.4.2 Let R be of inequicharacteristic type. R is said to be *unramified* if $p \in \mathfrak{m} - \mathfrak{m}^2$. Otherwise R is said to be *ramified*.

We propose to obtain the Structure Theorem for complete regular local rings. This theorem will be a consequence of the following theorem on the existence of a coefficient ring.

Theorem 9.4.3 (Cohen's Theorem). *Let R be a complete local ring with maximal ideal $\mathfrak{m}$, $k = R/\mathfrak{m}$ and $\operatorname{ch} k = p \geq 0$. Then R contains a subring S satisfying*

(i) $\mathfrak{m} \cap S = (p)$,

(ii) *S is a complete local ring and* $\dfrac{S}{\mathfrak{m} \cap S} \simeq k$.

Proof We first consider the case when $p = 0$. Let $\phi : R \to R/\mathfrak{m}$ be the natural projection. Since $\operatorname{ch} k = 0$, we have $\operatorname{ch} R = 0$ and $R \supset \mathbb{Z}$, $\mathbb{Z} \cap \mathfrak{m} = 0$. Then $R \supset \mathbb{Q}$, the rational number field and by Zorn's Lemma, R contains a maximal subfield L. We claim that L is the required subring, *i.e.*, ϕ maps L isomorphically onto $k = R/\mathfrak{m}$. Clearly, $\phi(L)$ is a subfield of k and if k is not algebraic over $\phi(L)$, there exists some $\alpha \in R - L$ such that $\phi(\alpha)$ is transcendental over $\phi(L)$. This implies, $\mathfrak{m} \cap L[\alpha] = 0$, *i.e.*, $R \supset L(\alpha)$, a contradiction to the maximality of L. Hence k is algebraic and separable over $\phi(L)$. If $\beta \in k$, then β is separable over $\phi(L)$, *i.e.*, there exists some monic polynomial $f(X) \in R[X]$ such that β is a simple root of the irreducible polynomial $\overline{f(X)} \in \phi(L)[X]$. By Hensel's Lemma, $f(X)$ has a simple root $\xi \in R$ such that $\phi(\xi) = \beta$. The isomorphism $\phi : L \to \phi(L)$ can be extended to an isomorphism $\overline{\phi} : L(\xi) \to \phi(L)(\beta)$. By maximality, $\xi \in L$ and then $\beta \in \phi(L)$, *i.e.*, $k = \phi(L)$. □

Before we consider the case when $\operatorname{ch} k = p > 0$, we prove some preliminary lemmas.

Definition 9.4.4 Let $\operatorname{ch} k = p > 0$. A p-basis for k is a subset $B \subset k$ such that

(i) $k^p(B) = k$,

(ii) all monomials of the type $b_1^{e_1} b_2^{e_2} \cdots b_r^{e_r}$, $b_i \in B$, $0 \leq e_i \leq p - 1$ are linearly independent over k^p.

Lemma 9.4.5 *Any field k with $\operatorname{ch} k = p > 0$ has a p-basis B satisfying the conditions*

(i) $k^{p^n}(B) = k$ *for all* $n \geq 1$,

(ii) $[k^{p^n}(b_l, b_2, \ldots, b_r) : k^{p^n}] = p^{nr}, \ b_i \in B \ (1 \leq i \leq r).$

Proof By Zorn's Lemma, there exists a maximal p-independent set, *i.e.*, a set satisfying condition (ii) of Definition 9.4.4. It will also satisfy condition (i) in view of the maximality. Hence k has a p-basis B. Let $\phi : k \to k^p$ be the isomorphism given by $\phi(a) = a^p$. Assume by induction that $k = k^{p^{n-1}}(B)$. Applying ϕ, we have $k^p = k^{p^n}(B^p)$. Hence $k = k^p(B) = k^{p^n}(B)$. This proves (i).

The relation (ii) is also proved by induction on n. It is clear for $n = 1$. Now,

$$\begin{aligned} p^{nr} &\geq [k^{p^n}(b_1, \ldots, b_r) : k^{p^n}] \\ &= [k^{p^n}(b_l, \ldots, b_r) : k^{p^n}(b_1^p, \ldots, b_r^p)][k^{p^n}(b_1^p, \ldots, b_r^p) : k^{p^n}] \\ &\geq p^r . p^{(n-1)r} \end{aligned}$$

as $[k^{p^n}(b_1^p, \ldots, b_r^p) : k^{p^n}] = [k^{p^{n-1}}(b_1, \ldots, b_r) : k^{p^{n-1}}] = p^{(n-1)r}$ by induction. Hence $[k^{p^n}(b_1, \ldots, b_r) : k^n] = p^{nr}$. □

Corollary 9.4.6 *If B is a p-basis of k, any $a \in k$ can be expressed uniquely as a linear combination over k^{p^n} of monomials of the type*

$$b_1^{e_1} b_2^{e_2} \cdots b_r^{e_r}, \quad 0 \leq e_i \leq p^n - 1, \ b_i \in B \ (1 \leq i \leq r).$$

Proof Follows from the relations (i) and (ii) of Lemma 9.4.5. □

Lemma 9.4.7 *Let $\mathfrak{m}$ be an ideal of R and $p \in \mathfrak{m}$, p prime. If $a - b \in \mathfrak{m}$, $a, b \in R$, then $a^{p^n} - b^{p^n} \in \mathfrak{m}^{n+1}$, for all $n \geq 0$.*

Proof The proof is by induction on n. We prove first for $n = 1$. Let $a = b + c$, $c \in \mathfrak{m}$. Then

$$a^p - b^p = \sum_{i>0} \binom{p}{i} b^{p-i} c^i \in \mathfrak{m}^2, \quad \text{as } p \mid \binom{p}{i}, \ i > 0.$$

Assume now that $a^{p^n} - b^{p^n} \in \mathfrak{m}^{n+1}$. Then $a^{p^n} - b^{p^n} = d, \quad d \in \mathfrak{m}^{n+1}$. Now, $a^{p^{n+1}} - b^{p^{n+1}} = \Sigma_{i>0} \binom{p}{i} (b^{p^n})^{p-i} d^i \in \mathfrak{m}^{n+2}, \ \text{as } p \mid \binom{p}{i}$. □

Corollary 9.4.8 *If $\mathfrak{m}$ is a maximal ideal of R and $\mathfrak{m}^t = 0$, then the mapping $a \mapsto a^{p^n}$ $(n + 1 \geq t)$ induces a well defined injective map of $R/\mathfrak{m}$ into R.*

Proof The mapping $a \mapsto a^{p^n}$ vanishes on $\mathfrak{m}$ as by Lemma 9.4.7, $a \in \mathfrak{m}$ implies $a^{p^n} \in \mathfrak{m}^{n+1} \subseteq \mathfrak{m}^t = (0)$. Hence there is an induced map $R/\mathfrak{m} \to R$ which is injective as it is injective when composed with the natural projection $R \to R/\mathfrak{m}$. □

We now return to the proof of Theorem 9.4.3 in the case $\operatorname{ch} k = p > 0$. Choose a p-basis B^* of k and let B be the set of representatives of B^* modulo $\mathfrak{m}$. Let $\phi_n : R/\mathfrak{m}^n \to R/\mathfrak{m}^n$ be the map given by $\phi_n(a) = a^{p^{2n}}$.

By Corollary 9.4.8, ϕ_n induces a well defined injective map

$$\sigma_n : k \simeq \frac{R/\mathfrak{m}^n}{\mathfrak{m}/\mathfrak{m}^n} \to R/\mathfrak{m}^n.$$

We denote the image $\sigma_n(k)$ by A_n. We note that A_n is closed under multiplication but not addition. Let

$$\eta : R/\mathfrak{m}^n \to \frac{R/\mathfrak{m}^n}{\mathfrak{m}/\mathfrak{m}^n} \simeq k$$

be the natural projection. Then $\eta(A_n) = k^{p^{2n}}$ and by Corollary 9.4.8, $\eta \mid A_n : A_n \to k^{p^{2n}}$ is bijective. Let $\overline{B}$ be the set of residue classes of elements of B modulo $\mathfrak{m}^n$ and T_n the set of polynomials in $\overline{B}$ with coefficients in A_n such that the degree in each variable of $\overline{B}$ is at most p^{2n}. The bijection $\eta : A_n \to k^{p^{2n}}$ can be extended to a bijection

$$\eta : T_n = A_n[\overline{B}] \to k^{p^{2n}}(B^*) = k.$$

Let $S_n = T_n + pT_n + \cdots + p^{n-1}T_n$.

We claim that S_n is a subring of $R/\mathfrak{m}^n$. Clearly the product of two elements of T_n is an element T_n and thus S_n is closed under multiplication. To show that S_n is closed under addition, it is sufficient to show that the sum of two elements of T_n is in S_n. Let G be the additive subgroup generated by T_n in $R/\mathfrak{m}^n$. We claim that $G \subset T_n + pG$. Each element ξ of T_n is of the type $\Sigma a_M M$, $a_M \in A_n$ and M a monomial of the type $\overline{b}_1^{e_1} \cdots \overline{b}_r^{e_r}$, $0 \le e_i \le p^{2n}$. Write $a_M = \overline{c}_M^{p^{2n}}$, $\overline{c}_M \in R/\mathfrak{m}^n$. By taking the image of $\overline{c}_M$ under the natural map $R/\mathfrak{m}^n \to R/\mathfrak{m} = k$ and taking its inverse image under η we have $\overline{c}_M{}^{p^{2n}} = \overline{d}_M{}^{p^{2n}}$, $\overline{d}_M \in T_n$. We may assume without loss of generality that $\overline{c}_M \in T_n$. Let $\xi' \in T_n$, where

$$\xi' = \Sigma \overline{d}_M{}^{p^{2n}} \overline{b}_1{}^{e_1} \cdots \overline{b}_r{}^{e_r}, \quad \overline{d}_M \in T_n.$$

Then

$$\xi + \xi' = \Sigma(\overline{c}_M + \overline{d}_M)^{p^{2n}} \overline{b}_1{}^{e_1} \cdots \overline{b}_r{}^{e_r} + p(\qquad)$$

where the expression in the bracket is a sum of products of elements of T_n, *i.e.*, an element of G. By iteration, we have

$$G \subset T_n + pT_n + p^2 G \subset T_n + pT_n + \cdots + p^{n-1}T_n = S_n$$

as $p^n G = 0$. Thus the sum of two elements of T_n is in S_n and S_n is a subring of $R/\mathfrak{m}^n$.

Clearly, $pS_n \subset \overline{\mathfrak{m}} \cap S_n$ where $\overline{\mathfrak{m}} = \mathfrak{m}/\mathfrak{m}^n$. If $\xi \in \overline{\mathfrak{m}} \cap S_n$, then $\xi = \xi_0 + p\xi'$, $\xi_0 \in \overline{\mathfrak{m}} \cap T_n$. Since η is an injection on T_n, we have $\mathfrak{m} \cap T_n = (0)$. Hence $\xi_0 = 0$ and $pS_n = \overline{\mathfrak{m}} \cap S_n$.

Since the mapping η restricted to T_n is a surjective map of T_n onto k, $S_n/\overline{\mathfrak{m}} \cap S_n \simeq k$. Thus pS_n is a maximal ideal of S_n and $S_n/pS_n \simeq k$. Let $\pi_n : R/\mathfrak{m}^n \to R/\mathfrak{m}^{n-1}$ be the natural map. Then $\pi_n(T_n) \subset S_{n-1}$, so that π_n defines a homomorphism $\pi_n : S_n \to S_{n-1}$. By definition of T_n, it is easy to see that each element of T_{n-1} is in the image $\pi_n(T_n)$, so that $\pi_n : S_n \to S_{n-1}$ is surjective. Thus (S_n, π_n) is an inverse system of rings with π_n surjective for each n. Let $S = \varprojlim_n S_n$. We claim that S is the required ring.

Since each S_n is a subring of $R/\mathfrak{m}^n$ and R is $\mathfrak{m}$-adically complete, by passing to the inverse limit, S can be identified with a subring of R. Since $(\mathfrak{m}/\mathfrak{m}^n) \cap S_n = pS_n$, by passing to the inverse limit we have $pS = \mathfrak{m} \cap S$.

Consider the exact sequence

$$0 \longrightarrow pS_n \longrightarrow S_n \longrightarrow S_n/pS_n \longrightarrow 0.$$

By passing to the inverse limit we have (Proposition 6.2.8) an exact sequence

$$0 \longrightarrow pS \longrightarrow S \longrightarrow k \longrightarrow 0,$$

showing that $pS = \mathfrak{m} \cap S$ is a maximal ideal of S and $S/\mathfrak{m} \cap S \simeq k$.

We claim that S is a local ring which is complete under the (p)-adic topology.

In fact the (p)-adic completion of S

$$= \varprojlim_k \frac{S}{p^k S} \simeq \varprojlim_k \varprojlim_n \frac{S_n}{p^k S_n}$$

$$\simeq \varprojlim_n \varprojlim_k \frac{S_n}{p^k S_n}.$$

Now, S_n is (p)-adically complete as pS_n is nilpotent.

Hence the right hand side is isomorphic to $\varprojlim_n S_n \simeq S$, showing that S is complete. Moreover, since S is (p)-adically complete, pS is contained in the Jacobson radical of S. Hence pS is the unique maximal ideal of S. Thus S is a complete local ring with $pS = \mathfrak{m} \cap S$ and $S/\mathfrak{m} \cap S \simeq k$. □

Definition 9.4.9 The ring S is called a *coefficient ring* of R.

Corollary 9.4.10 *The coefficient ring S is Noetherian.*

Proof Since $\cap p^n S \subseteq \cap \mathfrak{m}^n = (0)$, we have by Exercise 5, Section 5.1, every non-zero ideal of S is of the type (p^n), $n \geq 1$. Hence S is Noetherian. □

Corollary 9.4.11 *The coefficient ring S is a field or a* DVR *when it is a domain.*

Proof If R is of equicharacteristic type, then $\mathfrak{m} \cap S = (0)$ and $S \simeq k$. In the other case, S is a Noetherian local domain with maximal ideal principal. Hence S is a DVR (Theorem 5.2.8). □

Corollary 9.4.12 *Let R be a complete local domain. Then R is isomorphic to a quotient of the power series ring $S[[X_1, X_2, \ldots, X_d]]$, where S is either a field or a* DVR.

Proof Let S be a coefficient ring of R. Choose a generating set $a_1, a_2, \ldots, a_d$ for $\mathfrak{m}$ and consider the ring homomorphism given by

$$\phi : S[X_1, X_2, \ldots, X_d] \to R$$

$$\phi(X_i) = a_i \ \ (1 \leq i \leq d)$$

and $\phi|_S = \text{Id}$. Since R is complete, ϕ can be extended to a ring homomorphism

$$\phi : S[[X_1, X_2, \ldots, X_d]] \longrightarrow R.$$

Now, $\text{gr}(\phi) : S[X_1, \ldots, X_d] \to \text{gr}_{\mathfrak{m}}(R)$ is surjective.

Since $S[[X_1, \ldots, X_d]]$ is complete and $\cap \mathfrak{m}^n = (0)$, ϕ is surjective by Proposition 6.4.2.

Hence R is a quotient of $S[[X_1, X_2, \ldots, X_d]]$ where S is a field or a DVR. □

Proposition 9.4.13 *Let R be a complete local ring of equicharacteristic type, $k = R/\mathfrak{m}$ and $a_1, a_2, \ldots, a_d$ a system of parameters for R. Then R is a module of finite type over $R' = k[[a_1, \ldots, a_d]]$.*

Proof R', being the quotient of a power series ring, is a complete local ring with maximal ideal $\mathfrak{m}' = (a_1, \ldots, a_d)$. If $I = (a_1, \ldots, a_d)R$, then I is an $\mathfrak{m}$-primary ideal. To show that R is of finite type over R', it is sufficient to show (Theorem 6.4.3) that $\text{gr}_I(R)$ is a module of finite type over $\text{gr}_{\mathfrak{m}'} R'$. Since $\text{gr}_I(R)$ is naturally isomorphic to a quotient of the polynomial ring $(R/I)[X_1, \ldots, X_d]$, it is sufficient to show that R/I is a module of finite type over $R'/\mathfrak{m}' = k$. Since I contains a power of $\mathfrak{m}$ and $R/\mathfrak{m}$, $\mathfrak{m}/\mathfrak{m}^2$, $\mathfrak{m}^2/\mathfrak{m}^3, \ldots$, are finite dimensional k-spaces, R/I is a finite dimensional k-space. □

Corollary 9.4.14 *Let R be an equicharacteristic complete local ring and $k = R/\mathfrak{m}$. If $a_1, a_2, \ldots, a_d$ is a system of parameters for R, then $a_1, a_2, \ldots, a_d$ are analytically independent over k.*

Proof By Proposition 9.4.13, R is a finite type module over $R' = k[[a_1, a_2, \ldots, a_d]]$. Hence by Theorem 4.1.3, R is integral over R'. This implies $\dim R' = \dim R = d$. Consider the ring homomorphism

$$\phi : k[[X_1, X_2, \ldots, X_d]] \to R'$$

given by

$$\phi(X_i) = a_i \ \ (1 \leq i \leq d), \quad \phi\,|_k = \text{Id}.$$

ϕ is surjective and if

$$\text{Ker } \phi = I \neq 0, \quad \dim \frac{k[[X_1, \dots, X_d]]}{I} < d,$$

i.e., $\dim R' < d$ a contradiction. Hence ϕ is an isomorphism showing that $a_1, a_2, \dots, a_d$ are analytically independent over k. □

Theorem 9.4.15 *Let R be an equicharacteristic complete local ring and $k = R/\mathfrak{m}$. Let K be an extension field of k. Then there exists a complete local ring R' with residue field $R'/\mathfrak{m}' = K$ such that,*

(i) *R' is faithfully flat over R,*

(ii) $\mathfrak{m}' = \mathfrak{m}R'$.

(iii) $\dim R' = \dim R$.

Proof Express R as the quotient of a power series ring $k[[X_1, \dots, X_n]]$ with kernel I so that we have an exact sequence

$$0 \to I \to k[[X_1, X_2, \dots, X_n]] \to R \to 0.$$

If I' be the ideal generated by I in $K[[X_l, X_2, \dots, X_n]]$, then we have a commutative diagram

$$\begin{array}{ccccccccc}
0 & \longrightarrow & I & \longrightarrow & k[[X_1, \dots, X_n]] & \longrightarrow & R & \longrightarrow & 0 \\
 & & \downarrow & & \downarrow f & & \downarrow \overline{f} & & \\
0 & \longrightarrow & I' & \longrightarrow & K[[X_1, \dots, X_n]] & \longrightarrow & R' & \longrightarrow & 0
\end{array}$$

where $R' = \dfrac{K[[X_1, X_2, \dots X_n]]}{I'}$ and f is induced by the inclusion $k \subset K$. Clearly R' is a local ring with maximal ideal $\mathfrak{m}' = \mathfrak{m}R'$ which proves (ii). Now, $\dim R' \leq \dim R$ as any $\mathfrak{m}$-primary ideal of R generates an $\mathfrak{m}'$-primary ideal of R'.

We now prove (i). Since $\overline{f}$ is a local homomorphism, it is sufficient to show that R' is R flat. If $A = k[[X_1, \dots, X_n]]$ and $B = K[[X_1, \dots, X_n]]$, it is sufficient to show that B is A-flat, for then it will follow that $B \otimes_A A/I \simeq R'$ will be $A \otimes_A A/I \simeq R$-flat. By Theorem 7.3.33, to show that B is A-flat it is sufficient to show that $\text{Tor}_1^A(B, k) = 0$. Consider the Koszul resolution of k as an A-module. By tensoring this resolution with B over A, we get the Koszul resolution of k as a B module. In particular the tensored complex is acyclic showing that $\text{Tor}_1^A(B, k) = 0$. This proves (i).

Since R' is faithfully flat over R by Corollary 8.2.21, $\dim R' \geq \dim R$ showing (iii). The following theorem gives the structure of complete regular local rings. □

Theorem 9.4.16 (Cohen's Structure Theorem). *Let R be a complete regular local ring of dimension d. Then R is exactly one of the following types.*

(i) $R \simeq k[[X_1, \ldots, X_d]]$, *k field* ($R$ *equicharacteristic*).

(ii) $R \simeq S[[X_2, \ldots, X_d]]$, S *complete unramified* DVR (R *inequicharacteristic unramified*).

(iii) $R \simeq \dfrac{R_0[X]}{(f(X))}$, R_0 *is of type* (ii) *and* $f(X)$ *is an Eisenstein polynomial over* R_0 i.e., $f(X) = X^r + c_1 X^{r-1} + \cdots + c_r$ with $c_i \in \mathfrak{m}_0$, $c_r \notin \mathfrak{m}_0^2$ (R *inequicharacteristic ramified*).

Before proving the theorem, we first prove the following lemma.

Lemma 9.4.17 *Let R be a complete local ring with maximal ideal $\mathfrak{m}$, M an R-module such that $M/\mathfrak{m}M$ is finitely generated over $k = R/\mathfrak{m}$ and $\cap \mathfrak{m}^i M = 0$. Then M is finitely generated over R.*

Proof Choose a basis of $M/\mathfrak{m}M$ over $k = R/\mathfrak{m}$ and lift the basis elements to $x_1, x_2, \ldots, x_t \in M$. If N is the submodule of M generated by $\{x_1, x_2, \ldots, x_t\}$, then we show that $M = N$.

$$\text{Now,} \qquad M = N + \mathfrak{m}M = N + \mathfrak{m}N + \mathfrak{m}^2 M = \cdots$$
$$= N + \mathfrak{m}N + \cdots + \mathfrak{m}^{s-1}N + \mathfrak{m}^s M.$$

If $x \in M$, write $x = y_0 + y_1 + \cdots + y^{s-1} + z$ with $y_j \in \mathfrak{m}^j N$ and $z \in \mathfrak{m}^s M$. We can express $y_j = \Sigma_{i=1}^t a_{ji} x_i$, $a_{ji} \in \mathfrak{m}^j$. Since R is complete, for each fixed i, $\Sigma_{j=0}^{\infty} a_{ji}$ converges to $b_i \in R$ (say). Let $w = \Sigma_{i=1}^t b_i x_i$. Then $w - x \in \mathfrak{m}^s M$ for all s and hence $w = x$. Thus $M = N$. □

We now return to the proof of Theorem 9.4.16.

Case (i): Let R be of equicharacteristic type. By Corollary 9.4.11, $S \simeq k$. Choose a regular system of parameters $x_1, x_2, \ldots, x_d$ for R and consider the natural k-algebra homomorphism

$$\phi : k[X_1, \ldots, X_d] \to R$$

given by $\phi(X_i) = x_i$, $1 \leq i \leq d$, $\phi\,|_k = \text{Id}$. Since R is complete, ϕ can be extended to a k-algebra homomorphism

$$\overline{\phi} : k[[X_1, \ldots, X_d]] \to R.$$

Now, $\text{gr}(\overline{\phi}) : k[X_1, \ldots, X_d] \to \text{gr}_{\mathfrak{m}}(R)$ is surjective. By Proposition 6.4.2, $\overline{\phi}$ is surjective as $\text{gr}(\overline{\phi})$ is surjective. Since $\dim R = \dim k[[X_1, \ldots, X_d]] = d$, $\overline{\phi}$ is an isomorphism.

Case (ii)**:** Let R be inequicharacteristic unramified type. Since $p \in \mathfrak{m} - \mathfrak{m}^2$, we can choose a regular system of parameters $x_1, x_2, \ldots, x_d$ with $x_1 = p$. Let S be the coefficients ring. By Corollary 9.4.11, S is a complete unramified DVR. Consider the S-algebra homomorphism $\phi : S[X_2, ..., X_d] \to R$ given by $\phi(X_i) = x_i$, $2 \leq i \leq d$, $\phi \mid_S = \text{Id}$ and extend it to an S-algebra homomorphism $\overline{\phi} : S[[X_2, \ldots, X_d]] \to R$. Then $\overline{\phi}$ is an isomorphism as in case (i).

Case (iii)**:** R is inequicharacteristic ramified type. We claim that there exist $x_2, x_3, \ldots, x_d \in \mathfrak{m}$ which is a part of a regular system of parameters for R such that $(p, x_2, \ldots, x_d)$ is an $\mathfrak{m}$ primary ideal. This is proved by induction on d. If $d = 1$, there is nothing to prove as (p) is $\mathfrak{m}$-primary. Assume $d > 1$. Since R is C.M. (Corollary 9.1.8), every $P \in \text{Ass}(p)$ is minimal of height one and hence $\mathfrak{m} \notin \text{Ass}(p)$. This implies $\mathfrak{m} \not\subset \mathfrak{m}^2 \cup P$ where the union is taken over all $P \in \text{Ass}(p)$; for if $\mathfrak{m} \subset \mathfrak{m}^2 \cup P$, then $\mathfrak{m} = \mathfrak{m}^2$ or $\mathfrak{m} = P$ for some $P \in \text{Ass}(p)$. This implies $\mathfrak{m} = 0$ or $d = 1$, a contradiction. Choose $x_2 \in \mathfrak{m} - \mathfrak{m}^2$ such that $x_2 \notin P$ for all $P \in \text{Ass}(p)$ and let $\overline{R} = R/(x_2)$. By induction, there exist $\overline{x}_3, \overline{x}_4, \ldots, \overline{x}_d$ which is a part of a regular system of parameters for $\overline{R}$ such that $(\overline{p}, \overline{x}_3, \ldots, \overline{x}_d)$ is an $\overline{\mathfrak{m}} = \mathfrak{m}/(x_2)$ primary ideal. Then $\{p, x_2, \ldots, x_d\}$ has the required properties.

Let S be a coefficient ring of R and $R_0 = S[[x_2, \ldots, x_d]]$. Since R is complete, R_0 is a complete local subring of R with maximal ideal $\mathfrak{m}_0 = (p, x_2, \ldots, x_d)$. By Lemma 9.4.17, R is a finite R_0-module as $\mathfrak{m}_0 R$ is $\mathfrak{m}$-primary and $R/\mathfrak{m}_0 R$ is a finite $R_0/\mathfrak{m}_0 \simeq k$-module. In particular, R is integral over R_0, $\dim R_0 = d$ and R_0 is isomorphic to the power series ring in d variables $X_1, X_2, \ldots, X_d$, *i.e.*, to $S[[X_2, \ldots, X_d]]$. Thus R_0 is a complete unramified local ring of dimension d.

We now show that R is an Eisenstein extension of R_0. Since $x_2, \ldots, x_d$ is a part of a regular system of parameters for R, choose a regular system of parameters $x_1, x_2, \ldots, x_d$ for R. We claim that $R = R_0[x_1]$. Clearly,

$$\begin{aligned}
R = R_0 + \mathfrak{m} &= R_0 + \sum_{i=1}^{d} R x_i \\
&= R_0 + \sum_{i=1}^{d} (R_0 + \mathfrak{m}) x_i \\
&\subset R_0 + \sum_{i=1}^{d} R_0 x_i + \mathfrak{m}^2 \\
&\subset R_0 + \sum_{i=1}^{d} R_0 x_i + \sum_{i,j=1}^{d} R_0 x_i x_j + \mathfrak{m}^3 \\
&\cdots\cdots\cdots \\
&\subset R_0 + R_0 x_1 + \cdots + R_0 x_1^{t-1} + \mathfrak{m}_0 R,
\end{aligned}$$

where t is chosen to be the smallest positive integer with $x_1^t \in \mathfrak{m}_0 R$ which exists in view of the fact that $\mathfrak{m}_0 R = (p, x_2, \ldots, x_d)$ is $\mathfrak{m}$-primary. By Nakayama Lemma,

$$R = R_0 + R_0 x_1 + \cdots + R_0 x_1^{t-1},$$

i.e., $R = R_0[x_1]$. Since R is integral over R_0, x_1 satisfies a monic polynomial $f(X)$ over R_0 of least degree r, say

$$f(X) = X^r + c_1 X^{r-1} + \cdots + c_r, \quad c_i \in R_0.$$

We claim that $f(X)$ is an Eisenstein polynomial over R_0. Clearly, $c_r \in R_0 \cap \mathfrak{m} = \mathfrak{m}_0$ as $x_1 \in \mathfrak{m}$. Suppose some $c_i \notin \mathfrak{m}_0$. Choose the largest i with this property. Then

$$\begin{aligned} \overline{f(X)} &= X^r + \overline{c}_1 X^{r-1} + \cdots + \overline{c}_i X^i \quad (\text{modulo } \mathfrak{m}_0) \\ &= X^i (X^{r-i} + \overline{c}_1 X^{r-i-1} + \cdots + \overline{c}_i). \quad (\overline{c}_i \neq 0) \end{aligned}$$

Since $\overline{f(X)}$ is a product of two coprime monic polynomials over $R_0/\mathfrak{m}_0$, $f(X)$ is reducible over R_0 by Hensel's Lemma. This is a contradiction. It remains to show now that $c_r \notin \mathfrak{m}_0^2$. Suppose $c_r \in \mathfrak{m}_0^2$. The relation $x_1^r + c_1 x_1^{r-1} + \cdots + c_r = 0$ gives, in view of the fact that $c_r \in \mathfrak{m}_0^2$, the relation

$$x_1^r = -(c_1 x_1^{r-1} + \cdots + c_r) \in \mathfrak{m}_0 \mathfrak{m} \subseteq Rpx_1 + (x_2, \ldots, x_d)R.$$

Going modulo $(x_2, \ldots, x_d)R$, we have $\overline{x}_1^r \in \overline{R}\overline{p}\overline{x}_0$.
Since $x_1^t \in R$, write

$$x_1^t = b_0 + b_1 x_1 + \cdots + b_{t-1} x_1^{t-1}, \quad b_i \in R_0.$$

By minimality of the choice of r, $t \geq r$ and hence $\overline{x}_1^t \in \overline{R}\, \overline{p}\, \overline{x}_1$. Since $\overline{x}_1$ is a non-zero-divisor of $\overline{R}$, we have $\overline{x}_1^{t-1} \in \overline{Rp}$, *i.e.*, $x_1^{t-1} \in \mathfrak{m}_0 R$, a contradiction to the choice of t. This completes the proof that $f(X)$ is an Eisenstein polynomial over R_0.
Consider the ring homomorphism $\phi : R_0[X] \to R$ given by

$$\phi(X) = x_1, \ \ \phi\,|_{R_0} = \text{Id}.$$

It is surjective and induces a surjective ring homomorphism

$$\overline{\phi} : \frac{R_0[X]}{(f(X))} \to R.$$

Now, $R_0[X]/(f(X))$ is an integral extension of R_0 and hence has dimension equal to $\dim R_0 = \dim R$. Thus $\overline{\phi}$ is an isomorphism. □

9.4. EXERCISES

1. Let $(R, \mathfrak{m}) \subset (R', \mathfrak{m}')$ be complete local rings such that R' is integral over R and $R'/\mathfrak{m}'$ is a separable extension of $R/\mathfrak{m}$. Show that for any coefficient ring S of R there exists a coefficient ring S' of R' containing S and that it is uniquely determined by S.

2. Let R be a complete regular local ring with quotient field K and L a finite field extension of K. Show that the integral closure R' of R in L is a finite R-module. Deduce that if R is a local domain whose completion has no non-zero nilpotent elements (analytically unramified), the integral closure R' of R in its quotient field is a finite R-module.

CHAPTER

10

SOME CONJECTURES

The homology theory developed in Chapter 7 was applied in the subsequent chapters for the study of dimension and regular local rings. We gave a homological charecterisation of regular local rings and as a consequence obtained its UFD property. We also showed that R_P is a regular local ring when R is a regular local and P is a prime ideal of R. In this chapter we give some more applications of homology in the solution of some outstanding conjectures made around the nineteen seventies. Further developments on these conjectures is not dealt with in this chapter.

10.1 Big Cohen–Macaulay modules conjecture

M. Hochster announced (*Bull. Am. Math. Soc.* Vol. 80, 1974) the following conjecture.

Big Cohen–Macaulay modules conjecture: If R is a local ring and $a_1, a_2, \ldots, a_n$ a system of parameters of R, then there exists an R-module M (not necessarily finitely generated) such that $a_1, a_2, \ldots, a_n$ is an M-sequence.

The solution was given by M. Hochster in the case of *equicharacteristic* local rings. (*Deep Local Rings*: Preprint, Arhus University Preprint Series, 1973 or M. Hochster: Topics in the homological theory of modules over commutative rings, *Regional Conference Series in Maths.* AMS, Vol. 24 (1975)).

We outline the proof in the case $\mathrm{ch}\, R = p > 0$. The idea of the proof is to start with the given ring R and modify it successively to kill all bad relations between $a_1, a_2, \ldots, a_n$. By passing to the direct limit, we get the required module.

We start with some preliminaries. By a local ring we mean a Noetherian local ring and $\mathfrak{m}$ always denotes its maximal ideal. Let R be a local ring and $a_1, a_2, \ldots, a_n$ a system of parameters (s.o.p.) for R, M an R-module (not necessarily of finite type) and $x \in M$.

Definition 10.1.1 A *type k-relation* for $a_1, a_2, \ldots, a_n$ on M is a tuple $\rho = (x_1, x_2, \ldots, x_{k+1}) \in M^{k+1}$ such that $\sum_{i=1}^{k+1} a_i x_i = 0 \ (0 \le k < n)$.

Given such a relation on M, we define a modification of (M, x) as follows. Let $y = x_{k+1} \oplus (a_1, a_2, \ldots, a_k) \in M \oplus R^k$ and $M' = (M \oplus R^k)/Ry$.

If x' is the image of x under the natural projection $M \longrightarrow M'$, then (M', x') is called a *modification* of (M, x) for $a_1, a_2, \ldots, a_n$ of type k $(0 \le k < n)$, and is denoted by the symbol $(M, x) \to (M', x')$.

Suppose now we have a sequence

$$(M, x) = (M_0, x_0) \to (M_1, x_1) \to \cdots \to (M_r, x_r)$$

where (M_{i+1}, x_{i+1}) is a modification of (M_i, x_i) with respect to a relation P_i on M_i of type k_{i+1} $(0 \le i < r)$. Then (M_r, x_r) is called an *r-th modification of (M, x) of type $(k_1, k_2, \ldots, k_r)$*.

Lemma 10.1.2 *Let R be a local ring and $a_1, a_2, \ldots, a_n$ an s.o.p. for R. If for every modification (M, x) of $(R, 1)$, $x \notin (a_1, a_2, \ldots, a_n)M$, then R has some a-regular module M, i.e., an R-module M for which $a_1, a_2, \ldots, a_n$ is an M-sequence.*

Proof Choose a sequence of modifications of $(R, 1)$

$$(R, 1) = (M_0, x_0) \to (M_1, x_1) \to \cdots \to (M_i, x_i) \to \cdots$$

satisfying the following conditions.

(i) For each i, there is a finite generating set for type k relation on M_i $(0 \le k < n)$ (R is Noetherian).

(ii) For every i and k, $(0 \le k < n)$ and each generator ρ for the type k-relation on M_i, there exists $j > i$ such that the relation used in the passage $(M_j, x_j) \to (M_{j+1}, x_{j+1})$ is the image of the relation ρ under the natural map $M_i^{k+1} \longrightarrow M_j^{k+1}$.

Let $M = \varinjlim_i M_i$ and y the image of 1 in M under the natural map $M_0 = R \longrightarrow M = \varinjlim_i M_i$.

It is clear that for each i, there exists $j \ge i$ such that the image of $(a_1, \ldots, a_t)M_t : Ra_{t+1}$ in M_j is contained in $(a_1, a_2, \ldots, a_t)M_j$ $\ (0 \le t < n)$. Hence

$$(a_1, \ldots, a_t)M : Ra_{t+1} = (a_1, \ldots, a_t)M, \ (0 \le t < n).$$

To show that $a_1, \ldots, a_n$ is an M-sequence, it is now sufficient to show that $y \notin (a_1, a_2, \ldots, a_n)M$. Since by assumption $x_i \notin (a_1, a_2, \ldots, a_n)M_i$ for each i, we have by passing to the limit $y \notin (a_1, a_2, \ldots, a_n)M$. □

Definition 10.1.3 An *s.o.p.*, $a_1, \ldots, a_n$ for R is called *amiable* if there exists some $c \in R$, non-nilpotent, such that for all i $(0 \leq i < n)$ and positive integer t,

$$c((a_1^t, a_2^t, \ldots, a_i^t) : Ra_{i+1}^t) \subset (a_1^t, a_2^t, \ldots, a_i^t)R.$$

Lemma 10.1.4 *Let R_0 be an integrally closed,* C.M. *local domain and $a_1, a_2, \ldots, a_n$, s.o.p. for R_0. If there exists a local domain $R \supset R_0$ such that (i) R is a finitely generated R_0-module and (ii) the quotient field K of R is a separable extension of the quotient field K_0 of R_0, then $a_1, a_2, \ldots, a_n$ is a an amiable s.o.p. for R.*

Proof : Since K/K_0 is separable, $K = K_0(\alpha)$, where α can be chosen to be in R. If $f(X)$ is the minimum polynomial of α over K_0, then $f(X)$ has coefficients in R_0, as R_0 is integrally closed (Corollary 4.3.8) and $R_0[\alpha] = \dfrac{R_0[X]}{(f(X))}$. Since R_0 is C.M., $R_0[X]$ is C.M. (Theorem 8.5.23) and since f is a non-zero-divisor of $R[X]$, $R_0[\alpha]$ is also C.M. By (i), R is integral over R_0 (Theorem 4.1.3) and hence over $R_0[\alpha]$. Thus $R_0[\alpha]$ is a local ring. Clearly, $a_1, a_2, \ldots, a_n$ is an *s.o.p.* for $R_0[\alpha]$ as well as for R.
Since R is a *finitely generated* $R_0[\alpha]$-module and has the same quotient field as $R_0[\alpha]$, there exists some $c \in R - \{0\}$ such that $cR \subset R_0[\alpha]$. We claim that $a_1, a_2, \ldots, a_n$ is an amiable *s.o.p.* for R with respect to c. Let

$$b \in (a_1^t, a_2^t, \ldots, a_i^t) : Ra_i^{t+1} \quad (0 \leq i < n) \quad \text{so that}$$

$$cba_{i+1}^t = \sum_{j=1}^{i} (cb_j)a_j^t \quad \text{and} \quad cb, cb_1, \ldots, cb_i \in R_0[\alpha].$$

Since $R_0[\alpha]$ is C.M. and $a_1, a_2, \ldots, a_n$ is an *s.o.p.* for $R_0[\alpha]$, we have

$$cb \in (a_1^t, a_2^t, \ldots, a_i^t)R_0[\alpha] \subset (a_1^t, \ldots, a_i^t)R$$

and the proof is complete. □

Before proving the next lemma, we recall some preliminaries on the Frobenius functor. Let R be a local ring of $\operatorname{ch} p > 0$, p prime and e a positive integer. The ring homomorphism $R \to R$, given by $a \mapsto a^{p^e}$, is called the *Frobenius map*. R considered as an R-algebra through this map is denoted by ${}^eR$. For any R-module M the module $M' = M \otimes_R {}^eR$ is also an R-module for the operation $a(x \otimes b) = x \otimes ab$, $x \in M$, $a, b \in R$. The correspondence $M \to M'$ is a functor called *Frobenius functor*. It is right exact, carries R to R and changes the entries of the matrix of a map of free modules by raising them to the p^e-th power. Such a functor is considered by C. Peskine and L. Szpiro (*Dimension projective finite et cohomologie locale*, I.H.E.S. Pub. Math, Vol. 42, (1973)).

Lemma 10.1.5 *Let R be a local ring of $\operatorname{ch} p > 0$, p prime and $a_1, a_2, \ldots, a_n$ an amiable s.o.p. for R. Then there exists an a-regular module M.*

Proof Suppose there does not exist any a-regular module. By Lemma 10.1.2, there exists a modification (M_r, x_r) of $(R, 1)$ such that $x_r \in (a_1, \ldots, a_n)M_r$. We obtain a contradiction by applying the Frobenius functor to the sequence of modifications from $(R, 1)$ to (M_r, x_r). Since $a_1, a_2, \ldots, a_n$ is an amiable *s.o.p.* for R, there exists some $c \in R$, non-nilpotent, such that for all i and positive integers t

$$c((a_1^t, a_2^t, \ldots, a_i^t) : Ra_{i+1}^t) \subset (a_1^t, a_2^t, \ldots, a_i^t)R.$$

Since $\bigcap\limits_e (a_1^{p^e}, a_2^{p^e}, \ldots, a_n^{p^e})R \subset \bigcap\limits_e \mathfrak{m}^{p^e} = (0)$ and $c^r \neq 0$, there exists some sufficiently large e such that $c^r \notin (a_1^{p^e}, a_2^{p^e}, \ldots, a_n^{p^e})$.

Consider the sequence of modifications

$$(R, 1) = (M_0, x_0) \to (M_0, x_1) \to \cdots \to (M_i, x_i) \to \cdots \to (M_r, x_r)$$

and apply the Frobenius functor F with respect to the e chosen above.
F is right exact and the natural map $M \to M' = FM$, given by $x \to 1 \otimes x$, is quasi-linear with respect to the Frobenius map $a \to a^{p^e}$. If (M_i', x_i') is the image of (M_i, x_i), where x_i' is the image of x_i, under the natural map $M_i \to FM_i = M_i'$, we have a sequence of modifications

$$(R, 1) = (R', 1') = (M_0', x_0') \to \cdots \to (M_i', x_i') \to \cdots \to (M_r', x_r'),$$

where (M_{i+1}', x_{i+1}') is a modification of (M_i', x_i') with respect to the *s.o.p.* $a_1^{p^e}, a_2^{p^e}, \ldots, a_n^{p^e}$. Since $x_r \in (a_1, ..., a_n)M_r$, we have $x_r' \in (a_1^{p^e}, \ldots, a_n^{p^e})M_r'$. We shall show that this leads to a contradiction by proving the existence of the following commutative diagram

$$\begin{array}{ccccccc}
(M_0', x_0') & \longrightarrow & (M_1', x_1') & \longrightarrow & \cdots & \longrightarrow & (M_r', x_r') \\
\downarrow{\scriptstyle \phi_0} & & \downarrow{\scriptstyle \phi_1} & & & & \downarrow{\scriptstyle \phi_r} \\
(R, 1) & \xrightarrow{\ c\ } & (R, 1) & \xrightarrow{\ c\ } & \cdots & \xrightarrow{\ c\ } & (R, 1)
\end{array} \tag{10.1}$$

with $\phi_0 = id$.
Clearly, if such a diagram exists, we have a contradiction as

$$x_r' \in (a_1^{p^e}, \ldots, a_n^{p^e})M_r'$$

implies $c^r = \phi_r(x'_r) \in (a_1^{p^e}, \ldots, a_n^{p^e})\phi_r(M'_r) \subset (a_1^{p^e}, \ldots, a_n^{p^e})R$, contradicting the choice of c.

The existence of ϕ_r, making (10.1) commutative, is obtained by induction on i. Suppose we have a commutative diagram ($i < r$).

$$\begin{array}{ccc} (M'_0, x'_0) & \longrightarrow \cdots \longrightarrow & (M'_i, x'_i) \\ \downarrow \phi_0 & & \downarrow \phi_i \\ (R, 1) & \xrightarrow{c} \cdots \xrightarrow{c} & (R, 1) \end{array} \tag{10.2}$$

We have to define $\phi_{i+1} : (M_{i+1}, x_{i+1}) \longrightarrow (R, 1)$ such that

$$\begin{array}{ccc} (M'_i, x'_i) & \longrightarrow & (M'_{i+1}, x'_{i+1}) \\ \downarrow \phi_i & & \downarrow \phi_{i+1} \\ (R, 1) & \xrightarrow{\quad c \quad} & (R, 1) \end{array} \tag{10.3}$$

is commutative.

Since (M'_{i+1}, x'_{i+1}) is a modification of (M'_i, x'_i) of type k (say), with respect to $a_1^{p^e}, \ldots, a_n^{p^e}$, we have a relation

$$\sum_{j=1}^{k+1} y_j a_j^{p^e} = 0, \quad y_j \in M'_i. \tag{10.4}$$

If $y = y_{k+1} \oplus (a_1^{p^e}, \ldots, a_k^{p^e})$, then $M'_{i+1} = \dfrac{M'_i \oplus R^k}{Ry}$. To define ϕ_{i+1} making the diagram (10.3) commutative, it is necessary to define an R-linear map $\sigma : R^k \longrightarrow R$ such that $\phi_i \oplus \sigma$ vanishes on y. From the expression for y, this is possible provided $c\phi_i(y_{k+l}) \in (a_1^{p^e}, \ldots, a_k^{p^e})R$. The relation (10.4) above gives

$$a_{k+1}^{p^e} \phi_i(y_{k+1}) \in (a_1^{p^e}, \ldots, a_k^{p^e})R,$$

i.e.,

$$\phi_i(y_{k+1}) \in (a_1^{p^e}, \ldots, a_k^{p^e}) : Ra_{k+1}^{p^e}.$$

By the choice of c for amiability, we have $c\phi_i(y_{k+1}) \in (a_1^{p^e}, \ldots, a_k^{p^e})R$ and the proof is complete. □

Theorem 10.1.6 *Let R be a local ring of* ch $p > 0$, *p prime and $a_1, a_2, \ldots, a_n$ an s.o.p. for R. Then there exists an a-regular R-module M.*

Proof We first note that R can be replaced by R' whenever R' is a local ring equipped with a homomorphism $R \to R'$ which maps an *s.o.p.* of R into an *s.o.p.* of R'. This is so because $\text{ch } R = p$ implies $\text{ch } R' = p$ and any regular R'-module for the image *s.o.p.* is also a regular R-module for the given *s.o.p.*

Complete R and enlarge the residue field $k = R/\mathfrak{m}$ to a perfect field K (Theorem 9.4.15). Taking quotient by a prime ideal of $\text{coht } n$ ($= \dim R$), we may assume that R is a complete local domain with perfect residue field K. Since $\text{ch } R = p > 0$, p prime, by Corollary 9.4.8, R contains an isomorphic image of K. Since $a_1, a_2, \ldots, a_n$ is an s.o.p. for R, they are analytically independent over K and R is a module of finite type over $K[[a_1, a_2, \ldots, a_n]]$ (Proposition 9.4.13). The quotient field of R may not be a separable extension of the quotient field of $K[[a_1, a_2, \ldots, a_n]]$. To achieve this, we modify R further. If $b_i = a_i^{1/p^e}$ $(1 \leq i \leq n)$, then $K[[b_1, b_2, \ldots, b_n]]$ will contain p^e-th roots of all elements of $K[[a_1, a_2, \ldots, a_n]]$ as K is perfect. For sufficiently large e, $R' = R[[b_1, b_2, \ldots, b_n]]$ is a separable and finite type module over $K[[b_1, \ldots, b_n]]$ and R' is local. If we construct $a_1, a_2, \ldots, a_n, b_1, b_2, \ldots, b_n$-regular module M over R', it will be an a-regular module over R.

Thus, it is sufficient to prove the theorem in the case when R is a separable module finite extension domain of a power series ring $K[\![a_1, \ldots, a_n]\!]$. The theorem in this case follows from Lemmas 10.1.4 and 10.1.5. □

We now state another conjecture which is closely related to the Big Cohen–Macaulay modules conjecture.

A module M of finite type over a local ring R is called a *maximal Cohen–Macaulay module* if $\text{depth }(M) = \dim(M) = \dim R$.

Small Cohen–Macaulay modules conjecture: If R is a complete local ring, then R has a maximal Cohen–Macaulay module.

Clearly, the above conjecture implies the Big Cohen–Macaulay modules conjecture; for if M is a maximal C.M. module over $\widehat{R}$, then the completion of R, then any s.o.p. $a_1, a_2, \ldots, a_n$ of R is an M-sequence.

Consequently, the above conjecture will give all the implications of the Big Cohen–Macaulay modules conjecture to be obtained in the subsequent sections. However, it has implications in multiplicity theory that do not follow from the Big Cohen–Macaulay modules conjecture.

10.2 Intersection conjecture

Conjecture: (Intersection conjecture). Let R be a local ring $M \neq 0$ and N finitely generated R-modules such that $l_R(M \otimes_R N) < \infty$. Then $\dim N \leq \text{pd}_R M$.

C. Peskine and L. Szpiro have proved the above conjecture (C. Peskine and L. Szpiro: *Dimension projective finie et cohomologie locale*, IHES Publ. Math. No. 42 (1973)) for a class of rings which includes the class of local rings of characteristic $p > 0$, p prime.

The proof uses Frobenius functor and local cohomology theory. We will give an alternative proof due to M. Hochster, deducing it from the Big Cohen–Macaulay modules conjecture.

Lemma 10.2.1 *Let R be a ring, M a finitely generated R-module, $a_1, a_2, \ldots, a_k \in I = \text{Ann}(M)$ and E an R-module satisfying following conditions.*

(i) $\text{Tor}_i^R(M, N) = 0$ *for all R-modules N and $i > s$, and for some fixed s.*

(ii) $M \otimes_R E \neq 0$.

(iii) $(a_1, a_2, \ldots, a_i)E : Ra_{i+1} = (a_1, \ldots, a_i)E, \quad 1 \leq i \leq k.$

Then $a_1, a_2, \ldots, a_k$ is an E-sequence and if d is the largest integer with $\text{Tor}_d^R(M, E) \neq 0$, then $d + k \leq s$.

Proof If $J \subset I$,

$$M \otimes_R \frac{E}{JE} \cong M \otimes_R \frac{R}{J} \otimes E \cong \frac{M}{JM} \otimes_R E \cong M \otimes_R E \neq 0.$$

This implies $(a_1, a_2, \ldots, a_k)E \neq E$, which together with the condition (iii) implies that $a_1, \ldots, a_k$ is an E-sequence. The second implication is proved by induction on k. It is clear for $k = 0$ by (i). Assume $k > 0$ and consider $E' = E/a_1E$. Then E' satisfies the conditions (ii) and (iii) with respect to $a_2, a_3, \ldots, a_k$, and by induction, if d' is the largest integer for which $\text{Tor}_{d'}^R(M, E') \neq 0$, then $d' + k - 1 \leq s$. The proof will be complete if we show that $d' = d + 1$.

Consider the exact sequence

$$0 \to E \xrightarrow{\lambda a_1} E \xrightarrow{f} E' \to 0,$$

where λ_{a_1} is scalar multiplication by a_1. The induced map

$$\text{Tor}_i^R(M, E) \to \text{Tor}_i^R(M, E)$$

is also scalar multiplication by a_1 (Exercise 7, Section 7.2) and it is the zero map as a_1 annihilates M. Consider the exact sequence

$$\to \text{Tor}_i^R(M, E) \xrightarrow{0} \text{Tor}_i^R(M, E) \xrightarrow{f^*} \text{Tor}_i^R(M, E') \to \text{Tor}_{i-1}^R(M, E) \xrightarrow{0} \text{Tor}_{i-1}^R(M, E)$$

i.e., $\quad 0 \to \text{Tor}_i^R(M, E) \xrightarrow{f^*} \text{Tor}_i^R(M, E') \to \text{Tor}_{i-1}^R(M, E) \to 0.$

Since $\text{Tor}_i^R(M, E) = 0$, $i > d$ and $\text{Tor}_d^R(M, E) \neq 0$, the above exact sequence shows that

$$\text{Tor}_i^R(M, E') = 0 \text{ for } i > d + 1 \text{ and } \text{Tor}_{d+1}^R(M, E') \neq 0.$$

□

Theorem 10.2.2 *If R is a local ring having the property that for every prime ideal P of R there is an s.o.p. $a_1, a_2, \ldots, a_k$ for R/P and a regular R/P-module E, then the intersection conjecture is true for R.*

Proof If $N = 0$, or $\operatorname{pd}_R M = \infty$, then the result is trivial. Assume therefore that $N \neq 0$ and $\operatorname{pd}_R M < \infty$. Let $I = \operatorname{Ann}(M)$, $J = \operatorname{Ann}(N)$ so that $I + J$ is $\mathfrak{m}$-primary as $\ell_R(M \otimes_R N) < \infty$.

Choose a prime ideal $P \supset J$ such that $\dim R/P = \dim R/J = \dim N$. By assumption, there exists an s.o.p. $b_1, b_2, \ldots, b_k$ for R/P and an R-module E for which $b_1, b_2, \ldots, b_k$ is an E-sequence. Lift b_i to $c_i \in \mathfrak{m}$ with $c_i + P = b_i$ $(1 \leq i \leq k)$. Since $I + J$ is $\mathfrak{m}$-primary, we may, after replacing c_i and b_i by sufficiently large powers, assume that $c_1, c_2, \ldots, c_k \in I + J$ and that it is an E-sequence. Write $c_i = a_i + t_i, a_i \in I, t_i \in J \subset P$, so that $a_i + P = b_i$ $(1 \leq i \leq k)$ and $a_1, a_2, \ldots, a_k$ is an E-sequence. We claim that $M \otimes_R E \neq 0$. It is sufficient to show that $R/\mathfrak{m} \otimes E \neq 0$, for then

$$R/\mathfrak{m} \underset{R}{\otimes} (M \underset{R}{\otimes} E) \simeq R/\mathfrak{m} \underset{R}{\otimes} M \underset{R/\mathfrak{m}}{\otimes} R/\mathfrak{m} \underset{R}{\otimes} E \neq 0,$$

being the tensor product of two non-zero $R/\mathfrak{m}$ spaces. This will imply that $M \otimes_R E \neq 0$. Since $b_1, b_2, \ldots, b_k$ is an E-sequence, $(b_1, b_2, \ldots, b_k)E \neq E$. This implies $P_0 E \neq E$, where P_0 is the inverse image of $(b_1, \ldots, b_k)$ under the natural projection $R \to R/P$. Since $b_1, \ldots, b_k$ is an s.o.p. for R/P, P_0 is $\mathfrak{m}$-primary, i.e. $\mathfrak{m}^t \subset P_0$ for some t. We claim that $\mathfrak{m}E \neq E$, for otherwise $\mathfrak{m}^t E = E$ and $P_0 E = E$, a contradiction. Hence $R/\mathfrak{m} \otimes E \neq 0$. All the conditions of Lemma 10.2.1 are now satisfied with $s = \operatorname{pd}_R M$. Hence $d + k \leq s = \operatorname{pd}_R M$. Since k is the length of an s.o.p., for R/P and $d > 0$, we have

$$k = \dim R/P = \dim R/J = \dim N \leq \operatorname{pd}_R M. \qquad \square$$

Corollary 10.2.3 *If the Big Cohen–Macaulay modules conjecture is valid for a class of local rings closed under passage to homomorphic images, then the intersection conjecture is also valid for the same class.*

Proof Follows from the above theorem. □

Corollary 10.2.4 *If R is a local ring of $\operatorname{ch} p > 0$, p prime, intersection conjecture is valid for R.*

Proof The class of local rings of $\operatorname{ch} p > 0$ satisfies conditions of Corollary 10.2.3 by Theorem 10.1.6. □

10.3 Zero-divisor conjecture

The following conjecture is due to M. Auslander.

Conjecture: (Zero-divisor conjecture). Let R be a local ring and M a finitely generated R-module with $\text{pd}_R M < \infty$. If $a \in R$ is a non-zero-divisor of M, it is a non-zero-divisor of R.

C. Peskine and L. Szpiro have proved that for any class of local rings closed under localisations at prime ideals, the intersection conjecture implies the zero-divisor conjecture. Before outlining this proof, we prove some preliminary results on homological dimension.

Lemma 10.3.1 *Let R be a local ring and M a finitely generated R-module. If $\mathfrak{m} \in \text{Ass}(R)$, then $\text{pd}_R M = 0$ or ∞.*

Proof Let $\text{pd}_R M = n > 0$. If $n < \infty$, we can always get an R-module M with $\text{pd}_R M = 1$, by constructing a projective resolution of M as in Proposition 7.1.10 and stopping at the $(n-1)$th stage. Hence assume that $\text{pd}_R M = 1$. Choose a minimal generating set $x_1, \ldots, x_t$ for M and express it as a quotient of a free module F of rank t with kernel K so that we have an exact sequence

$$0 \longrightarrow K \longrightarrow F \longrightarrow M \longrightarrow 0,$$

with $K \subset \mathfrak{m}F$ (Theorem 2.2.13).

Since $\text{pd}_R M = 1$, $\text{pd}_R K = 0$, i.e., K is free over R. Since $\mathfrak{m} \in \text{Ass}(R)$, $\mathfrak{m}$ can be expressed as $\mathfrak{m} = 0 : (a)$ for some $a \in R$, $a \neq 0$. Since $K \subset \mathfrak{m}F$, $aK = 0$, $a \neq 0$, which is a contradiction as K is R-free. □

Lemma 10.3.2 *Let R be a local ring, $a \in R$, a non-unit and a non-zero-divisor of R. Let $R^* = R/(a)$ and M a finitely generated R-module such that a is also a non-zero-divisor of M and $\text{pd}_R M < \infty$. Then*

$$\text{pd}_R M = \text{pd}_{R^*} \frac{M}{aM}.$$

Proof We prove by induction on $n = \text{pd}_R M$. If $n = 0$, then M is free, i.e., it is a direct sum of copies of R. Then M/aM is a direct sum of copies of R^*, i.e., it is free over R^*. Assume $n > 0$ and consider an exact sequence $0 \to K \to F \to M \to 0$ with F-free. Since a is a non-zero-divisor of M, we have an exact sequence of R^*-modules

$$0 \to \frac{K}{aK} \to \frac{F}{aF} \to \frac{M}{aM} \to 0. \tag{10.5}$$

We claim that if M/aM is free over R^*, then M is free over R. Choose a basis $y_1, y_2, \ldots, y_n \in M/aM$ over R^* and lift it to elements $x_1, \ldots, x_n \in M$. Then by Nakayama Lemma, $x_1, \ldots, x_n$ span M. To show that $x_1, \ldots, x_n$ are linearly independent over R, consider a linear relation $\Sigma_{i=1}^n \lambda_i x_i = 0$, $\lambda_i \in R$. Then $\Sigma_{i=1}^n \overline{\lambda}_i y_i = 0$ in M/aM

and hence $\overline{\lambda}_i = 0$ $(1 \le i \le n)$. Write $\lambda_i = \mu_i a$, $(1 \le i \le n)$, so that $\Sigma \mu_i a x_i = 0$. Since a is a non-zero-divisor of M, $\Sigma \mu_i x_i = 0$. Continuing the same argument, we observe that a divides each μ_i, so that if $\mu_i = \nu_i a$, $\Sigma \nu_i x_i = 0$. Consider for each i the chain of ideals of R, $(\lambda_i) \subset (\mu_i) \subset (\nu_i) \subset \cdots$. Since R is Noetherian, the chain is stationary for each i. We may therefore assume without changing the notation that $(\lambda_i) = (\mu_i)$, for each i. Then $\mu_i = \theta_i \lambda_i$, $\theta_i \in R$ and since $\lambda_i = \mu_i a$, we have $\mu_i(1 - \theta_i a) = 0$. Since $a \in \mathfrak{m}$, $(1 - \theta_i a)$ is a unit and $\mu_i = 0$, i.e., $\lambda_i = 0$ $(1 \le i \le n)$, showing that $x_1, x_2, \ldots, x_n$ are linearly independent over R. Hence M/aM is not free over R^*. By Exercise 10, Section 7.3, we have $\operatorname{pd}_R K = n - 1 < \infty$ and by induction, $\operatorname{pd}_{R^*} K/aK = n - 1$. This implies by the exact sequence (10.5)

$$\operatorname{pd}_{R^*} M/aM = n = \operatorname{pd}_R M. \qquad \square$$

The following theorem is a generalisation of the result proved for regular local rings (Theorem 9.2.1) to an arbitrary local ring.

Theorem 10.3.3 (M. Auslander and D. Buchsbaum) *Let R be a local ring, M a finitely generated R-module with $\operatorname{pd}_R M < \infty$. Then $\operatorname{pd}_R M + \operatorname{depth}_R(M) = \operatorname{depth} R$.*

Proof We prove the theorem by induction on $\operatorname{depth}(M)$. Assume $\operatorname{depth}(M) = 0$. We prove the result in this case by induction on $\operatorname{depth}(R)$. If $\operatorname{depth} R = 0$, $\mathfrak{m} \in \operatorname{Ass}(R)$ and by Lemma 10.3.1, $\operatorname{pd}_R M = 0$ and the theorem is true. Assume $\operatorname{depth} R > 0$. We may also assume that $\operatorname{pd}_R M > 0$; for if $\operatorname{pd}_R M = 0$, then M is free and hence $\operatorname{depth} M = \operatorname{depth} R$ and the theorem is proved. Since $\operatorname{depth}(M) = 0$, $\mathfrak{m} \in \operatorname{Ass}(M)$ and there exists $y \in M$, $y \neq 0$ with $\mathfrak{m} y = 0$. Choose an exact sequence of R-modules

$$0 \to K \to F \xrightarrow{\phi} M \to 0,$$

with F-free and an element $x \in F$ with $\phi(x) = y$. Then $x \notin K$ and $\mathfrak{m}x \subset K$. Since $\operatorname{depth} R > 0$, choose $a \in \mathfrak{m}$, which is not a zero-divisor of R. Since F is free, a is also a non-zero-divisor of F as well as K. If $R^* = R/(a)$ and $\mathfrak{m}^* = \mathfrak{m}/(a)$, then $\mathfrak{m}^* \in \operatorname{Ass}(K/aK)$ as $\mathfrak{m}^*$ is the annihilator of the non-zero element $ax + aK \in K/aK$. Hence $\operatorname{depth}_{R^*}(K/aK) = 0$. Since $\operatorname{pd}_R K = \operatorname{pd}_R M - 1 < \infty$, by Lemma 10.3.2, $\operatorname{pd}_{R^*}(K/aK) = \operatorname{pd}_R K < \infty$. By Exercise 4, Section 8.4, $\operatorname{depth}(R^*) = \operatorname{depth}(R) - 1$ and the theorem is true by induction for R^*. Since K/aK is a finitely generated non-zero module over R^* with depth 0, we have $\operatorname{pd}_{R^*}(K/aK) = \operatorname{depth}(R^*)$.

Now, $\operatorname{pd}_{R^*}(K/aK) = \operatorname{pd}_R K = \operatorname{pd}_R M - 1$. Hence $\operatorname{pd}_R M = \operatorname{depth} R$, completing the proof when $\operatorname{depth} M = 0$. Assume now that $\operatorname{depth}(M) > 0$. We prove the theorem by induction on $\operatorname{depth}(M)$. We may also assume that $\operatorname{depth} R > 0$, for if $\operatorname{depth} R = 0$, $\mathfrak{m} \in \operatorname{Ass}(R)$ and by Lemma 10.3.1, $\operatorname{pd}_R M = 0$, i.e., M is free and $\operatorname{depth} M = \operatorname{depth} R$. Hence

$$\mathfrak{m} \not\subset \bigcup_{P \in \operatorname{Ass}(M)} P \text{ and } \mathfrak{m} \not\subset \bigcup_{P' \in \operatorname{Ass}(R)} P' \text{ i.e., } \mathfrak{m} \not\subset \bigcup_{P} P \cup \bigcup_{P'} P'.$$

Choose $a \in \mathfrak{m}$ which is not a zero-divisor of R as well as M. By Exercise 4, Section 8.4, $\operatorname{depth}_{R/(a)} M/aM = \operatorname{depth} M - 1$ and by induction

$$\operatorname{pd}_{R/(a)} M/aM + \operatorname{depth}_{R/(a)} M/aM = \operatorname{depth} R/(a).$$

By Lemma 10.3.2, $\operatorname{pd}_{R/(a)} M/aM = \operatorname{pd}_R M$ and hence

$$\operatorname{pd}_R M + \operatorname{depth}_R M = 1 + \operatorname{depth} R/(a) = \operatorname{depth} R.$$

□

Theorem 10.3.4 For any class of local rings closed under localisations at prime ideals, the intersection conjecture implies the zero-divisor conjecture.

Proof Let R be a local ring in the class and M a finitely generated R-module with $\operatorname{pd}_R M < \infty$. It is sufficient to show that every prime ideal associated to R is contained in a prime ideal associated to M. We prove this by induction on $\dim M$. The result is clear when $\dim M = 0$. Assume $\dim M > 0$ and let P be any prime ideal associated to R. If there exists a prime ideal $Q \in \operatorname{Supp}(M)$, $Q \neq \mathfrak{m}$ and $Q \supset P$, then $\dim M_Q < \dim M$. Since R_Q belongs to the given class of local rings and $\operatorname{pd}_{R_Q} M_Q \leq \operatorname{pd}_R M < \infty$, by induction, there exists $Q'R_Q \in \operatorname{Ass}(M_Q)$ with $PR_Q \subset Q'R_Q$. Hence $P \subset Q'$, $Q' \in \operatorname{Ass}(M)$ and the proof is complete. Suppose now that $\mathfrak{m}$ is the only prime ideal in $\operatorname{Supp}(M)$ with $\mathfrak{m} \supset P$. Then

$$\operatorname{Supp}((R/P) \otimes M) = \mathfrak{m} \quad \text{and} \quad \ell_R((R/P) \otimes M) < \infty.$$

By intersection theorem, $\dim R/P \leq \operatorname{pd}_R M$. Since $P \in \operatorname{Ass}(R)$, depth $(R) \leq \dim R/P \leq \operatorname{pd}_R M$. By Theorem 10.3.3,

$$\operatorname{depth}(R) = \operatorname{pd}_R M + \operatorname{depth}(M) \geq \operatorname{pd}_R M$$

and hence $\operatorname{depth}(R) = \operatorname{pd}_R M$, i.e., $\operatorname{depth}(M) = 0$. This implies that $\mathfrak{m} \in \operatorname{Ass}(M)$. □

Corollary 10.3.5 *For any local ring R of* $\operatorname{ch} p > 0$, *p prime, zero-divisor conjecture is true.*

Proof The class of local rings of $\operatorname{ch} p > 0$ satisfies conditions of Theorem 10.3.4. The result now follows from Corollary 10.2.4. □

Theorem 10.3.6 *For any class of local rings closed under the passage to homomorphic images and localisations at prime ideals, the intersection conjecture implies the following:*

If M is a finitely generated R-module with $\operatorname{pd}_R M < \infty$, then any M-sequence is an R-sequence.

Proof The proof is by induction on the length of an M-sequence. Let $a_1, a_2, \ldots, a_k$ be an M-sequence. If $k = 1$, a_1 is a non-zero-divisor on M and by Theorem 10.3.4, it is a non-zero-divisor of R and the result is proved. Assume $k > 1$ and consider $M' = M/a_1M$, as an $R/(a_1)$-module. By assumption, the intersection conjecture is true for the $R/(a_1)$-module M'. Since by Lemma 10.3.2, $\mathrm{pd}_{R'} M' < \infty$ and $a_2, \ldots, a_k$ is an M'-sequence, by induction on k, it is an $R/(a_1)$-sequence. Hence $a_1, a_2, \ldots, a_k$ is an R-sequence. □

Corollary 10.3.7 *Let R be a local ring of* ch $p > 0$, *p prime and M a finitely generated R-module with* $\mathrm{pd}_R M < \infty$. *Then any M-sequence is an R-sequence.*

Proof Follows from the above theorem (Theorem 10.3.6) and Corollary 10.2.4. □

10.4 Bass's conjecture

Before stating the main conjecture we recall some preliminaries on injective modules.

Definition 10.4.1 Let $M \subset N$ be R-modules. N is called an *essential extension* of M if for every non-zero submodule K of N, $K \cap M \neq 0$.

Given any R-module M, there exists an injective R-module Q such that $M \subset Q$ (Proposition 7.3.13). Choose by Zorn's Lemma a submodule E of Q such that E is a maximal essential extension of M in Q. We claim that E is a direct summand of Q.

Let L be the submodule of Q maximal with respect to the property that $L \cap E = (0)$. We will prove that $L + E = Q$, showing that E is a direct summand of Q and hence E is injective.

The composite of maps $E \xrightarrow{i} Q \xrightarrow{p} Q/L$ is injective as $L \cap E = (0)$. Consider the diagram

$$\begin{array}{ccc} E & \xrightarrow{pi} & Q/L \\ {\scriptstyle i}\big\downarrow & & \\ Q & & \end{array}$$

Since Q is injective, there exists an R-linear map $\phi : Q/L \to Q$ with $\phi pi = i$. Now, $\mathrm{Ker}\,\phi = 0$ for if Ker $\phi = K/L \neq 0$, then $L \subsetneqq K \subset Q$. and by maximality of L, $K \cap E \neq 0$.

If $x \in K \cap E, x \neq 0, x = i(x) = \phi pi(x) = 0$, a contradiction. Hence $M \subset E \xrightarrow{pi} Q/L \xrightarrow{\phi} Q$ are injective maps. Since E is an essential extension of M in Q, so is Q/L and by maximality pi is surjective, i.e., $L + E = Q$.

Definition 10.4.2 Let $M \subset E$ be an essential extension with E injective. Then E is called an *injective envelope* of M.

The injective envelope of M exists as shown above. It is unique up to isomorphism, for let $M \subset E_1$ and $M \subset E_2$ be two essential extensions of M, with E_1, E_2, injective. Consider the diagram

$$\begin{array}{ccccc} 0 & \longrightarrow & M & \xrightarrow{i_1} & E_1 \\ & & \downarrow{\scriptstyle i_2} & & \\ & & E_2 & & \end{array}$$

Since E_2 is injective, there exists an R-linear map $f : E_1 \to E_2$ with $fi_1 = i_2$. Now, f is injective, for if $K = \operatorname{Ker} f$, $K \neq 0$ would imply $K \cap M \neq 0$, i.e., there exists $x \in K \cap M$, $x \neq 0$ with

$$x = i_2(x) = fi_1(x) = f(x) = 0,$$

a contradiction. Hence $f(E_1)$ is a non-zero injective submodule of E_2. This implies that $f(E_1)$ is a direct summand of E_2. As E_2 is an essential extension of M, $f(E_1) = E_2$, i.e., f is an isomorphism.

The injective envelope of M is denoted by $E(M)$. It occurs naturally in the structure theory of injective modules over Noetherian rings developed by E. Matlis (E. Matlis: Injective modules over Noetherian rings: *Pacific Jr. Math.* Vol. 8 (1958)).

We recall some preliminaries from this structure theory.

(i) Every injective module over a Noetherian ring is a direct sum of indecomposable injectives and the decomposition is unique in the sense of Krull-Schmidt.

(ii) There is a bijection between non-zero indecomposable injectives and $E(R/P)$, where P runs over the set of prime ideals of R.

(iii) If $0 = \cap_i N_i$ is an irredundant decomposition of 0 as an intersection of irreducible submodules of M, then the monomorphism $M \to \oplus\Sigma M/N_i$ induces an isomorphism $E(M) \simeq \oplus E(M/N_i)$, where $E(M/N_i) \simeq E(R/P)$, with M/N_i being P-primary.

(iv) If R is a local ring with maximal ideal $\mathfrak{m}$ and $E = E(R/\mathfrak{m})$, then the contravariant exact functor $\operatorname{Hom}_R(-, E)$ defines a duality for R-modules of finite length.

(v) If R is a complete Noetherian local ring with maximal ideal $\mathfrak{m}$ and $E = E(R/\mathfrak{m})$, the contravariant exact functor $\operatorname{Hom}_R(-, E)$ establishes a bijection between the set of Noetherian R-modules and Artinian R-modules such that $\operatorname{Hom}_R(Hom_R(M, E), E) \simeq M$, for an R-module M which is of either type.

(vi) If R is a Noetherian local ring, then $\operatorname{Hom}_R(E, E) \simeq \widehat{R}$.

Definition 10.4.3 A minimal injective resolution of M is an exact sequence $0 \longrightarrow M \longrightarrow E_0 \xrightarrow{d_0} E_1 \longrightarrow \cdots \longrightarrow E_i \xrightarrow{d_i} E_{i+1} \longrightarrow \cdots$, where E_i is the injective envelope of Ker d_i $(i \geq 0)$.

Such a minimal injective resolution exists. Clearly, $E_0 = E(M)$ and E_i are uniquely determined by M up to isomorphism. We denote E_i by $E_i(M)$. By the structure theorem, we can write

$$E_i(M) = \oplus \Sigma \mu_i(P, M) E(R/P),$$

where μT denotes the sum of μ copies (finite or infinite) of T.

Then the $\mu_i(P, M)$ (finite or infinite) are uniquely determined by P and M.

H. Bass has proved the following. (H. Bass: Injective dimensions in Noetherian rings: *Tr. Am. Math. Soc.* Vol. 102 (1962)).

(i) If R is Noetherian and S a multiplicatively closed set of R, $E(M)_S$ is the injective envelope $E_{R_S}(M_S)$. In particular, localisation under S preserves minimal injective resolutions.

(ii) $id_{R_S} M_S \leq id_R M$.

(iii) $id_R M = \operatorname{Sup}_P id_{R_P} M_P$, where P runs over the prime ideals of R.

(iv) $\mu_i(P, M) = \mu_i(P_S, M_S)$ if $P \cap S = \emptyset$.

He has also proved (H. Bass: On the ubiquity of Gorenstein Rings, *Math. Zeitschrift*, Vol. 82 (1963)) that if a Noetherian local ring has a finitely generated module of finite injective dimension r, then r is equal to the depth of R. We outline this proof here.

Proposition 10.4.4

$$\begin{aligned} \mu_i(P, M) &= \dim_{k(P)} \operatorname{Ext}^i_{R_P}(k(P), M_P) \\ &= \dim_{k(P)} \operatorname{Ext}^i_R(R/P, M)_P, \end{aligned}$$

where $k(P) = R_P/PR_P$.

Proof By (iv), above it is sufficient to prove the result when R is a local ring with maximal ideal P. Let

$$0 \longrightarrow M \longrightarrow E_0 \longrightarrow E_1 \longrightarrow \cdots \longrightarrow E_i \xrightarrow{d_i} E_{i+1} \longrightarrow \cdots$$

be a minimal injective resolution of M. Let T_i, be the submodule of E_i annihilated by P. Then $\mu_i(P, M) = \dim_k T_i$, where $k = R/P$. Since the resolution is minimal, $d_i(T_i) = 0$.

Hence the complex

$$0 \longrightarrow \mathrm{Hom}_R(k, E_0) \longrightarrow \mathrm{Hom}_R(k, E_1) \longrightarrow \cdots$$

has zero differentiation. In particular,

$$\mathrm{Ext}^i_R(k, M) \simeq \mathrm{Hom}_R(k, E_i) \simeq T_i$$

and

$$\mu_i(P, M) = \dim_k T_i = \dim_k \mathrm{Ext}^i_R(k, M).$$

□

Corollary 10.4.5 *If M is a finitely generated R-module over a Noetherian ring R, then $\mu_i(P, M) < \infty$ for all i and all prime ideals P of R.*

Proof Assume R is Noetherian local. Since k and M are finitely generated over R, $\mathrm{Ext}^i_R(k, M)$ is finitely generated over R and since it is annihilated by P, it is of finite dimension over k.

Proposition 10.4.6 *Let R be a Noetherian ring, $Q \subsetneq P$ prime ideals of R such that there is no prime ideal strictly lying between Q and P. If M is a finitely generated R-module, then $\mu_i(Q, M) \neq 0$ implies $\mu_{i+1}(P, M) \neq 0$.*

Proof By localising at P, we may assume without loss of generality that R is a local ring with maximal ideal P. Choose $a \in P, a \notin Q$ and let $S = R/Q, S' = R/(Q, a)$. Since P is the only prime ideal of R containing (Q, a) and is maximal, $\ell_R(S') < \infty$. Since a is a non-zero-divisor of S, we have an exact sequence $0 \longrightarrow S \xrightarrow{\lambda_a^*} S \longrightarrow S' \longrightarrow 0$ where λ_a is scalar multiplication by a. This gives an exact sequence

$$\mathrm{Ext}^i_R(S, M) \xrightarrow{\lambda_a^*} \mathrm{Ext}^i_R(S, M) \longrightarrow \mathrm{Ext}^{i+1}_R(S', M),$$

where λ_a^* is the scalar multiplication by a. Now, $\mu_i(Q, M) \neq 0$ implies, by Proposition 10.4.4, $\mathrm{Ext}^i_R(S, M) \neq 0$. If $\mathrm{Ext}^{i+1}_R(S', M) = 0$, Coker $\lambda_a^* = 0$ and by Nakayama Lemma this will imply $\mathrm{Ext}^i_R(S, M) = 0$. Hence $\mathrm{Ext}^{i+1}_R(S', M) \neq 0$.

Now, $P \in \mathrm{Ass}(S')$ so that there is an injection of $k = R/P$ in S' and $\ell_R(S') < \infty$. The exact sequence

$$0 \longrightarrow k \longrightarrow S' \longrightarrow S'/k \longrightarrow 0$$

induces an exact sequence

$$\mathrm{Ext}^{i+1}_R(S'/k, M) \longrightarrow \mathrm{Ext}^{i+1}_R(S', M) \longrightarrow \mathrm{Ext}^{i+1}_R(k, M).$$

If $\mathrm{Ext}^{i+1}_R(k, M) = 0$, by induction on $\ell_R(S')$, it follows that

$$\mathrm{Ext}_R^{i+1}(S', M) = 0,$$

a contradiction. Hence $\mathrm{Ext}_R^{i+1}(k, M) \neq 0$ and $\mu_{i+1}(P, M) \neq 0$. □

Corollary 10.4.7 *If M is finitely generated R-module and $id_R M = r < \infty$, then*

(i) $r \geq \dim \mathrm{Supp}(M)$, and

(ii) $\mu_r(P, M) \neq 0$ implies that P is maximal.

Proof Let Q_0 be a minimal element of Supp (M) so that $Q_0 \in \mathrm{Ass}(M)$, i.e., R/Q_0 is isomorphic to a submodule of M. This implies $\mu_0(Q_0, M) \neq 0$. Choose $Q_1 \in \mathrm{Supp}(M)$, $Q_0 \subsetneqq Q_1$ such that there is no prime ideal strictly lying between Q_0 and Q_1. By Proposition 10.4.6, $\mu_1(Q_1, M) \neq 0$. Continue the process s times where $s = \dim \mathrm{Supp}(M)$. Then $\mu_s(Q_s, M) \neq 0$ for some $Q_s \in \mathrm{Supp}(M)$. This implies that $\mathrm{Ext}_R^s(R/Q_s, M) \neq 0$, i.e., $id_R M \geq s$.

To prove the second relation, assume that $\mu_r(P, M) \neq 0$. If P is not maximal, there exists a prime ideal Q with $P \subsetneqq Q$ such that there is no prime ideal strictly lying between them. Then

$$\mu_{r+1}(Q, M) \neq 0,$$

i.e.,

$$\mathrm{Ext}_R^{r+1}(R/Q, M) \neq 0,$$

contradicting $id_R M = r$. □

Corollary 10.4.8 *If $id_R R = r < \infty$, then $r \geq \dim R$.*

Proof Take $M = R$ in Corollary 10.4.7.

Proposition 10.4.9 *Let R be a Noetherian local ring, M a finitely generated R-module with $\mathrm{pd}_R\, M = r$. Then $\mathrm{Ext}_R^r(M, N) \neq 0$ for all finitely generated R-modules $N \neq 0$.*

Proof The proof is by induction on r. If $r = 0$, then M is free and the result is clear. If $r = 1$, express M as the quotient of a free module F with kernel K so that there is an exact sequence

$$0 \longrightarrow K \xrightarrow{i} F \longrightarrow M \longrightarrow 0.$$

K is free and we may also assume that $K \subset \mathfrak{m}F$. Consider the exact sequence

$$\mathrm{Hom}_R(F, N) \xrightarrow{i*} \mathrm{Hom}_R(K, N) \longrightarrow \mathrm{Ext}_R^1(M, N).$$

It is sufficient to show that i^* is not surjective. Since $N \neq 0$, $\mathfrak{m}N \neq N$ and since K is free, there exists an R-linear map $f : K \to N$ such that $\mathrm{Im}(f) \not\subset \mathfrak{m}N$. If i^* is surjective, then $f = i^*(g)$ for some $g \in \mathrm{Hom}_R(F, N)$ and then $\mathrm{Im}(f) \subset g(K) \subset \mathfrak{m}N$, a contradiction. Assume $r > 1$ and consider the exact sequence

$$0 \longrightarrow K \longrightarrow F \longrightarrow M \longrightarrow 0$$

with F free. By Exercise 10, Section 7.3, $\mathrm{pd}_R K = r - 1$ and by induction

$$\mathrm{Ext}_R^{r-1}(K, N) \neq 0$$

for all finitely generated $N \neq 0$. Consider the exact sequence $(r > 1)$

$$0 = \mathrm{Ext}_R^{r-1}(F, N) \longrightarrow \mathrm{Ext}_R^{r-1}(K, N) \longrightarrow \mathrm{Ext}_R^{r}(M, N) \longrightarrow \mathrm{Ext}_R^{r}(F, N) = 0.$$

Thus $\mathrm{Ext}_R^r(M, N) \simeq \mathrm{Ext}_R^{r-1}(K, N) \neq 0$. □

Theorem 10.4.10 *Let R be a Noetherian local ring, M a finitely generated R-module with $id_R M = r < \infty$. Then* $\mathrm{depth}(R) = id_R(M)$.

Proof Since $id_R(M) = r$, $E_r \neq 0$ in a minimal injective resolution of M and for some prime ideal P of R

$$\mu_r(P, M) \neq 0.$$

By Corollary 10.4.7, P is maximal and if $k = R/P$,

$$\mathrm{Ext}_R^r(k, M) \neq 0, \quad k = R/P.$$

Let $s = \mathrm{depth}\, R$ and choose a maximal R-sequence $a_1, a_2, \ldots, a_s$ in R. If $N = R/(a_1, a_2, \ldots, a_s)$, $P \in \mathrm{Ass}(N)$ and there is an injection of k in N. Since $\mathrm{Ext}_R^r(k, M) \neq 0$ and $id_R M = r$, it follows by the right exactness of $\mathrm{Ext}_R^r(\ , M)$ that $\mathrm{Ext}_R^r(N, M) \neq 0$. Hence by Corollary 8.4.22, $\mathrm{pd}_R N = s \geq r$. However, from Proposition 10.4.9, $\mathrm{Ext}_R^s(N, M) \neq 0$. Hence $id_R M = r \geq s$, i.e., $id_R M = r = s = \mathrm{depth}(R)$. □

Corollary 10.4.11 *Let R be a Noetherian local ring and M a finitely generated R-module with $id_R M = r < \infty$. If $\widehat{M}$ is the $\mathfrak{m}$-adic completion of M, then $id_R M = id_{\widehat{R}} \widehat{M}$.*

Proof We claim that $id_R M = \sup\ \{i \mid \mathrm{Ext}_R^i(k, M) \neq 0\}$. This is clear from the above theorem if $id_R M = r < \infty$. If $id_R M = \infty$, then any minimal injective resolution of M has infinite length. Hence for any i, there exists $j \geq i$ such that $\mu_j(\mathfrak{m}, M) \neq 0$, i.e.,

$$\mathrm{Ext}_R^i(k, M) \neq 0$$

and the result follows. Since $\hat{R}$ is faithfully flat over R,

$$\mathrm{Ext}^i_R(k, M) \simeq \mathrm{Ext}^i_{\widehat{R}}(k, \widehat{M})$$

and it follows that $id_R M = id_{\widehat{R}} \widehat{M}$. □

Corollary 10.4.12 *Let R be a Noetherian local ring with $id_R R = r < \infty$. Then $id_R R = \dim R$ and R is* C.M.

Proof By Theorem 10.4.10, depth $R = id_R R = r \leq \dim R$. By Corollary 10.4.8, $r \geq \dim R$. Hence $r =$ depth $R = \dim R$ and R is C.M. □

We now state Bass' conjecture.

Bass' conjecture: Let R be a Noetherian local ring such that there exists a finitely generated R-module M with $id_R M < \infty$. Then R is C.M.

We first show that the condition stated above is necessary for R to be C.M. i.e., if R is a C. M. local ring, then there exists a finitely generated R-module M with $id_R M < \infty$. For proving this we need some preliminary results on the behaviour of injective dimension under change of rings.

Proposition 10.4.13 (Kaplansky). *Let R be a ring, $a \in R$ a non-unit and a non-zero divisor and $R^* = R/(a)$. If M is a non-zero R^*-module with $id_{R^*} M < \infty$, then $id_R M = 1 + id_{R^*} M$.*

We first prove the following lemmas.

Lemma 10.4.14 *Let $a \in R$ be a non-unit and non-zero-divisor and $R^* = R/(a)$. If M is an R-injective module and N an R-module for which a is a non-zero-divisor, then* $\mathrm{Ext}^1_R(N, M) = 0$.

Proof Express N as the quotient of a free R-module F with kernel K, so that we have an exact sequence of R-modules

$$0 \longrightarrow \mathrm{K} \longrightarrow \mathrm{F} \longrightarrow \mathrm{N} \longrightarrow 0 \tag{10.6}$$

Since a is a non-zero-divisor of N, this induces an exact sequence of R^*-modules

$$0 \longrightarrow K/aK \longrightarrow F/aF \longrightarrow N/aN \rightarrow 0.$$

Then 10.6 induces an exact sequence

$$\longrightarrow \mathrm{Hom}_R(F, M) \xrightarrow{f^*} \mathrm{Hom}_R(K, M) \longrightarrow \mathrm{Ext}^1_R(N, M).$$

To show that $\mathrm{Ext}^1_R(N, M) = 0$, it is sufficient to show that f^* is surjective. Let $\phi : K \longrightarrow M$ be an R-linear map. It induces an R^*-linear map $\phi^* : K/aK \longrightarrow M$ and since M is R^*-injective, ϕ^* can be extended to an R^*-linear map $\overline{\phi} : F/aF \longrightarrow M$. If $\psi = \overline{\phi}p$, where p is the natural projection $p : F \longrightarrow F/aF$, then $\psi : F \to M$ is R-linear and $f^*(\psi) = \phi$. □

Lemma 10.4.15 *Let M and N be R-modules. Express N as the quotient of a projective R-module X_0 with it kernel, K_1 and M as a submodule of an injective R-module Q_0 with cokernel T_1. Then $\mathrm{Ext}^1_R(K_1, M) = 0$ if and only if $\mathrm{Ext}^1_R(N, T_1) = 0$.*

Assume $\mathrm{Ext}^1_R(K_1, M) = 0$ and consider the diagram

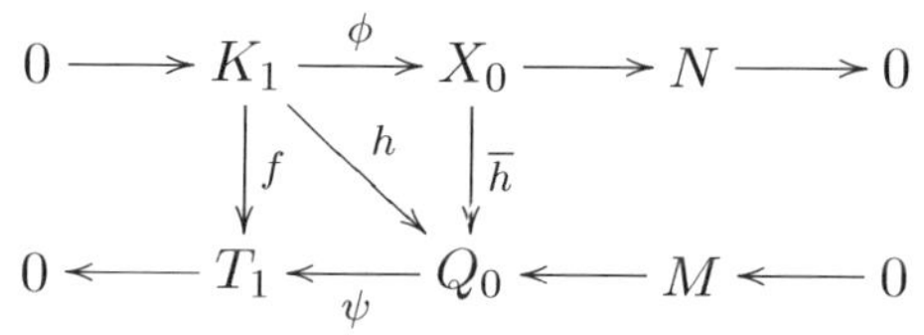

To show that $\mathrm{Ext}^1_R(N, T_1) = 0$, it is sufficient to show that for any R-homomorphism $f : K_1 \to T_1$, there exists an R-homomorphism $g : X_0 \to T_1$ with $g\phi = f$. Consider the exact sequence

$$\mathrm{Hom}_R(K_1, Q_0) \longrightarrow \mathrm{Hom}_R(K_1, T_1) \longrightarrow \mathrm{Ext}^1_R(K_1, M).$$

Since $\mathrm{Ext}^1_R(K_1, M) = 0$, there exists some map $h : K_1 \to Q_0$ with $\psi h = f$. Since Q_0 is injective, h can be extended to a map $\overline{h} : X_0 \to Q_0$ with $\overline{h}\phi = h$. Then $g = \psi\overline{h}$ satisfies the condition $g\phi = f$. This shows that $\mathrm{Ext}^1_R(N, T_1) = 0$. The converse can be proved similarly. □

Lemma 10.4.16 *Let M be an R-module and I an ideal of R. Let $M' = \{x \in M | Ix = 0\}$. If M is R-injective, then M' is R/I-injective.*

In particular, if $I = (a)$ and M is R-injective, then $M' = 0 : (a)$ is $R^* = R/(a)$-injective.

Proof Let $0 \to X \to Y$ be an exact sequence of R/I-modules and $f : X \to M'$ an R/I-linear map. Since $M' \subset M$ and M is R-injective, f can be extended to an R-linear map $g : Y \to M$. Since Y is annihilated by I, $\mathrm{Im}(g) \subset M'$, i.e., $g : Y \to M'$ is R/I-linear and is an extension of f. Hence M' is R/I-injective. □

Proof of Proposition 10.4.13 The proof is by induction on $id_{R^*}M = n < \infty$. It is sufficient to prove the proposition only for $n = 0$ and 1, for if $n > 1$, express M as a submodule of an injective R^*-module Q with Coker T to obtain an exact sequence

$$0 \longrightarrow M \longrightarrow Q \longrightarrow T \longrightarrow 0.$$

Now, $id_{R^*}M = n$ implies $id_{R^*}T = n - 1$, (Exercise 12, Section 7.3) and by induction, $id_R T = n$. But $id_R Q = 1$ (by the case $n = 0$) and if

$$n > 1, \quad id_R M = 1 + id_R T = n + 1.$$

(Exercise 12, Section 7.3)

Case (i)**:** $n = 0$. Then M is R^*-injective. We claim that M is not R-injective, for if it were so, M will be divisible by a, as any homomorphism from $(a) \to M$ can be extended to R. This will imply that $M = 0$, as a annihilates M. Hence it is sufficient to show that $id_R M \leq 1$. Consider an exact sequence of R-modules $0 \to M \to Q_0 \to T_1 \to 0$ where Q_0 is R-injective. It is sufficient to show that T_1 is R-injective, i.e., $\text{Ext}^1_R(N, T_1) = 0$ for all R-modules, N. Consider an exact sequence

$$0 \longrightarrow K_1 \longrightarrow X_0 \longrightarrow N \longrightarrow 0 \text{ with } X_0 \text{ free.}$$

By Lemma 10.4.15, it is sufficient to show that $\text{Ext}^1_R(K_1, M) = 0$. This follows from Lemma 10.4.14, as M is R^*-injective and a is a non-zero-divisor of K_1, being a submodule of the free module X_0.

Case (ii)**:** $id_{R^*}M = 1$. In this case, $id_{R^*}T = 0$. By case (i), $id_R T = 1$ and $id_R Q = 1$. It follows from Exercise 12, Section 7.3 that $id_R M \leq 1 + id_R Q = 2$. We have already seen in case (i) that M cannot be R-injective. It is therefore sufficient to show that $id_R M = 1$ is impossible. Suppose $id_R M = 1$. Then we have an exact sequence $0 \to M \to E \to E' \to 0$ with E and E', R-injective. Consider the commutative diagram

$$\begin{array}{ccccccccc} 0 & \longrightarrow & M & \longrightarrow & E & \longrightarrow & E' & \longrightarrow & 0 \\ & & \downarrow \lambda_a & & \downarrow \lambda_a & & \downarrow \lambda_a & & \\ 0 & \longrightarrow & M & \longrightarrow & E & \longrightarrow & E' & \longrightarrow & 0 \end{array}$$

where the vertical maps are scalar multiplications by a. By Proposition 2.2.15 we have an exact sequence

$$0 \to M \to E_0 \xrightarrow{\theta} E'_0 \to M \to 0,$$

where E_0 and E'_0 are the submodules of E and E' annihilated by a. By Lemma 10.4.16, E_0 and E'_0 are R^*-injective. Split the above exact sequence as $0 \to M \to E_0 \to \text{Im}\,\theta \to 0$, and

$$0 \to \text{Im}\ \theta \to E'_0 \to M \to 0.$$

Since $id_{R^*}M = 1$, $\text{Im}(\theta)$ is R^*-injective, which implies that M is R^*-injective, a contradiction to the assumption. □

Proposition 10.4.17 *Let R be a C.M. local ring. Then there exists a finitely generated R-module E with $id_R E < \infty$.*

Proof Choose a maximal R-sequence $a_1, a_2, \ldots, a_n$ for R ($n = \dim R$) and let $S = \dfrac{R}{(a_1, \ldots, a_n)}$. Then S is an Artinian local ring and $\ell_S(S) < \infty$. If E is the injective envelope of the residue field of S, we have by Matlis duality, $\mathrm{Hom}_S(S, E) \simeq E$ is a module of finite length over S, i.e., E is a finitely generated injective module over S. E is also finitely generated over R, and by a repeated application of Proposition 10.4.13, we have $id_R E = n + id_S E = n < \infty$. □

C. Peskine and L. Szpiro have proved that the intersection conjecture implies Bass' conjecture for a class of local rings which is closed under completions and localisations at prime ideals. Before outlining the proof, we prove some preliminary results.

Definition 10.4.18 Let M be an R-module. Then the grade M

$$\mathrm{gr}(M) = \mathrm{Inf}\{\mathrm{depth}\, R_P, P \in \mathrm{Supp}(M)\}.$$

Theorem 10.4.19 *Let R be a Noetherian local ring, M a finitely generated R-module. Then* $\mathrm{depth}\, R \leq \mathrm{gr}(M) + \dim M \leq \dim R$.

Proof Choose $P \in \mathrm{Supp}(M)$ with $\dim M = \dim R/P$. We have $\mathrm{gr}(M) \leq \mathrm{depth}\, R_P \leq \dim R_P$ and hence

$$\dim M + \mathrm{gr}(M) \leq \dim R/P + \dim R_P \leq \dim R.$$

The other inequality is proved by induction on $\mathrm{gr}(M)$. If $\mathrm{gr}(M) = 0$, then there exists some $P \in \mathrm{Supp}(M)$ with $\mathrm{depth}\, R_P = 0$ i.e., $P \in \mathrm{Ass}(R)$. Then $\dim M \geq \dim R/P \geq \mathrm{depth}\, R$ and the result follows. Assume that $\mathrm{gr}(M) > 0$ and let $I = \mathrm{Ann}(M)$. Then there exists some $a \in I$ which is not a zero-divisor of R. Then M is an $R' = R/(a)$-module and $\mathrm{gr}_{R'}(M) = n - 1$. By induction, $\mathrm{gr}_{R'}(M) + \dim_{R'} M \geq \mathrm{depth}\, R'$. But $\dim_{R'} M = \dim_R M$. Hence $\mathrm{gr}_R(M) + \dim_R M \geq 1 + \mathrm{depth}\, R' = \mathrm{depth}\, R$. □

Theorem 10.4.20 *Let R be a Noetherian local ring, M a finitely generated R-module with $id_R M < \infty$. Then for any $P \in \mathrm{Supp}(M)$, we have*

$$\mathrm{depth}\, R_P + \dim R/P = \mathrm{depth}\, R.$$

Proof Now, $M_P \neq 0$ and $id_{R_P} M_P \leq id_R M < \infty$. By Theorem 10.4.10, $id_{R_P} M_P = \mathrm{depth}\, R_P = s$, say, so that $\mathrm{Ext}^s_{R_P}(k(P), M_P) \neq 0$, where $k(P) = R_P/PR_P$. Hence $\mu_s(P, M) \neq 0$. Let $\mathfrak{m}$ be the unique, maximal ideal of R and $d = \dim R/P$. By repeated application of Proposition 10.4.6, we have $\mu_{s+d}(\mathfrak{m}, M) \neq 0$, i.e., $\mathrm{Ext}^{s+d}_R(k, M) \neq 0$. This implies $id_R M \geq s+d = \operatorname{depth} R_P + \dim R/P$. By Theorem 10.4.10, $id_R M = \operatorname{depth} R$, so that $\operatorname{depth} R \geq \operatorname{depth} R_P + \dim R/P$. To prove the reverse inequality, observe that

$$\operatorname{depth} R_P \geq \operatorname{gr}_R R/P \geq \operatorname{depth} R - \dim R/P,$$

(Theorem 10.4.19), so that $\operatorname{depth} R_P + \dim R/P \geq \operatorname{depth} R$. □

Corollary 10.4.21 *Let R be a Noetherian local ring, M a finitely generated R-module with $id_R M < \infty$. Then*

$$\operatorname{gr}(M) + \dim M = \operatorname{depth} R.$$

Proof Choose $P \in \operatorname{Supp}(M)$ with $\dim M = \dim R/P$. Since $\operatorname{gr}(M) \leq \operatorname{depth} R_P$, we have $\operatorname{gr}(M) + \dim M \leq \operatorname{depth} R_P + \dim R_P = \operatorname{depth} R$. The result now follows from Theorem 10.4.19. □

Theorem 10.4.22 *Let R be a Noetherian local ring, T a finitely generated R-module with $id_R T = r < \infty$ and E the injective envelope of $k = R/\mathfrak{m}$. Then*

(i) $M = \operatorname{Ext}^r_R(E,T)$ is a finitely generated $\widehat{R}$-module of finite projective dimension,

(ii) M has the same support as $\widehat{T}$.

Proof Consider a minimal injective resolution of T

$$0 \to T \to I^0 \to I \cdots \to I^r \to 0 \tag{10.7}$$

where $I^i = \oplus \sum \mu_i(P,T)E(R/P)$.

If $P \neq \mathfrak{m}$, $\operatorname{Hom}_R(E, E(R/P)) = 0$, for if $f : E \to E(R/P)$ is an R-linear map and $x \in E$, then Rx is an R-module of finite length and hence $Rf(x)$ is a submodule of $E(R/P)$ of finite length. Since $E(R/P)$ is an essential extension of R/P, it does not contain any non-trivial module of finite length. Hence $f(x) = 0$, i.e.,

$$\operatorname{Hom}_R(E, E(R/P)) = 0, \quad P \neq \mathfrak{m}.$$

We compute $\operatorname{Hom}_R(E, I^i)$.

Now, $\operatorname{Hom}_R(E,E) \simeq \widehat{R}$ and this implies $\operatorname{Hom}_R(E, I^i)$ is a free $\widehat{R}$-module of rank $\mu_i(\mathfrak{m},T) = \mu_i$ (say). Applying the functor $\operatorname{Hom}_R(E, \ )$ to the deleted complex (10.7), we have the complex

$$0 \to \widehat{R}^{\mu_0} \to \widehat{R}^{\mu_1} \to \cdots \to \widehat{R}^{\mu_r} \to 0 \tag{10.8}$$

whose i-th homology is $\operatorname{Ext}^i_R(E,T)$. If we show that $\operatorname{Ext}^i_R(E,T) = 0$, $i < r$, it will follow that $\operatorname{Ext}^r_R(E,T)$ is an $\widehat{R}$-module of finite type and of finite projective dimension over $\widehat{R}$.

E is also the injective envelope of the residue field k of $\widehat{R}$ and by Corollary 10.4.11, $id_R T = id_{\widehat{R}} \widehat{T} = r$. If we start with a minimal injective $\widehat{R}$-resolution of $\widehat{T}$ and apply the functor $\mathrm{Hom}_{\widehat{R}}(E, \ )$, we get the same complex as (10.8). It is therefore sufficient to show that $\mathrm{Ext}^i_{\widehat{R}}(E, \widehat{T}) = 0$ for $i < r$. Since $\widehat{T}$ is finitely generated $\widehat{R}$-module, we have by Matlis duality, $\widehat{T} \simeq \mathrm{Hom}_{\widehat{R}}(\mathrm{Hom}_{\widehat{R}}(\widehat{T}, E), E)$. Let

$$T' = \mathrm{Hom}_{\widehat{R}}(\widehat{T}, E),$$

so that $\widehat{T} = \mathrm{Hom}_{\widehat{R}}(T', E)$. Now,

$$\begin{aligned} T' = \mathrm{Hom}_{\widehat{R}}(\widehat{T}, E) &= \mathrm{Hom}_{\widehat{R}}\left(\varprojlim \frac{\widehat{T}}{\mathfrak{m}^n \widehat{T}}, E\right) \\ &\simeq \varinjlim \mathrm{Hom}_{\widehat{R}}\left(\frac{\widehat{T}}{\mathfrak{m}^n \widehat{T}}, E\right) \simeq \varinjlim T'_n \end{aligned}$$

where $T'_n = \mathrm{Hom}_{\widehat{R}}\left(\frac{\widehat{T}}{\mathfrak{m}^n \widehat{T}}, E\right)$.

By the duality isomorphism (Exercise 13, Section 7.3), we have

$$\begin{aligned} \mathrm{Ext}^i_{\widehat{R}}(E, \widehat{T}) &\simeq \mathrm{Ext}^i_{\widehat{R}}(E, \mathrm{Hom}_{\widehat{R}}(T', E)) \\ &\simeq \mathrm{Hom}_{\widehat{R}}(\mathrm{Tor}^{\widehat{R}}_i(E, T'), E) \\ &\simeq \mathrm{Hom}_{\widehat{R}}\left(\mathrm{Tor}^{\widehat{R}}_i(E, \varinjlim T'_n), E\right) \\ &\simeq \mathrm{Hom}_{\widehat{R}}\left(\varinjlim \ \mathrm{Tor}^{\widehat{R}}_i(E, T'_n), E\right) \\ &\simeq \varprojlim \mathrm{Hom}_{\widehat{R}}\left(\mathrm{Tor}^{\widehat{R}}_i(E, T'_n), E\right) \\ &\simeq \varprojlim \mathrm{Hom}_{\widehat{R}}\left(\mathrm{Tor}^{\widehat{R}}_i(T'_n, E), E\right) \simeq \varprojlim \mathrm{Ext}^i_{\widehat{R}}\left(T'_n, \mathrm{Hom}_{\widehat{R}}(E, E)\right) \\ &\simeq \varprojlim \mathrm{Ext}^i_{\widehat{R}}(T'_n, \widehat{R}). \end{aligned}$$

Now, depth $\widehat{R} =$ depth $R = r$ and for each n, T'_n is an $\widehat{R}$ module of finite length. We have $\mathrm{Ext}^i_{\widehat{R}}(T'_n, \widehat{R}) = 0 \ i < r$ by induction on length of T'_n as $\mathrm{Ext}^i_{\widehat{R}}(k, R) = 0, i < r$. Hence $\mathrm{Ext}^i_{\widehat{R}}(E, \widehat{T}) = 0, i < r$ and the proof of (i) is complete.

To prove the relation (ii), consider the duality isomorphism

$$\mathrm{Ext}^i_{\widehat{R}}(M, \widehat{R}) \simeq \mathrm{Ext}^{n-(r-1)}_A(\widehat{T}, A)$$

([C. Peskine and L. Szpiro], Page 67), where A is an n dimensional Gorenstein local ring whose quotient is $\widehat{R}$.

$$\begin{aligned}\text{Supp }(M) &= \bigcup_{i\geq 0} \text{Supp }\left(\text{Ext}^{i}_{\widehat{R}}(M,\widehat{R})\right)\\ &= \bigcup_{i\geq 0} \text{Supp }\left(\text{Ext}^{n-(r-i)}_{A}(\widehat{T},A)\right)\\ &= \text{Supp }(\widehat{T})\end{aligned}$$ □

Theorem 10.4.23 *For a class of local rings closed under completions and localisations at prime ideals, the intersection conjecture implies Bass's conjecture.*

Proof Suppose for some ring in the class Bass's conjecture is false. Choose such a ring of least Krull dimension, i.e., there exists a non C.M. local ring R with $\dim R = n$, (n least) and a finitely generated R-module T with $id_R T < \infty$. We shall obtain a contradiction by showing that R is C.M. By replacing R by $\widehat{R}$ we may assume that R is complete. Choose a prime ideal P of R with $\dim R/P = n$.

Case (i)**:** Suppose there exists some $Q \in \text{Supp }(T)$ such that $\mathfrak{m} \supsetneqq Q \supset P$. Then we may choose such a Q satisfying $\dim R/Q = 1$. Now, $T_Q \neq 0$ and $id_{R_Q} T_Q < \infty$. Since $\dim R_Q < \dim R$, R_Q is C.M. Since R/P is a complete local domain, by Corollary 9.4.12, R/P is the quotient ring of a power series ring over a DVR, i.e., it is the quotient ring of a regular local ring. Since R/P is a local domain which is the quotient of a C.M. ring, by Corollary 8.5.17, we have $\text{ht}\, Q/P = \dim R/P - \dim R/Q = n-1$. Now, $\text{depth}\, R_Q = \dim R_Q = \text{ht}\, Q \geq \text{ht}\, Q/P = n-1$. By Theorem 10.4.20, $\text{depth}\, R = \dim R/Q + \text{depth}\, R_Q \geq 1 + (n-1) = n = \dim R$. Hence $\text{depth}\, R = \dim R$, i.e., R is C.M., which is a contradiction.

Case (ii)**:** Assume that there is no $Q \in \text{Supp }(T)$ except $\mathfrak{m}$ containing P, i.e., $\text{Supp }(T \otimes R/P) = \{\mathfrak{m}\}$. Let E be the injective envelope of $k = R/\mathfrak{m}$. If $r = \text{depth}\, R$, then by Theorem 10.4.22, $M = \text{Ext}^r_R(E,T)$ is a finitely generated R-module of finite projective dimension and it has the same support as T. Thus $\text{Supp }(M \otimes R/P) = \{\mathfrak{m}\}$, i.e., $\ell_R(M \otimes R/P) < \infty$. By the intersection conjecture, this implies $\dim R/P \leq \text{pd}_R M$. Hence $\dim R = n = \dim R/P \leq \text{pd}_R M \leq \text{depth}\, R$ (Theorem 10.3.3). Hence R is C. M., which is a contradiction. □

Corollary 10.4.24 *If* R *is a local ring of* ch $p > 0$, p *prime, Bass's conjecture is true for* R.

Proof Follows from the above theorem and Corollary 10.2.4. □

Final remark

We have thus obtained the validity of the intersection conjecture, zero-divisors conjecture and Bass's conjecture for any local ring of $\mathrm{ch}\, p > 0$ as a consequence of the Big Cohen–Macaulay modules conjecture.

The solution of the Big Cohen–Macaulay modules conjecture in the $\mathrm{ch}\, 0$ (equi-ch.) case is more involved as it uses Artin's approximation theorem, Chevalley's constructibility, properties of descent, generic flatness and reduces to the case of $\mathrm{ch}\, p > 0$. For a proof of this and some more applications of homological methods in commutative algebra, we refer you to M. Hochster: Topics in homological theory of modules over commutative rings, *Regional Conference Series in Maths*. AMS. Vol. 24 (1975).